W9-ACU-975

CONTROL AND
DYNAMIC SYSTEMS

*Advances in Theory
and Applications*

Volume 35

CONTRIBUTORS TO THIS VOLUME

DARÉ AFOLABI
M. D. ARDEMA
GEORGE BOJADZIEV
DONG DA
FAYEZ M. DAMRA
HENRYK FLASHNER
JONATHAN E. GAYEK
RAMESH S. GUTTALU
SHIV P. JOSHI
JAE H. KIM
YEONG CHING LIN
OSITA D. I. NWOKAH
ANDRZEJ OLAS
H. ALI PAK
W. E. SCHMITENDORF
KAVEH SHAMSA
G. F. SHANNON
ROWMAU SHIEH
ROBERT E. SKELTON
J. M. SKOWRONSKI
R. J. STONIER
THOMAS L. VINCENT
C. N. WHEELER
C. WILMERS

CONTROL AND DYNAMIC SYSTEMS

ADVANCES IN THEORY AND APPLICATIONS

Edited by

C. T. LEONDES

School of Engineering and Applied Science
University of California, Los Angeles
Los Angeles, California
and
College of Engineering
University of Washington
Seattle, Washington

VOLUME 35: ADVANCES IN CONTROL MECHANICS
PART 2 OF 2

ACADEMIC PRESS, INC.
Harcourt Brace Jovanovich, Publishers
San Diego New York Boston
London Sydney Tokyo Toronto

This book is printed on acid-free paper. ∞

Academic Press, Inc.
San Diego, California 92101

United Kingdom Edition published by
Academic Press Limited
24-28 Oval Road, London NW1 7DX

Library of Congress Catalog Card Number: 64-8027

ISBN 0-12-012735-0 (alk. paper)

Printed in the United States of America
90 91 92 93 9 8 7 6 5 4 3 2 1

Dedicated to
Drs. George Leitmann and Angelo Miele
in appreciation for their many contributions over the years

CONTENTS

CONTRIBUTORS

Numbers in parentheses indicate the pages on which the authors' contributions begin.

Daré Afolabi (137), *School of Engineering and Technology, Purdue University, Indianapolis, Indiana 46202*

M. D. Ardema (235), *Santa Clara University, Santa Clara, California*

George Bojadziev (215), *Mathematics and Statistics Department, Simon Fraser University, Burnaby, B. C., Canada V5A 1S6*

Dong Da (65), *School of Aeronautics and Astronautics, Purdue University, West Lafayette, Indiana 47907*

Fayez M. Damra (137), *School of Aeronautics an Astronautics, Purdue University, West Lafayette, Indiana 47907*

Henryk Flashner (xiii, 43), *Department of Mechanical Engineering, University of Southern California, Los Angeles, California 90089-1453*

Jonathan E. Gayek (31), *Department of Mathematics, Trinity University, San Antonio, Texas 78284*

Ramesh S. Guttalu (xiii), *Department of Mechanical Engineering, University of Southern California, Los Angeles, California 90089*

Shiv P. Joshi (87), *Aerospace and Mechanical Engineering, University of Arizona, Tucson, Arizona*

Jae H. Kim (65), *School of Aeronautics and Astronautics, Purdue University, West Lafayette, Indiana 47907*

Yeong Ching Lin (87), *Aerospace and Mechanical Engineering, University of Arizona, Tucson, Arizona*

Osita D. I. Nwokah (137), *School of Mechanical Engineering, Purdue University, West Lafayette, Indiana 47907*

Andrzej Olas (1), *Oregon State University, Department of Mechanial Engineering, Corvallis, Oregon 97331*

H. Ali Pak (273), *Department of Mechanial Engineering, University of Southern California, Los Angeles, California 90089-1453*

W. E. Schmitendorf (165), *Mechanical Engineering, University of California, Irvine, Irvine, California 92714*

Kaveh Shamsa (43), *Department of Mechanical Engineering, University of Southern California, Los Angeles, California 90089-1453*

G. F. Shannon (295), *Associate Professor of Electrical Engineering, University of Queensland, St Lucia, Queensland, Australia*

Rowmau Shieh (273), *Department of Mechanial Engineering, University of Southern California, Los Angeles, California 90089-1453*

Robert E. Skelton (65), *School of Aeronautics and Astronautics, Purdue University, West Lafayette, Indiana 47907*

J. M. Skowronski (xiii, 235), *Department of Mechanical Engineering, University of Southern California, Los Angeles, California 90089*

R. J. Stonier (185, 255), *Department of Mathematics & Computing, Capricornia Institute of Advanced Education, Rockhampton, Queensland, Australia 4702*

Thomas L. Vincent (87), *Aerospace and Mechanical Engineering, University of Arizona, Tucson, Arizona*

C. N. Wheeler (255), *Tarong Power Station, Nanango, Queensland, Australia 4610*

C. Wilmers (165), *Institute for System Dynamics and Control, University of Stuttgart, Stuttgart, Federal Republic of Germany*

PREFACE

Modern technology makes possible amazing things in control and dynamic systems and will continue to do so to an increasing extent with the passage of time. It is a fact, however, that control systems go back over the millennia, and while rudimentary in earlier centuries, they were, generally, enormously effective. A classic example of this is the Dutch windmill. In any event, this is true today, and, as just noted, amazing things are now possible. As a result, this volume and the previous one, Volumes 35 and 34, respectively, in the series *Control and Dynamic Systems,* are based on a National Science Foundation–Sponsored Workshop on Control Mechanics, i.e., control system development using analytical methods of mechanics and active control of mechanical systems. Publication of the presentations at this workshop in this Academic Press series has made it possible to expand them into a format which will facilitate the study and utilization of their significant results by working professionals and research workers on the international scene.

Ordinarily, the preface to the individual volumes in this series presents a summary of the individual contributions in the respective volumes. In the case of these two volumes, this is provided by the Introduction, which follows immediately.

INTRODUCTION

The articles in this volume were presented at the National Science Foundation–Sponsored Second Workshop on Control Mechanics, held at the University of Southern California, Los Angeles, January 23–25, 1989. This workshop is the second in a series devoted to promoting control mechanics, i.e., control system development using analytical methods of mechanics and active control of mechanical systems.

Research in control mechanics is motivated by the demands imposed on modern control systems. The tasks of modern industry in areas like high technology manufacturing, construction and control of large space structures, and aircraft control can be accomplished only by precisely controlled and highly autonomous mechanical systems. These systems are inherently nonlinear due to simultaneous high speed large motions of multiple interconnected bodies and their complex interactions with the environment. In addition, weight limits imposed by space-based systems and power constraints lead to highly flexible structures. High speed operation combined with structural flexibility necessitates inclusion of both nonlinear effects and vibrational modes in control law development. Consequently, one needs to apply methods of analytical mechanics to develop an adequate mathematical representation of the system. Then the underlying global characteristics of the equations of motion must be employed to develop sophisticated multivariable, possibly nonlinear, control laws.

The theme of the above-mentioned promotion of control mechanics is covered by a wide scope of papers in mechanical systems control theory presented at the workshop. The topics are arranged in two volumes—seven papers in the first and thirteen papers in the second. The first two chapters of Part 1 deal with microburst, a severe meteorological condition pertinent to aircraft control. These two chapters consider the problem of stable control of an aircraft subjected to windshear caused by microburst. New results concerning control of an aircraft under windshear conditions are given by G. Leitmann and S. Pandey in the first chapter.

The problem of control during windshear was first posed and has been extensively studied by A. Miele, who kindly accepted the invitation of C. T. Leondes to present a review of his results in the second chapter. In the third chapter, M. Corless addresses the issue of designing controllers for uncertain mechanical systems which are robust against unmodelled flexibility. Results concerning robust control design without matching conditions are presented by M. S. Chen and M. Tomizuka in the fourth chapter. Their results are applicable to SISO systems with disturbances or modelling errors that are either bounded or that have a cone-bounded growth rate. In the fifth chapter, S. Hui and S. H. Zak deal with the robust control problem using the variable structure control method. They present a methodology to design controllers and state observers and analyze their stability properties.

W. G. Grantham and A. M. Athalye study the control of chaotic systems in the sixth chapter. This article is concerned with the numerical chaotic behavior which can occur under feedback control, even with a stabilizing control law. Part 1 concludes with an article by R. S. Guttalu and P. J. Zufria that considers the problem of finding zeros of a nonlinear vector function, using methods of dynamical systems analysis. The role of singularities and their effect on the global behavior of dynamical systems is studied in detail.

Liapunov design method, an approach often used in control mechanics, is the first topic considered in Part 2. In the first chapter of Part 2, A. Olas studies the question of finding recursive Liapunov functions for autonomous systems. This is a sequel to the novel method of "converging series" presented by him at the First Control Mechanics Workshop in 1988. In the second chapter, J. E. Gayek discusses a new approach to verifying the existence of stabilizing feedback control laws for systems with time-delay by using Liapunov functionals. In the third chapter, K. Shamsa and H. Flashner use the notions of passivity and Liapunov stability to define a class of discrete-time control laws for mechanical systems. Their approach is based on the Hamiltonian structure of the equations of motion, a characteristic common to a wide class of mechanical systems. Reduction in dimensionality of models is discussed by R. E. Skelton, J. H. Kim, and D. Da in the fourth chapter. This problem is of interest for controlling high-order mechanical systems such as large space structures and robotic manipulators with structural flexibility. Regarding the systems with structural flexibility, a new method of control via active damping augmentation is introduced by T. L. Vincent, Y. C. Lin, and S. P. Joshi in the fifth chapter. This study is an extension to two-dimensional structures that the authors presented for beams in the First Control Mechanics Workshop. O. D. I. Nwokah, D. Afolabi, and F. M. Damra discuss the modal stability of imperfect cyclic systems in the sixth chapter. The paper is of particular value in resolving some of the disagreement in the literature concerning the qualitative behavior of cyclic systems. In the seventh chapter, W. E. Schmitendorf and C. Wilmers investigate the problem of developing reduced-order stabilizing controllers. They present a numerical algorithm for designing a minimum-order compensator to stabilize a given plant.

Most of the papers presented in this workshop are applicable to the analysis and design of robotic manipulators. This field has recently attracted attention in developing strategies for coordination control of multi-arm systems, adaptive control of robots, and control of manipulators with varying loads. A numerical step-by-step collision avoidance technique is proposed and demonstrated by R. J. Stonier in the eighth chapter. A theoretical basis for collision avoidance using Liapunov stability theory is investigated by G. Bojadziev in the ninth chapter. In the tenth chapter, a new method of using differential game approach to coordination is proposed by M. Ardema and J. M. Skowronski. A single arm problem using nonlinear Model Reference Adaptive Control studied by R. J. Stonier and C. N. Wheeler appears in the next chapter. A path-tracking method for control of mechanical systems is presented by H. A. Pak and R. Shieh in the twelfth chapter. In this chapter, a class of optimal feedforward tracking controllers have been proposed using preview and feedback control actions. Finally, our Australian participants presented a number of applications with this field, specifically on how to use the robotic manipulators in cane-sugar production analysis, sheep shearing, and the mining industry. The last-mentioned application is discussed in the final chapter by G. F. Shannon.

The participants of the workshop are in debt to Professor G. Leitmann for initiating this series of meetings; to Professor L. M. Silverman, Dean of the School of Engineering at the University of Southern California, for supporting and opening the workshop; the administrative staff of the Mechanical Engineering Department, Ms. G. Acosta and Ms. J. Givens, for their invaluable help; and to the editor of *Control and Dynamic Systems,* Professor C. T. Leondes, for inviting these proceedings for publication. The organizing committee gratefully acknowledges a grant from the National Science Foundation.

Janislaw M. Skowronski
Ramesh S. Guttalu
Henryk Flashner

RECURSIVE LYAPUNOV FUNCTIONS: PROPERTIES, LINEAR SYSTEMS

ANDRZEJ OLAS

Oregon State University
Department of Mechanical Engineering
Corvallis, OR 97331

I. PROBLEM STATEMENT

We discuss the procedure of designing the recursive Lyapunov function for autonomous asymptotically stable systems. At each step of the procedure a new and better, in the below defined sense, Lyapunov function is obtained.

The autonomous system

$$\dot{x} = f(x) \quad , \quad x \in R^n \tag{1}$$

defined on the set $Z = \{||x|| < H > 0\}$ and such that $f(0) = 0$ is considered under the assumption that $f \in C^1(Z)$ satisfies the condition for existence and uniqueness of solutions, which are denoted by $p(t,x_0)$, $p(0,x_0) = x_0$. Together with the system (1) the positive-definite function $V_1(x)$ is considered with an assumption that it is of a class $C^2(Z)$; this assumption will be utilized only when considering the second derivative of V_1. For other considerations the assumption $V \in C^1(Z)$ is sufficient.

The classic efforts to construct a Lyapunov function for a given system are well reviewed by Hahn [1]. Recent years have brought out a number of papers on this subject and the closely related subject of an estimation of the domain of attraction. Brayton and Tong [2,3], in their series of papers, introduced computer generated Lyapunov functions and considering the Aizerman conjecture obtained the new, better results. Leipholz [4], Olas [5] considered a generalization of Lyapunov Direct Method. Vanelli and Vidyasagar [6] introduced maximal Lyapunov function in the form of a rational function for estimation of the domain of attraction. They derived the partial differential equation characterizing the maximal Lyapunov function and proposed an iterative method for solving this equation. The estimates obtained using their method to two- and three-dimensional examples are in many cases substantially better than previous results.

The properties of two sequences are analyzed: the sequence of recursive Lyapunov functions and the sequence of their performance measures. The recursive Lyapunov function concept was introduced in [7]. The performance measure of a Lyapunov function which is defined as

$$\lambda = \sup_{x \in Z/0} \dot{V}(x)/V(x)$$

allows the estimate

$$V(p(t,x)) \leq V(x)\exp(\lambda t)$$

and the value $(-2/\lambda)$ corresponds, as defined by Ogata [8], to the largest time constant of the system, relating to changes in the Lyapunov function $V(p(t,x))$.

Finally the procedure is applied to linear systems. The recursive Lyapunov equation is introduced and the properties of the recursive Lyapunov function for linear systems analyzed.

II. RECURSIVE ALGORITHM AND LEMMAS

In [7] the recursive algorithm for design of Lyapunov function was introduced by defining the sequence of functions

$$V_{i+1}(x) = \int_0^T V_i(p(t,x))dt \ , \qquad i = 1,2,... \tag{2}$$

where $T > 0$ is some constant.

It was proved that the Lyapunov derivative of $V_i(x)$ was given by the formula

$$\dot{V}_{i+1}(x) = \int_0^T \dot{V}_i(p(t,x))dt = V_i(p(T,x)) - V_i(x) \ , \qquad i = 1,2,... \tag{3}$$

Similarly the second Lyapunov derivative of $V_i(x)$ may be expressed as

$$\ddot{V}_{i+1}(x) = \int_0^T \ddot{V}_i(p(t,x))dt = \dot{V}_i(p(T,x)) - \dot{V}_i(x) \ , \qquad i = 1,2,... \tag{4}$$

Here we assume that the solution to the system (1) exists on the interval $[0,T]$ for $x \in Z_1$, where Z_1 is a certain compact subset of Z containing the equilibrium point. This is enough to have the above functions well-defined on Z_1.

In [7] the following Lemma was proved

Lemma 1. Let $a_i, b_i \in R$, $b_i > 0$, $i = 1,...,n$. Then the inequality holds

$$\frac{\sum_{i=1}^{m} a_i}{\sum_{i=1}^{m} b_i} \leq \max_i \frac{a_i}{b_i}$$

Moreover if for some j's we have

$$\frac{a_j}{b_j} < \max_i \frac{a_i}{b_i} = \frac{a_i^*}{b_i^*}$$

then the inequality is a strong one.

We need also Lemma 2, which deals with the lower bound.

Lemma 2. Let $a_i, b_i \in R$, $b_i > 0$, $i = 1,\ldots,n$. Then the inequality holds

$$\frac{\sum_{i=1}^{m} a_i}{\sum_{i=1}^{m} b_i} \geq \min_i \frac{a_i}{b_i}$$

Moreover if for some j's we have

$$\frac{a_j}{b_j} > \min_i \frac{a_i}{b_i} = \frac{a_i^*}{b_i^*}$$

then the inequality is a strong one.

The proof of the Lemma 2 is given in Section I of the Appendix.

III. FUNCTION $\Lambda_i(x)$ AND PERFORMANCE MEASURE

Assume that the function $V_1(x)$ is selected in such a manner that the function

$$\Lambda_1(x) = \dot{V}_1(x)/V_1(x) \tag{5}$$

exists and is bounded on the set $Z_1\backslash 0$. Define the functions $\Lambda_i(x)$, $i = 1,2,\ldots$

$$\Lambda_i(x) = \dot{V}_i(x)/V_i(x) \tag{6}$$

The function $\Lambda_i(x)$ measures the Lyapunov function performance at the point x. We define the performance measure on the set $Z_1\backslash 0$ by introducing

$$\lambda_i = \sup_{x \in Z1\backslash 0} \Lambda_i(x)$$

We have

$$\Lambda_{i+1}(x) = \frac{\int_0^T \dot{V}_i(p(t,x))dt}{\int_0^T V_i(p(t,x))dt}$$

Selecting the same Δt for both integrals we may write

$$\Lambda_{i+1}(x) = \lim_{\substack{\Delta t \to 0 \\ m \to \infty}} \frac{\sum_{i=1}^{m} \dot{V}_i(p(t,x))\Delta t}{\sum_{i=1}^{m} V_i(p(t,x))\Delta t} \tag{7}$$

By the definition of the upper bound we have for all x

$$\lim_{\substack{\Delta t \to 0 \\ m \to \infty}} \frac{\sum_{i=1}^{m} \dot{V}_i(p(t,x))\Delta t}{\sum_{i=1}^{m} V_i(p(t,x))\Delta t} \le \lambda_i$$

i.e.

$$\sup_{x \in Z1 \setminus 0} \Lambda_{i+1}(x) = \lambda_{i+1} \le \lambda_i$$

Similar reasoning leads to introduction of

$$\nu_i = \inf_{x \in Z1 \setminus 0} \Lambda_i(x)$$

The relation

$$\nu_{i+1} \ge \nu_i \tag{8}$$

may be easily proven by utilizing the Lemma 2. Thus the sequence $\{\lambda_i\}$ is monotonically decreasing and by virtue of Eq. (8) it is also bounded below. As such the sequence is convergent. Denote its limit by λ_{lim}. Similar reasoning leads to to the concept of ν_{lim}.

The estimations

$$V_i(p(t,x)) \geq V_i(x)\exp(\nu_i t) \tag{9}$$

$$V_i(p(t,x)) \leq V_i(x)\exp(\lambda_i t) \tag{10}$$

hold for all i's.

IV. FUNCTION Λ_i AS THE FUNCTION OF TIME

Consider an arbitrary non-zero solution $p(t,x)$, $x \in Z_1$, and the function $\Lambda_i(p(t,x))$ on the interval $[0,T]$. Denote

$$\omega_i = \max_{t\in[0,T]}\Lambda_i(p(t,x))$$

and

$$\gamma_i = \min_{t\in[0,T]}\Lambda_i(p(t,x))$$

The following estimations hold

$$\max_{t\in[0,T]}V_i(p(t,x)) \leq K_1V_i(x) \tag{11}$$

$$\min_{t\in[0,T]}V_i(p(t,\dot{x})) \geq K_2V_i(x) \tag{12}$$

where

$$K_1 = \max\ [1,\exp\omega_i T]$$

$$K_2 = \min\ [1,\exp\gamma_i T]$$

Since as previously we have

$$\omega_{i+1} \leq \omega_i$$

$$\gamma_{i+1} \geq \gamma_i \tag{13}$$

the estimations (11), (12) remain valid for all i's larger than this for which they were found.

Section II of the Appendix contains the proof that the time derivative of $\Lambda_i(p(t,x))$ is uniformly bounded for all i. We denote

$$\sup_i \max_{t\in[0,T]} \left|\frac{d\Lambda_i}{dt}\right| = b \tag{14}$$

In Section III of the Appendix the estimation of the upper bound of the function $\Lambda_{i+1}(x)$ has been derived. We have at the point x at which the maximum value of Λ_i is attained

$$\Lambda_{i+1}(x) \leq \Lambda_i(x) - K(\Lambda_i(x) - \gamma_i)^2 \tag{15}$$

where K is some positive constant independent of i. To estimate the upper bound of the function $\Lambda_{i+1}(p(t,x))$ at the neighborhood of the point x we utilize the estimation (14) of the derivative, obtaining

$$\Lambda_{i+1}(p(\varphi,x)) \leq \Lambda_{i+1}(x) + |b\varphi| \tag{16}$$

V. THEOREM

We close the discussion of properties of the functions $\Lambda_i(x)$, i =1,2,... along an arbitrary solution p(t,x) by proving the respective theorem. We assume definitions of the system (1) and of the function V_1 as in Sections 1 and 4. We introduce the set Z_2 requiring that for $x \in Z_2$ the solution exists for the interval [0,2T].

Theorem. Let the system (1), function V_1, a set Z_2 and number T be given. For any $\epsilon > 0$ there is an i^* such that for all $i \geq i^*$ and $x \in Z_2$ the relation

$$\max_{t\in[0,T]} \Lambda_i(p(t,x)) - \min_{t\in[0,T]} \Lambda_i(p(t,x)) < \epsilon$$

holds.

Proof. Assume contrary, i.e., that it is possible to select the infinite, increasing sequence $\{j\}$, $j \geq i^*$ and corresponding sequences of instants $\{t'_j\}$, $\{t''_j\}$ and sequence $\{x_j\}$, $x_j \in Z_2$ such that for each j we have

$$\Lambda_j(p(t'_j, x_j)) - \Lambda_j(p(t''_j, x_j)) \geq \epsilon \tag{17}$$

Denote by τ_j the instant such, that

$$\max_{t \in [0,T]} \Lambda_j(p(t, x_j)) = \Lambda_j(p(\tau_j, x_j))$$

and

$$x_j^* = p(\tau_j, x_j)$$

Then the relation (17) implies that

$$\max_{t \in [0,T]} \Lambda_j(p(t, x_j)) - \Lambda\ (p(t''_j, x_j)) \geq \epsilon$$

i.e.

$$\Lambda_j(p(\tau_j, x_j)) - \Lambda_j(p(t''_j, x_j)) \geq \epsilon$$

or

$$\Lambda_j(x_j^*) - \Lambda_j(p(t''_j, x_j)) \geq \epsilon$$

Separate the sequence $\{j\}$ onto two subsequences $\{k\}$, $\{\ell\}$ such that for any number out of the sequence $\{k\}$ we have

$$\tau_k < t''_k$$

and for any number out of the sequence $\{\ell\}$ we have

$$\tau_\ell > t''_\ell$$

At least one of the sequences $\{k\}$, $\{\ell\}$ must be infinite. The proof may be done separately for each sequence. For the sake of simplicity we prove the theorem only for the case of the sequence $\{k\}$, it means that we assume further that $\tau_j < t''_j$ for all j's. The proof for the case of the sequence $\{\ell\}$ results from the proof for $\{k\}$ by reversing the time, namely introducing $x'_j = p(T, x_j)$ and $t_1 = -t$.

The function $\Lambda_j(p(t,x_j))$ is defined on Z_2, i.e. for $t \in [0,2T]$. Therefore the function $\Lambda_j(p(t,x_j^*))$ while considered for $t \in [0,T]$ is well defined.

We shall utilize further the fact that $\Lambda_j(x)$ is lower bounded on Z_2. Denote the lower bound by L. On the basis of (8) taking into account that T is an arbitrary positive number we may state that

$$\Lambda_j(x) \geq L \quad \text{for all } j \geq i^* , \; x \in Z_2$$

Use the estimation (15) inserting

$$\Lambda_j(x_j^*) - \gamma_j = \epsilon$$

We obtain

$$\Lambda_{j+1}(x_j^*) \leq \Lambda_j(x_j^*) - K\epsilon^2$$

Adding $|b\varphi|$ to both sides of the inequality yields

$$\Lambda_{j+1}(x_j^*) + |b\varphi| \leq \Lambda_j(x_j^*) - K\epsilon^2 + |b\varphi|$$

Utilizing (16) we get

$$\Lambda_{j+1}(p(\varphi,x_j^*)) \leq \Lambda_j(x_j^*) - K\epsilon^2 + |b\varphi|$$

Therefore if

$$|\varphi| \leq \frac{K\epsilon^2}{2b}$$

we have finally

$$\Lambda_{j+1}(p(\varphi,x_j^*)) \le \Lambda_j(x_j^*) - \frac{K\epsilon^2}{2}$$

This means that for any j and any point x_j such that (17) is satisfied, there is a finite, independent of j, interval of time $[\tau_j - \varphi, \tau_j + \varphi]$, with τ_j being the middle point of the interval, such that the value of $\Lambda_{j+1}(p(t,x_j))$ drops on this interval through the finite, independent of j, value for each step.

The value of the function

$$\Lambda_{j+1}(p(t,x_j)) - L \tag{18}$$

defined for $t \in [0,T]$ drops correspondingly. Therefore after finite number of steps out of the sequence $\{j\}$ the function (18) attains a negative value. But (18) must be non-negative which shows contradiction and proves the theorem.

The theorem leads to the following:

Corollary. Consider $x \in Z_2$ and the solution $p(t,x)$ for $t \in [0,T]$. For each fixed x, the sequence of functions of time $\{\Lambda_i(p(t,x))\}$ has a limit $\Lambda_{lim}(p(t,x))$ such that

$$\Lambda_{lim}(p(t,x)) = C \text{ for } t \in [0,T]$$

Remark. The constant C depends on the selection of x; $C = C(x)$.

Properties of the Sequence $\{V_i(x)\}$

Select $\epsilon > 0$ and the sufficiently large i such that

$$C - \epsilon \le \Lambda_i(p(t,x)) \le C + \epsilon \quad \text{for} \quad t \in [0,T]$$

Then

$$\begin{aligned} V_i(x)\exp(Ct) - V_i(x)\exp(\epsilon t) &\le V_i(p(t,x)) \\ &\le V_i(x)\exp(Ct) + V_i(x)\exp(\epsilon t) \end{aligned} \tag{19}$$

and $\epsilon \to 0$ while $i \to \infty$.

VI. COMPARING LYAPUNOV FUNCTIONS

Asking is it possible to compare different Lyapunov functions applied to the given system leads to the concept of the best Lyapunov function discussed in paper [9]. In this paper the class of Lyapunov functions were quadratic forms applied to linear system.

If the class of Lyapunov functions and systems is not limited, the fact that for some functions the performance measure is smaller than for the other does not necessarily mean that the first function is better than the other.

To compare two functions V_1^* and V_1^{**} we treat them as the generators of the sequences $\{V_i^*\}$, $\{V_i^{**}\}$. The functions will be compared by relating the limits C^*, C^{**} of the function Λ_i^*, Λ_i^{**}, see Corollary. Assume that the system considered has an asymptotically stable zero solution, and denote the attraction set of this solution by Z_3. Assume that for both functions the corresponding functions Λ_1^*, Λ_1^{**} are bounded on $Z_3\backslash 0$. Finally assume that

$$\inf_{i\in N}\ \inf_{x\in Z_3\backslash 0}\left(\frac{V_i^*}{V_i^{**}}\right) = K_1 > 0 \quad , \quad \sup_{i\in N}\ \sup_{x\in Z_3\backslash 0}\left(\frac{V_i^*}{V_i^{**}}\right) = K_2 < \infty$$

Under these assumptions the following Proposition holds.

Proposition. For any $x \in Z_3\backslash 0$ the constants C^*, C^{**} fulfill

$$C^* = C^{**}$$

Proof. Due to asymptotic stability the solution p(t,x) tends to zero and the recursive process may be introduced for any finite value of T. Assume contrary, namely that there is an $\epsilon > 0$ such that

$$C^* = C^{**} + \epsilon \tag{20}$$

Select

$$T = \frac{2}{\epsilon} \ln\left(\frac{2K_1}{K_2}\right)$$

and find $i = i^*$ such that for all $j \geq i^*$ we have for $t \in [0,T]$

$$C^* - \frac{\epsilon}{4} \leq \Lambda_j^*(p(t,x)) \leq C^* + \frac{\epsilon}{4}$$

$$C^{**} - \frac{\epsilon}{4} \leq \Lambda_j^{**}(p(t,x)) \leq C^{**} + \frac{\epsilon}{4}$$

Then utilizing the results of Sec. 11 and (20) we have for $t \in [0,T]$

$$V_i^*(x)\exp\left[\left(C^{**} + \frac{3\epsilon}{4}\right)t\right] \leq V_i^*(p(t,x)) \leq V_i^*(x)\exp\left[\left(C^{**} + \frac{5\epsilon}{4}\right)t\right]$$

$$V_i^{**}(x)\exp\left[\left(C^{**} - \frac{\epsilon}{4}\right)t\right] \leq V_i^{**}(p(t,x)) \leq V_i^{**}(x)\exp\left[\left(C^{**} + \frac{\epsilon}{4}\right)t\right]$$

and correspondingly

$$\frac{V_i^*(p(t,x))}{V_i^{**}(p(t,x))} \geq \frac{V_i^*(x)\exp\left[\left(C^{**} + \frac{3\epsilon}{4}\right)t\right]}{V_i^{**}(x)\exp\left[\left(C^{**} + \frac{\epsilon}{4}\right)t\right]} = \frac{V_i^*(x)}{V_i^{**}(x)} \exp\left(\frac{\epsilon}{2}\, t\right)$$

Using the constant K_2 and $t = T$ we have

$$\frac{V_i^*(p(t,x))}{V_i^{**}(p(t,x))} \geq K_2 \left(\frac{2K_1}{K_2}\right) = 2K_1$$

which contradicts the definition of K_1 and proves the Proposition.

Remark. It may be observed that if the asymptotic stability condition is fulfilled, any two recursive functions from the sequence $\{V_i\}$ satisfy the assumption of the Proposition.

Assume asymptotic stability of the system zero solution. Use the expressions (2) and (19) to relate V_i and V_{i+1}. Selecting an $x \in Z_3 \backslash 0$ we have

$$\frac{V_i(x)}{C} \{\exp[(C-\epsilon)T]-1\} \leq V_{i+1}(x) \leq \frac{V_i(x)}{C} \{\exp[(C+\epsilon)T]-1\}$$

For assumed asymptotic stability the constant C is negative. Selecting sufficiently large T we estimate

$$-\frac{V_i(x)}{C}(1-\epsilon_1) \leq V_{i+1}(x) \leq -\frac{V_i(x)}{C}(1+\epsilon_1)$$

where $\epsilon_1 > 0$ and $\epsilon_1 \to 0$ for $i \to \infty$. Therefore we have

$$-\frac{V_i(x)}{V_{i+1}(x)}(1+\epsilon_1) \leq C \leq -\frac{V_i(x)}{V_{i+1}(x)}(1-\epsilon_1) \tag{21}$$

The relation (21) implies the alternative procedure of asymptotic stability investigation, based on generating the sequence $\{V_i(x)\}$ and determining the sign of the ratio $V_i(x)/V_{i+1}(x)$.

VII. RECURSIVE FUNCTIONS FOR LINEAR SYSTEMS

Consider a linear stationary system

$$\dot{x} = Ax, \qquad x \in R^n \tag{22}$$

with the solution $p(t,x_0)$, $p(0,x_0) = x_0$. Together with (1) consider the quadratic Lyapunov function

$$V = x^T P x$$

where P is a symmetric, positive-definite matrix. We get

$$\dot{V}(x) = x^T(A^TP + PA)x$$

For $x \neq 0$ introduce the function

$$\Lambda(x) = \dot{V}(x)/V(x)$$

Due to the fact that the matrix P is positive-definite, the function $\Lambda(x)$ exists and is bounded for all $x \in R^n \backslash 0$. It attains its extrema on the set $R^n \backslash 0$.

Introducing the number

$$\sup_{x \in R^n \backslash 0} \Lambda(x) = \lambda$$

we get for $x_0 \neq 0$

$$V(p(t,x_0)) \leq V(x_0)\exp(\lambda t)$$

we call the number λ the performance measure of the function V.

A. Recursive Matrix Equation

Assume now that the system (22) is asymptotically stable. Then using the recursive Lyapunov function procedure we introduce the function V_{i+1} as the integral

$$V_{i+1}(x) = \int_0^T V_i(p(t,x))dt , \quad i = 1,2,... \tag{23}$$

The integral (23) is well defined and converges even when the upper integration limit T is infinite. Entering for the function V_i under the integral the corresponding quadratic form yields

$$V_{i+1}(x) = \int_0^T p^T(t,x)P_i p(t,x)dt$$

Integration of the above quadratic form, due to the fact that p(t,x) is the linear form of x, results in $V_{i+1}(x)$ being also the quadratic form with some positive-definite matrix P_i, i.e.

$$V_{i+1}(x) = x^T P_{i+1} x \tag{24}$$

In paper [7] the following relation was proven, relating the Lyapunov derivative $\dot{V}_{i+1}$ of the recursive function V_{i+1}, obtained by the formula (23), to the function V_i

$$\dot{V}_{i+1}(x) = V_i(p(T,x)) - V_i(x) \tag{25}$$

By virtue of (25), utilizing the fact that (1) is asymptotically stable, we have for $T = \infty$

$$\dot{V}_{i+1} = V_i(p(\infty,x)) - V_i(x) = -x^T P_i x$$

Differentiating V_{i+1} along the solutions of (1) we get

$$\dot{V}_{i+1} = x^T(A^T P_{i+1} + P_{i+1}A)x$$

for all x's and thus

$$A^T P_{i+1} + P_{i+1}A = -P_i \qquad i = 0,1,2,\ldots \tag{26}$$

This equation is a specific form of well known Lyapunov equation. It represents the recursive algorithm to determine the sequence of matrices P_i. The matrix P_o, introduced to generate P, is selected as positive-definite, usually as unit matrix. This and the fact that the system is asymptotically stable is enough [10] to ensure the positive-definiteness of all the matrices P_i, $i = 1,2,3,\ldots$. The equation was considered by Jury [11] for other purposes, namely for evaluating the integral square of signals.

Due to proven properties of the recursive Lyapunov function, while solving Eq. (26) at each step we obtain the Lyapunov function with the better performance measure

$$\lambda_{i+1} \leq \lambda_i \ .$$

B. Example

Consider 3D system $\dot{x} = Ax$ with the matrix

$$A = \begin{bmatrix} 0 & 1 & 0 \\ -1 & -3 & 1 \\ -2 & 0 & 0 \end{bmatrix}$$

The system eigenvalues are −0.0533557 ± 0.8298703i, −2.89335. Utilizing Eq. (26) the matrices $P_1, P_2, .., P_5$ have been found. Specifically

$$P_1 = \begin{bmatrix} 37.5000 & 11.5000 & -5.5000 \\ 11.5000 & 4.0000 & -.5000 \\ -5.5000 & -.5000 & 6.0000 \end{bmatrix}; \quad P_2 = \begin{bmatrix} 354.0001 & 111.2500 & -46.2500 \\ 111.2500 & 37.7500 & -3.0000 \\ -46.2500 & -3.0000 & 54.3750 \end{bmatrix}$$

The eigenvalues of matrices in sequence are listed in Table 1. As it is seen for i=5 the single precision procedure for linear equation solving is not good enough and the matrix P_5 is calculated as not being positive-definite.

Table 1.

Matrix	Three Eigenvalues		
$P_0 = I$	1	1	1
P_1	.172637	5.44583	41.8815
P_2	.0298653	50.9335	395.162
P_3	$.531006 \cdot 10^{-2}$	478.719	3692.95
P_4	$.488281 \cdot 10^{-3}$	4486.32	34602.1
P_5	−.015625	42041.4	324264

By checking the eigenvalues of the corresponding matrices the estimates have been found to be

$$\dot{V}_1 < -.02\ V_1$$

$$\dot{V}_2 < -.101\ V_2$$

$$\dot{V}_3 < -.1063\ V_3$$

The ratio of elements of consequent matrices $[P_i]_{k\ell}/[P_{i+1}]_{k\ell}$ converges quickly to the value 2(−0.0533557); for $i = 5$ this ratio varies from .1067105 to .1067115. However, during the iteration process the minimal eigenvalue of the matrix P_i quickly tends to zero (see Table 1), so that for P_5 it is less than round-up error and is shown as negative.

C. Properties of the Sequence $\{V_i(x)\}$ at the Limit

Consider properties of the functions $V_i(x)$ for large i's. With accuracy to arbitrarily small, positive ϵ we have

$$\Lambda_i(p(t,x)) = C_x$$

where the subscript x represents the fact that C_x depends on the solution. We then get

$$\dot{V}_i(p(t,x)) = C_x V_i(pH,x))$$

therefore

$$V_i(p(t,x)) = V_i(x)\exp(C_x t)$$

and

$$V_{i+1}(x) = \int_0^\infty V_i(p(t,x))dt = -(1/C_x)V_i(x)$$

or

$$V_i(x) = -C_x V_{i+1}(x)$$

and consequently

$$x^T(P_i + C_x P_{i+1})x = 0 \ .$$

D. Jordan Form of the System

For the sake of simplicity we consider the case where the real matrix A is simple, i.e. it is similar to a diagonal matrix Q composed from k real and 21 complex eigenvalues of A. We put the matrix Q into the block form

$$Q = \begin{bmatrix} Q^R & & & & \\ & Q^I_{k+1} & & & \\ & & \cdot & & \\ & & & \cdot & \\ & & & & \cdot \\ & & & & & Q^I_{k+\ell} \end{bmatrix}$$

where Q^R is the kxk diagonal matrix of real eigenvalues and

$$Q^I_j = \begin{bmatrix} \mu_j + i\nu_j & 0 \\ 0 & \mu_j - i\nu_j \end{bmatrix} \quad j = k+1, \ldots, k+\ell$$

contains the pair of complex conjugate eigenvalues.

Introduce the block matrix B containing the kxk identity matrix I_k and the sequence of ℓ matrices B_2,

$$B = \begin{bmatrix} I_k & & & & \\ & B_2 & & & \\ & & \cdot & & \\ & & & \cdot & \\ & & & & \cdot \\ & & & & & B_2 \end{bmatrix} \qquad \text{where } B_2 = \begin{bmatrix} 1/2 & 1/2 \\ -i/2 & i/2 \end{bmatrix}$$

It is easy to see that the product

$$M = BQB^{-1}$$

is the real matrix of the form

$$M = \begin{bmatrix} Q^R & & & & \\ & M_2^{(k+1)} & & & \\ & & \cdot & & \\ & & & \cdot & \\ & & & \cdot & M_2^{(k+\ell)} \end{bmatrix}$$

where

$$M_2^{(j)} = \begin{bmatrix} \mu_j & -\nu_j \\ \nu_j & \mu_j \end{bmatrix} \qquad j = k+1,\ldots,k+\ell$$

Observe that $M_2^{(j)}$ may be decomposed onto symmetric diagonal matrix $M_{2d}^{(j)}$ and skew–symmetric matrix $M_{2s}^{(j)}$. As it is known [12] the matrices V,U of correspondingly left and right eigenvectors of the matrix A can be normalized in such a manner that

$$V^T AU = Q \quad , \quad V^T = U^{-1} \tag{27}$$

But then utilizing (27) we have

$$M = BV^T AUB^{-1}$$

Therefore the transformation

$$y = BV^T x = E^x \quad , \quad \text{where } E \triangleq BV^T \tag{28}$$

transforms Eq. (22) to the equation

$$\dot{y} = My$$

It is easy to see that the matrix E is real.

E. Solution to Equation (26)

Denote

$$R^{(i)} = (E^T)^{-1}P_iE^{-1} \tag{29}$$

Then Eq. (26) transforms to the form

$$M^TR^{(i+1)} + R^{(i+1)}M = -R^{(i)} \tag{30}$$

where assuming that $R^{(i)}$ is symmetric we look for the symmetric matrix $R^{(i+1)}$. We consider here the case of real eigenvalues, i.e., the case when

$$M = Q^R$$

Inserting this expression into (30) yields

$$(Q^R)^TR^{(i+1)} + R^{(i+1)}Q^R = -R^{(i)}$$

giving

$$R_{jk}^{(i+1)} = -\frac{1}{\mu_j+\mu_k} R_{jk}^{(i)} \tag{31}$$

and finally using (29)

$$P_{i+1} = E^TR^{(i+1)}E$$

Formula (31) allows to explain why during iteration the smallest eigenvalue of P_i tends to zero. For example take $\mu_2 < \mu_1 < -1$. We have

$$\frac{R_{11}^{(0)}}{R_{11}^{(i)}} = \frac{1}{(-2\mu_1)^i}$$

and

$$\frac{R_{22}^{(0)}}{R_{22}^{(i)}} = \frac{1}{(-2\mu_2)^i}$$

Thus when $i \to \infty$ the element $R_{22}^{(i)}$ tends to zero quicker than the element $R_{11}^{(i)}$. It can be shown that such situation results in the matrix $R^{(i)}$ tending to become semi-definite when $i \to \infty$. Also it explains why the ratio of elements in the example tends to achieve the doubled value of the largest real part of the system eigenvalue, $\lambda = 2\mu$.

F. Applications

The recursive procedure applied to a design of state-variable feedback is the subject of the current interest of the author. The sequence of controls $u^{(i)}$ is considered with the constraints imposed on the length of the vector $u^{(i)}$. Each vector $u^{(i)}$ is designed to improve the performance measure. The convergence of the sequence of its limit vector $u^* = K^*x$ is expected.

The importance of the problem may be illustrated by reconsidering the Section B example, namely the 3-D systems

$$\dot{x} = Ax$$

with the matrix A defined previously. The iteration of the matrix P_i was terminated at the fifth step, when the ratio of elements of the consequent matrices converged close to the doubled value of the largest real part of the system eigenvalues. The matrix P_5 has the form

$$P_5 = 10^3 \times \begin{bmatrix} 290.4688 & 91.33466 & -37.91838 \\ 91.33466 & 30.99615 & -2.392497 \\ -37.91838 & -2.392497 & 44.84047 \end{bmatrix}$$

Consider the state-variable feedback applied to reconsidered example, i.e., the system

$$\dot{x} = Ax + Bu$$

where 3 × 3 matrix B is the identity matrix. Select the gain matrix $K = -\alpha B^T P_5$, obtaining the controller

$$u = -\alpha B^T P_5 x \tag{32}$$

Table 2 illustrates interesting properties of this controller. As may be seen, the controller (32) executes the shift of the system dominant eigenvalue to the left without significant change of its imaginary part, and without affecting the remaining eigenvalue. This pole placement result seems to be new and it may find important applications.

VIII. CONCLUSIONS

The paper creates the theoretical basis for applications of recursive Lyapunov functions.

The proposed algorithm is not the only one possible; the alternate convergent algorithms should be reviewed.

In the field of linear systems the Lyapunov equation seems to be a sufficient tool for Lyapunov function design. The author thinks that the recursive procedure may be used to establish the bound k of the disturbance e(x,t)

Table 2

Value of α	Eigenvalues of the Matrix $A-\alpha BB^TP_5$
0	−0.0533557 ± 0.829703i −2.89329
1×10^{-7}	−0.0716708 ± 0.829584i −2.89329
2×10^{-7}	−0.089986 ± 0.829223i −2.89329
3×10^{-7}	−0.108301 ± 0.828623i −2.89329
4×10^{-7}	−0.126616 ± 0.827781i −2.89329
5×10^{-7}	−0.144932 + 0.826698i −2.89329
6×10^{-7}	−0.163247 ± 0.825372i −2.8939
7×10^{-7}	−0.181562 ± 0.823803i −2.89329
8×10^{-7}	−0.198878 ± 0.821988i −2.89329
9×10^{-7}	−0.218193 ± 0.819926i −2.89329
10×10^{-7}	−0.236508 ± 0.817616i −2.89329

$$\|e(x,t)\| \le k\|x\|$$

such that the disturbed system remains asymptotically stable.
It seems that number of applications may be found, such as generation of Lyapunov functions for local and global asymptotic stability, generation of the best Lyapunov function, linear and nonlinear system time constant estimation etc.

The pole placement problem, particularly shifting of the dominant pole to the left has been demonstrated to be possible by applying recursive procedures.

In the field of nonlinear systems the author thinks that the procedure may be particularly useful for polynomial systems. The procedure can provide the Lyapunov functions for local and global asymptotic stability investigation, and also for Lypaunov-based controller design.

APPENDIX

I. PROOF OF LEMMA 2

Lemma 2. Let $a_i, b_i \in R$, $b_i > 0$, $i = 1,\ldots,n$. Then the inequality holds

$$\frac{\sum_{i=1}^{n} a_i}{\sum_{i=1}^{n} b_i} \geq \min_i \frac{a_i}{b_i} \tag{A1}$$

Moreover, if for some j's we have

$$\frac{a_j}{b_j} > \min_i \frac{a_i}{b_i} = \frac{a_i^* x}{b_i^*} \tag{A2}$$

then the inequality is a strong one.

Proof.

(i) If $\min_i \frac{a_i}{b_i} = 0$, then $a_i \geq 0$ for $i = 1,\ldots,n$ and $\sum a_i \geq 0$.

Thus (A1) holds.

(ii) If $\min\limits_{i} \frac{a_i}{b_i} = \frac{a_{i*}}{b_{i*}} < 0$, then $a_{i*} < 0$. We have

$$\frac{a_k}{b_k} \geq \frac{a_{i*}}{b_{i*}} \qquad k = 1,\ldots,n$$

i.e.,

$$\frac{a_k}{a_{i*}} \leq \frac{b_k}{b_{k*}} \qquad k = 1,\ldots,n$$

and

$$\sum_{k=1}^{n} \frac{a_k}{a_{i*}} \leq \sum_{k=1}^{n} \frac{b_k}{b_{i*}} \; .$$

Consequently, since both sides of the last inequality are positive we have

$$\frac{\sum\limits_{k=1}^{n} a_i}{\sum\limits_{k=1}^{n} b_i} = \frac{a_{i*}\left(\sum\limits_{i=1}^{n} \frac{a_i}{a_{i*}}\right)}{b_{i*}\left(\sum\limits_{i=1}^{n} \frac{b_i}{b_{i*}}\right)} \geq \frac{a_{i*}}{b_{i*}} = \min_{i} \frac{a_i}{b_i} \; .$$

(iii) if $\min\limits_{i} \frac{a_i}{b_i} = \frac{a_{i*}}{b_{i*}} > 0$, then $a_{i*} > 0$, and all $a_i > 0$. We have

$$\frac{a_k}{b_k} \geq \frac{a_{i*}}{b_{i*}} \qquad k = 1,\ldots,n$$

i.e.

$$\frac{a_k}{a_{i*}} \geq \frac{b_k}{b_{i*}} \qquad k = 1,\ldots,n$$

and

$$\sum_{k=1}^{n} \frac{a_k}{a_{i*}} \geq \sum_{k=1}^{n} \frac{b_k}{b_{i*}}$$

Consequently

$$\frac{\sum_{k=1}^{n} a_i}{\sum_{k=1}^{n} b_i} = \frac{a_{i*} \left(\sum_{i=1}^{n} \frac{a_i}{a_{i*}} \right)}{b_{i*} \left(\sum_{i=1}^{n} \frac{b_i}{b_{i*}} \right)} \geq \frac{a_{i*}}{b_{i*}} = \min_i \frac{a_i}{b_i} .$$

which ends the proof that (A1) holds.

We prove now that if (A2) holds then

$$\frac{\sum a_i}{\sum b_i} > \frac{a_{i*}}{b_{i*}} .$$

For the sake of simplicity, consider the case when there is only one j such that

$$\frac{a_j}{b_j} > \frac{a_i^*}{b_i^*} .$$

Denote

$$a_j^* = b \frac{a_i^*}{b_i^*} .$$

Then

$$a_{j*} - a_j < 0 .$$

Denote

$$a_{j*} - a_j = \xi < 0 \ .$$

Then

$$a_j = a_{j*} + \xi \ .$$

The initial sum takes the form and fulfills

$$\frac{\sum a_i}{\sum b_i} = \frac{a_1 + \ldots + a_j^* + \ldots + a_n - \xi}{\sum b_i} \geq \frac{a_i^*}{b_i^*} - \frac{\xi}{\sum b_i} > \frac{a_i^*}{b_i^*}$$

which ends the proof.

II. TIME DERIVATIVE OF $\Lambda(p(t,x))$

We prove that the time derivative of $\Lambda_i(p(t,x))$ is uniformly bounded for all i. We have

$$\frac{d\Lambda_i(p(t,x))}{dt} = \frac{\ddot{V}_i(p(t,x))}{V_i(p(t,x))} - \frac{\dot{V}_i(p(t,x))}{V_i(p(t,x))}$$

The first term of the right-hand-side of the above expression, by virtue of (4), fulfills the relation

$$\frac{\ddot{V}_i(p(t,x))}{V_i(p(t,x))} = \frac{\int_0^T \ddot{V}_{i-1}(p(t,x))dt}{\int_0^T V_{i-1}(p(t,x))dt}$$

Following the previous reasoning it is easy to get

$$\frac{\ddot{V}_i(P(t,x))}{V_i(p(t,x))} \leq \max_{t\in[0,T]} \frac{\ddot{V}_{i-1}(p(t,x))}{V_{i-1}(p(t,x))}$$

Since the second term of the right-hand-side is bounded we conclude that $d\Lambda_i/dt$ is uniformly bounded for all i.

III. ESTIMATION OF THE UPPER BOUND OF $\Lambda_{i+1}(x)$

We have

$$\Lambda_{i+1}(x) = \frac{\int_0^T \dot{V}_i(p(t,x))dt}{\int_0^T V_i(p(t,x))dt} \tag{A3}$$

Considering the function $\Lambda_i(p(t,x))$ denote the instant at which it attains its minimal value γ on $[0,T]$ as t'. Taking into account that the derivative $d\Lambda_i/dt$ is bounded by the number b we get

$$\Lambda_i(p(t,x)) \le \begin{cases} \gamma_i + b(t'-t) & \text{for} \quad t \in \left[t' - \frac{\omega_i - \gamma_i}{b}, t'\right] \overset{\Delta}{=} I_2 \\ \gamma_i - b(t'-t) & \text{for} \quad t \in \left[t', t' + \frac{\omega_i - \gamma_i}{b}\right] \overset{\Delta}{=} I_2 \\ \omega_i & \text{for} \quad t \in I \backslash [I_1 U I_2] \end{cases}$$

Correspondingly the following estimations hold

$$\dot{V}_i(p(t,x) \le [\gamma_i + b(t'-t)]V_i(p(t,x)) \qquad \text{for } t \in I_1$$

$$\dot{V}_i(p(t,x) \le [\gamma_i - b(t'-t)]V_i(p(t,x)) \qquad \text{for } t \in I_2$$

$$\dot{V}_i(p(t,x)) \le \omega_i V_i(p(t,x)) \qquad \text{for } t \in I \backslash [I, U I_2]$$

Then the integral $\int_0^T \dot{V}_i(p(t,x))dt$ may be estimated as the following

$$\int_0^T \dot{V}_i(p(t,x))dt \le \omega_i \int_0^T V_i(p(t,x)dt$$

$$+ \int_{t' + \frac{\omega_i - \gamma_i}{b}}^{t'} [\gamma_i - \omega_i + b(t'-t)]V_i(p(t,x))dt$$

$$+ \int_{t'}^{t' + \frac{\omega_i - \gamma_i}{b}} [\gamma_i - \omega_i - b(t'-t)]V_i(p(t,x))dt$$

Integrating and using the estimation (12) we get

$$\int_0^T \dot{V}_i(p(t,x))dt \le \omega_i \int_0^T V_i(p(t,x))dt - \frac{(\omega_i - \gamma_i)^2}{b} K_2 V_i(x)$$

Utilizing this expression to estimate $\Lambda_{i+1}(x)$ at (A3) we have

$$\Lambda_{i+1}(x) \le \omega_i - \frac{(\omega_i - \gamma_i)^2 K_2 V_i(x)}{b \int_0^T V_i(p(t,x))dt} \le \omega_i - \frac{(\omega_i - \gamma_i)^2 K_2}{bTK_1}$$

If the point x is the point at which the maximum of $\Lambda_i(x)$ is attained then we get

$$\Lambda_{i+1}(x) \le \Lambda_i(x) - \frac{(\Lambda_i(x) - \gamma_i)^2 K_2}{bTK_1}$$

Since all the constants b, T, K_1, K_2 once established for some $i=i^*$ remain solid we replace them by the constant $K > 0$, also independent of i and state that for all $i \ge i^*$

$$\Lambda_{i+1}(x) \le \Lambda_i(x) - K(\Lambda_i(x) - \gamma_i)^2 \qquad \text{(A4)}$$

REFERENCES

1. W. Hahn, "Stability of Motion," Springer-Verlag, New York, 1967.

2. R.K. Brayton and C.H. Tong, "Stability of Dynamical Systems: A Constructive Approach," *IEEE Trans. Circ. Syst.*, Vol. 26, No. 4, 224–234, 1979.
3. R.K. Brayton and C.H. Tong, "Constructive Stability and Asymptotic Stability of Dynamical Systems," *IEEE Trans. Circ. Syst.*, Vol. 27, No. 11, 1121–1130, 1980.
4. H. Leipholz, "Stability of Elastic Systems," Sijthoff et Noordhoff, Alphen aan den Rijn, The Netherlands, 1980, 85–88.
5. A. Olas, "A Criterion of Stability of the Trivial Solution of a Set of Ordinary Differential Equations," *Bull. de l'Acad. Polon. des Sci.*, Serie des Sci. Tech., Vol. 29, No. 5–6, 67–73, 1981.
6. A. Vannelli and M. Vidyasagar, "Maximal Lyapunov Function and Domains of Attraction for Autonomous Nonlinear Systems," *Automatica*, Vol. 21, No. 1, 69–80, 1985.
7. A. Olas, "Recursive Lyapunov Functions, Workshop on Control Mechanics," *USC*, January 1988.
8. K. Ogata, "Modern Control Engineering," Prentice-Hall, 1970.
9. B. Radziszewski, "On the Best Lyapunov Function," IFTR Reports, Polish Academy of Sciences, Warsaw, 1977.
10. M. Vidyasagar, "Nonlinear Systems Analysis," Prentice-Hall, 1978.
11. E.I. Jury, "Inners and Stability of Dynamic Systems," Robert E. Krieger Publ. Co., Malabar, 1982.
12. L. Meirovitch, "Elements of Vibration Analysis," McGraw-Hill, New York, 1986.

LYAPUNOV FUNCTIONAL APPROACH TO UNCERTAIN SYSTEMS GOVERNED BY FUNCTIONAL DIFFERENTIAL EQUATIONS WITH FINITE TIME–LAG

JONATHAN E. GAYEK[1]

Department of Mathematics
Trinity University
San Antonio, TX 78284 USA

I. INTRODUCTION

The design of controllers which lead to uniformly ultimately bounded solutions of systems governed by ordinary differential equations has been of continued interest to researchers since it was originally considered by Leitmann [1] – [3] and Corless and Leitmann [4]. Variations of the basic problem, where a set of matching conditions are satisfied, have been considered by Corless and Manela [5] and Magaña and Zak [6] for discrete-time problems, and by Thowsen [7] for models governed by retarded functional differential equations with finite time-lag (i.e., systems where the rate of change depends on the recent past). In each of these settings the determination of the uniform ultimate bound is arrived at using Lyapunov functions. There is, however, a peculiarity in doing this in all three settings, namely that in continuous-time and discrete-time problems Lyapunov functions employ the state of the system, whereas in the retarded functional differential equation case, Lyapunov functions do not use the state of the system. If the state is used in this case, then the proper formulation requires the use of Lyapunov *functionals* (see [8] for a contrast between Lyapunov functions and Lyapunov functionals in stability considerations of functional differential equations).

[1]Current mailing address: The Aerospace Corporation, Mail Station M4/971, P.O. Box 92957, Los Angeles, CA 90009-2957, USA.

In this paper we will use a modified version of a theorem by Burton and Zhang [9] to do a full state space analysis of nonlinear uncertain retarded differential equations with finite time-lag subject to matching conditions. The resulting Lyapunov functional approach is used to verify that the controller proposed by Thowsen [7] leads to uniformly ultimately bounded solutions. However, unlike the analysis found in [7], the ultimate bound is shown to be dependent on delay-size, an observation which can lead to tighter estimates of the ultimate bound in the case of small delay.

The paper is organized as follows. Section II contains a discussion of relevant terminology and the basic assumptions to be made throughout the paper (one of the assumptions being a set of matching conditions). In Section III is found the main result relating the use of Lyapunov functionals to verify the existence of controllers leading to uniformly ultimately bounded solutions. An example of the main result is presented in Section IV, one which indicates the state space approach advocated can have advantages over the Lyapunov function procedure. A summary of the paper is contained in Section V.

II. Preliminaries

Let $x \in \mathbf{R}^n$, let x' denote the transpose of x, and let $|x|$ denote a vector norm. Let $r > 0$ and let C denote the set of continuous functions $\phi : [-r, 0] \to \mathbf{R}^n$ with $\|\phi\| = \sup_{-r \leq s \leq 0} |\phi(s)|$, a function norm. For $\tau > 0$ and $x : [-r, \tau] \to \mathbf{R}^n$ continuous, then for fixed $t \in [0, \tau]$ we let x_t represent $x(\sigma)$ restricted to the interval $t - r \leq \sigma \leq t$ and shifted to the interval $[-r, 0]$; i.e., $x_t(s) = x(t + s)$ where $s \in [-r, 0]$. If $Z : \mathbf{R} \times C \to \mathbf{R}^{k \times k}$ is a continuous, real matrix and $|\cdot|_m$ is the matrix norm induced by $|\cdot|$, then $\|Z\|_m = \sup_{-r \leq s \leq 0} |Z(t, \phi(s))|_m$.

Consider the uncertain retarded functional differential equation

$$\dot{x}(t) = f(t, x_t) + \Delta f(t, x_t) + \left[B(t, x_t) + \Delta B(t, x_t)\right] u(t, x_t) + \Delta C(t, x_t) v(t) \quad (1)$$

where $t \geq 0, x, f, \Delta f \in \mathbf{R}^n$, $u \in \mathbf{R}^m$, $v \in \mathbf{R}^l$, B and ΔB are $n \times m$ matrices, and ΔC is a $n \times l$ matrix. In addition the right-hand side of Eq. 1 is continuous, locally Lipschitz in x_t, and quasibounded [10]. This is enough to guarantee the local existence and uniqueness of solutions to Eq. 1 from an initial condition $\phi \in C$. In posing Eq. 1 we assume that the exact form of each of the quantities $\Delta f(t, x_t)$, $\Delta B(t, x_t)$, and $\Delta C(t, x_t)v(t)$ is unknown but they satisfy the following *matching conditions*.

Assumption 1. There exists matrices $D(t, x_t)$, $E(t, x_t)$ and $F(t, x_t)$ of appropriate dimension such that

$$\Delta f(t, x_t) = B(t, x_t)D(t, x_t) \tag{2}$$

$$\Delta B(t, x_t) = B(t, x_t)E(t, x_t) \tag{3}$$

$$\Delta C(t, x_t) = B(t, x_t)F(t, x_t). \tag{4}$$

Furthermore, bounds on $\|D\|$, $\|E\|_m$, and $\|Fv\|$ are known and $\sup \|E\|_m < 1$.

Thowsen [7] has considered the uniform ultimate boundedness of Eqs. (1)—(4) through a Lyapunov function analysis (e.g., a Razumikhin method, see [8]). In making the following assumption we begin the analysis of the same system employing Lyapunov *functional* arguments.

Assumption 2. There exists continuous, strictly increasing functions u, v_0, v_1, $w : \mathbf{R}^+ \to \mathbf{R}^+$ with $u(0) = v_0(0) = v_1(0) = w(0) = 0$ and $u(s) \to \infty$ as $s \to \infty$; there exists a differentiable positive definite function $\nu(t, y(t))$ and a nonnegative functional $\eta(t, y_t)$ which is continuous and differentiable with respect to t and such that $\eta(t, 0) = 0$; and there exists a continuous $g(t, y_t) : \mathbf{R} \times C \to \mathbf{R}^m$ such that for any $y_t \in C$

$$\text{i)} \quad u\big(|y(t)|\big) \leq V(t, y_t) \leq v_0\big(|y(t)|\big) + v_1\left(\int_{t-r}^{t} w\big(|y(s)|\big)\, ds\right) \tag{5}$$

and

$$\text{ii)} \quad \frac{\partial \nu}{\partial t}(t, y) + \nabla_y' \nu(t, y)\big[f(t, y_t) + B(t, y_t)g(t, y_t)\big]$$

$$+ \dot{\eta}(t, y_t) \leq -w\big(|y(t)|\big) \tag{6}$$

where $V(t, y_t) = \nu(t, y(t)) + \eta(t, y_t)$, $\dot{\eta}(t, y_t)$ is the total derivative with respect to t, and $\nabla'_y \nu(t, y(t))$ is the transpose of the gradient of ν with respect to the vector y.

From Eq. 5 it follows that

$$u(|y(t)|) \leq V(t, y_t) \leq v_0(\|y_t\|) + v_1(rw(\|y_t\|)) = v(\|y_t\|) \tag{7}$$

where v is a continuous, strictly increasing function with $v : \mathbf{R}^+ \to \mathbf{R}^+$ and $v(0) = 0$. Furthermore, if we consider the nominal system of Eq. 1, i.e.,

$$\dot{y}(t) = f(t, y_t) + B(t, y_t)u(t, y_t) \tag{8}$$

and set $u(t, y_t) = g(t, y_t)$, then time derivatives of V along solutions of Eq. 8 satisfy

$$\dot{V}_{(8)}(t, y_t) = \frac{\partial \nu}{\partial t}(t, y) + \nabla'_y \nu(t, y)\left[f(t, y_t) + B(t, y_t)g(t, y_t)\right] + \dot{\eta}(t, y_t)$$

$$\leq -w(|y(t)|). \tag{9}$$

Thus Eq. 7 and Eq. 9 indicate that under the control law $u = g(t, y_t)$ the nominal system Eq. 8 is uniformly asymptotically stable to $y_t = 0$ ([8] and [10]).

III. Main Result

THEOREM. *Let* $q(t, x_t) = \|D(t, x_t)\| + \|E(t, x_t)g(t, x_t)\| + \|F(t, x_t)v(t)\|$. *Let* σ, δ, ρ *be given positive constants. Consider Eq. 1 subject to Assumptions 1 and 2, and the feedback control law* $u(t, x_t) = g(t, x_t) + h(t, x_t)$, *where*

$$h(t, x_t) = \begin{cases} \frac{-B'\nabla_x \nu q}{\|B'\nabla_x \nu\|(1-\max\|E\|_m)}, & \text{if } \|B'\nabla_x \nu q\| > \sigma \\ \frac{-B'\nabla_x \nu q^2}{\sigma(1-\max\|E\|_m)}, & \text{if } \|B'\nabla_x \nu q\| \leq \sigma. \end{cases} \tag{10}$$

Then solutions to Eq. 1 for any initial function $\phi \in C$ with $\|\phi\| < \beta$ are uniformly bounded and uniformly ultimately bounded with ultimate bound

$$\varepsilon = u^{-1}\left[v_0\left(w_0^{-1}\left(\delta + \frac{\sigma}{4}\right)\right) + v_1\left(\frac{(\rho + \sigma r)}{4}\right) + \rho\right]. \tag{11}$$

PROOF. Under Assumption 1 system Eq. 1 becomes

$$\begin{aligned}
\dot{x}(t) &= f(t, x_t) + B(t, x_t)\big[g(t, x_t) + h(t, x_t)\big] \\
&\quad + B(t, x_t)\big\{D(t, x_t) + E(t, x_t)\big[g(t, x_t) + h(t, x_t)\big] + F(t, x_t)v(t)\big\} \\
&= f(t, x_t) + B(t, x_t)g(t, x_t) + B(t, x_t)\big[I + E(t, x_t)\big]h(t, x_t) \\
&\quad + B(t, x_t)e(t, x_t)
\end{aligned} \tag{12}$$

where I is the $m \times m$ identity matrix and

$$e(t, x_t) = D(t, x_t) + E(t, x_t)g(t, x_t) + F(t, x_t)v(t) \tag{13}$$

with $\|e(t, x_t)\| \leq q(t, x_t)$. Let $V(t, x_t) = \nu\big(t, x(t)\big) + \eta(t, x_t)$ be the functional in Assumption 2. Differentiating V with respect to time along solutions of Eq. 12 leads to

$$\begin{aligned}
\dot{V}_{(12)}(t, x_t) &= \left\{\frac{\partial \nu}{\partial t}\big(t, x(t)\big) + \nabla_x' \nu\big(t, x(t)\big)\left[f(t, x_t) + B(t, x_t)g(t, x_t)\right]\right. \\
&\quad \left. + \dot{\eta}(t, x_t)\right\} \\
&\quad + \nabla_x' \nu\big(t, x(t)\big)B(t, x_t)\big\{\big[I + E(t, x_t)\big]h(t, x_t) + e(t, x_t)\big\} \\
&\leq -w\big(|x(t)|\big) \\
&\quad + \nabla_x' \nu\big(t, x(t)\big)B(t, x_t)\big\{\big[I + E(t, x_t)\big]h(t, x_t) + e(t, x_t)\big\}
\end{aligned} \tag{14}$$

by virtue of Eq. 6 of Assumption 2. If $\|B'(t,x_t)\nabla_x\nu(t,x(t))q(t,x_t)\| > \sigma$ we have upon substitution for $h(t,x_t)$

$$\dot{V}_{(12)}(t,x_t) \leq -w(|x|) - \frac{\|B'\nabla_x\nu\|q}{1-\max\|E\|_m}$$

$$- \frac{\nabla'_x\nu BEB'\nabla_x\nu q^2}{(1-\max\|E\|_m)\|B'\nabla_x\nu q\|} + \nabla'_x\nu Be$$

$$\leq -w(|x|) - \frac{\|B'\nabla_x\nu q\|}{1-\max\|E\|_m} + \frac{\|E\|\cdot\|B'\nabla_x\nu q\|}{1-\max\|E\|_m} + \|B'\nabla_x\nu q\|$$

$$\leq -w(|x|) + \frac{\|B'\nabla_x\nu q\|}{1-\max\|E\|_m}(\|E\|_m - \max\|E\|_m)$$

$$\leq -w(|x|). \tag{15}$$

However, if $\|B'(t,x_t)\nabla_x\nu(t,x(t))q(t,x_t)\| \leq \sigma$ we have

$$\dot{V}_{(12)}(t,x_t) \leq -w(|x|) - \frac{\|B'\nabla_x\nu q\|^2}{\sigma(1-\max\|E\|_m)} - \frac{\nabla'_x\nu BEB'\nabla_x\nu q^2}{\sigma(1-\max\|E\|_m)} + \nabla'_x\nu Be$$

$$\leq -w(|x|) - \frac{\|B'\nabla_x\nu q\|^2}{\sigma(1-\max\|E\|_m)} + \frac{\|B'\nabla_x\nu q\|^2\cdot\max\|E\|_m}{\sigma(1-\max\|E\|_m)}$$

$$+ \|B'\nabla_x\nu q\|$$

$$= -w(|x|) - \frac{\|B'\nabla_x\nu q\|^2}{\sigma} + \|B'\nabla_x\nu q\|$$

$$\leq -w(|x|) + \sigma/4. \tag{16}$$

Thus, under the control law $u(t,x_t) = g(t,x_t) + h(t,x_t)$ we have

$$\dot{V}_{(12)}(t,x_t) \leq -w(|x|) + \sigma/4 \tag{17}$$

for all $t \geq 0$ and $x_t \in C$. Combining Eq. 17 and Eq. 5 of Assumption 2 it follows that the solution to Eq. 1 under Assumptions 1 and 2 is uniformly bounded with bound $B = \max(B_1, B_2)$, where

$$B_1 = u^{-1}\left[v_0(\beta) + v_1\big(rw(\beta)\big) + \sigma r/4\right] \tag{18}$$

and

$$B_2 = u^{-1}\left[v_0\left(w^{-1}(\sigma/4) + w_1(\sigma r/4)\right]\right. \tag{19}$$

as given in Burton and Zhang [9]. Uniform ultimate boundedness also follows from [9] but it is a very conservative bound. Below we present a modified version of the Burton and Zhang proof, one which gives a tighter estimate.

Let $B_3 > 0$ be given. From Eq. 17 we find that $\dot{V}_{(12)}(t, x_t) \leq -\delta < 0$ provided $|x(t)| \geq w^{-1}(\delta + \sigma/4) = U$. Observe that by using the uniform boundedness result above we can find a $B_4 > U$ such that for $t_0 \geq 0$, $\|\phi\| \leq B_3$ and $t \geq t_0$ we have $|x(t, t_0, \phi)| < B_4$. Thus for any $\|\phi\| \leq B_3$ we have $0 \leq V(t, x_t) \leq v_0(B_4) + v_1\big(rw(B_4)\big)$ for $t \geq t_0$, i.e., $V(t, x_t)$ is bounded on $[t_0, \infty)$. Furthermore, there is a sufficiently large positive N such that on any interval $[t, t+Nr]$ with $t \geq t_0$, we have a time $\bar{t} \in [t, t+Nr]$ with $|x(\bar{t})| < U$. Since $\dot{V}_{(12)}(t, x_t) \leq -w(|x|) + \sigma/4$ we have

$$\int_{t-r}^{t} w\big(|x(s)|\big)\, ds \leq -\int_{t-r}^{t} \dot{V}(s, x_s)\, ds + \sigma r/4$$

$$= V(t-r, x_{t-r}) - V(t, x_t) + \sigma r/4. \tag{20}$$

Now consider intervals $I_1 = [t_0, t_0 + Nr]$, $I_2 = [t_0 + Nr, t_0 + 2Nr], \ldots$, $I_i = [t_0 + (i-1)Nr, t_0 + iNr], \ldots$, and take $t_i \in I_i$ such that $V(t_i) = V(t_i, x_{t_i})$ is a maximum on I_i (such a maximum exists since we have a continuous functional on a closed bounded set). If it turns out that $t_i = t_0 + (i-1)Nr$, i.e. t_i is at the left edge of I_i, and $|x(t_i)| > U$, then because the way the width of the interval I_i is chosen, there is a first time $\bar{t}_i \in I_i$ such that $|x(\bar{t}_i)| = U$. In such a case we will redefine $I_i = [\bar{t}_i, t_0 + iNr]$ and let $V(t_i) = \max_{s \in I_i} V(s)$. Whether or not the interval I_i was modified we have that $|x(t_i)| \leq U$ for $i = 1, 2, \cdots$.

Consider the intervals $L_2 = [t_2 - r, t_2]$, $L_3 = [t_3 - r, t_3], \cdots$, $L_i = [t_1 - r, t_i], \cdots$. For each $i = 2, 3, 4, \cdots$ we have two cases: (i) $V(t_i) + \rho \geq V(s)$ for all $s \in L_i$, or (ii) $V(t_i) + \rho < V(s_i)$ for some $s_i \in L_i$. Observe that if we are in case (ii) $s_i \notin I_i$ because $V(t_i)$ is the maximum of V in I_i. Note that if I_{i-1} and I_i are contiguous, then $s_i \in I_{i-1}$. However, if I_{i-1} and I_i are not contiguous and $s_i \in [t_0 + (i-1)Nr, \bar{t}_i]$ then $|x(t)| \geq U$ and $\dot{V}_{(12)}(t, x_t) \leq -\delta$ on $[t_0 + (i-1)Nr, \bar{t}_i]$. Hence, $V(t_0 + (i-1)Nr) \geq V(s_i) > V(t_i) + \rho$. So regardless of I_{i-1} and I_i touching we find that $V(t_i) + \rho < V(t_{i-1})$ because the maximum of V on I_{i-1} is $V(t_{i-1})$. This observation, together with the fact that $V(t)$ is bounded leads to the existence of an integer $N^* > 0$ such that case (ii) is true for no more than N^* consecutive intervals I_i. Thus on some interval L with $j \leq N^*$ we have $V(t_j) + \rho \geq V(s)$ for all $s \in L_j$, i.e. the system changes to case (i). Using Eq. 20 with $t = t_j$ we have

$$\int_{t_j - r}^{t_j} w\big(|x(s)|\big)\, ds \leq V(t_j - r) - V(t_j) + \sigma r/4 \leq \rho + \sigma r/4 \tag{21}$$

which, together with the inequality Eq. 5 leads to

$$\begin{aligned} V(t_j) &\leq v_0\big(|x(t_j)|\big) + v_1\left(\rho + \sigma r/4\right) \\ &\leq v_0(U) + v_1\left(\rho + \sigma r/4\right) \\ &= v_0\left[w^{-1}\left(\delta + \sigma/4\right)\right] + v_1\left(\rho + \sigma r/4\right) \end{aligned} \tag{22}$$

Let $\varepsilon = v_0\left[w^{-1}(\delta + \sigma/4)\right] + v_1(\rho + \sigma r/4) + \rho$. It follows that $V(t) < \varepsilon$ for all $t > t_j$. For assume the contrary: then there exists a first time $t_p > t_j$ with $V(t_p) = \varepsilon$. We have two possibilities:

a) $t_p - r > t_j$. Then $V(t_p - r) < V(t_p)$ and from Eq. 20 with $t = t_p$ we have $\int_{t_p - r}^{t} w\big(|x(s)|\big)\, ds \leq \sigma r/4$. Hence, Eq. 5 leads to $V(t_p) \leq v_0\big(|x(t_p)|\big) + v_1\left(\sigma r/4\right) \leq v_0(U) + v_1\left(\sigma r/4\right) < \varepsilon$, a contradiction.

b) $t_p - r \leq t_j < t_p$. The key thing to note in this situation is that t_j is such that $V(t_j) + \rho \geq V(s)$ for all $s \in L$ and that $t_p - r \in L_j$. Applying Eq. 20 with $t = t_p$ we have $\int_{t_p - r}^{t} w\big(|x(s)|\big)\, ds \leq V(t_p - r) - V(t_p) + \sigma r/4 \leq$

$\rho + V(t_j) - \varepsilon + \sigma r/4 \leq \rho + \sigma r/4$. Hence, $V(t_p) \leq v_0(U) + v_1(\rho + \sigma r/4) < \varepsilon$, a contradiction.

Consequently, $u(|x(t)|) \leq V(t) \leq v_0\left[w^{-1}(\delta + \sigma/4)\right] + v_1\left[(\rho + \sigma r/4)\right] + \rho$, $t \geq t_j$, results in $|x(t)| \leq u^{-1}\left\{v_0\left[w^{-1}(\delta + \sigma/4)\right] + v_1(\rho + \sigma r/4) + \rho\right\}$ for all $t \geq t_j = t_0 + N^* N r$. ∎

Since the parameters δ and ρ are arbitrary positive constants needed in the proof we can let $\delta, \rho \to 0^+$ with the tightest ultimate bound given by

$$\varepsilon^* = u^{-1}\left[v_0\left(w^{-1}\left(\frac{\sigma}{4}\right)\right) + v_1\left(\frac{\sigma r}{4}\right)\right] \tag{23}$$

which depends only on the delay time r and the switching parameter σ for $h(t, x_t)$. Note that we can achieve an arbitrary ultimate bound by taking σ sufficiently small. Furthermore, the estimate obtained in [4] in the case of uncertain systems governed by ordinary differential equations is a special case for when $r = 0$ we have $\varepsilon^* = u^{-1}\left[v_0\left(w_0^{-1}(\sigma/4)\right)\right]$.

A number of comparisons can be made between the current approach to uncertain retarded systems and that which employs Lyapunov *functions* [7]. First, the functional Eq. 10, in a slightly altered form, is common to both procedures. Second, the use of a positive definite, radially unbounded function $\nu(t, x(t))$ is essential to both methods. Typically, we can expect that the same $\nu(t, x(t))$ can be employed in both approaches. The major difference in the two techniques is the use of the entire state of the system in creating the Lyapunov functional $V(t, x_t)$ for the theorem above. Intuitively, we suspect that making use of as much information in the analysis as possible will lead to better estimates, at least for some values of delay. As we will see in the next section, this is precisely what happens.

IV. An Example

Thowsen [7] considered the following uncertain system:

$$\begin{aligned} \dot{x}(t) &= [3\cos(t) + \Delta a(t)]\, x(t) + [1 + 0.5\sin(t)]\, x(t - r) \\ &\quad + (0.5 + \Delta b(t))\, u(t) + v(t) \end{aligned} \tag{24}$$

where $|\Delta a(t)| \leq 0.28$, $|\Delta b(t)| \leq 0.02$, and $|v(t)| \leq 0.24$. In the context of the present paper $f(t, x_t) = 3\cos(t)x(t) + \left[1 + 0.5\sin(t)\right]x(t - r)$, $B = 0.5$

and the matching conditions become $\Delta f(t, x_t) = \Delta a(t)x(t) = 0.5D\big(t, x(t)\big)$, $\Delta B(t, x_t) = \Delta b(t) = 0.5E(t)$, and $\Delta C(t, x_t) = 1 = 0.5F$, where $\|D\| = 0.56|x(t)|$, $\|E\| = 0.04$, and $\|F\| = 2$.

To satisfy Assumption 2, let

$$V(t, x_t) = 0.5x^2(t) + 0.75 \int_{t-r}^{t} x^2(s)\, ds \tag{25}$$

with $\nu\big(t, x(t)\big) = 0.5x^2(t)$ and $\eta(t, x_t) = 0.75 \int_{t-r}^{t} x^2(s)\, ds$. Letting $g\big(x(t)\big) = -10x(t)$ we find that

$$\begin{aligned}
&\frac{\partial \nu}{\partial t}(t, x) + \nabla'_x \nu(t, x)\left[f(t, x_t) + B(t, x_t) g(t, x_t)\right] + \dot{\eta}(t, x_t) \\
&\quad = x(t)\left\{3\cos(t)x(t) + [1 + 0.5\sin(t)]\, x(t-r) + 0.5\big(-10x(t)\big)\right\} \\
&\qquad + 0.75x^2(t) - 0.75x^2(t-r) \\
&\quad \le -1.25x^2(t) + 0.5\,[1 + 0.5\sin(t)]\left[x^2(t) + x^2(t-r)\right] - 0.75x^2(t-r) \\
&\quad \le -0.5x^2(t).
\end{aligned} \tag{26}$$

Therefore, Assumption 2 is satisfied with $g(x) = -10x(t)$ if we let $u(s) = v_0(s) = w(s) = 0.5s^2$, and $v_1(s) = 1.5s^2$. Now we can compute $q\big(t, x(t)\big) = \|D\big(t, x(t)\big)\| + \|E(t) g\big(x(t)\big)\| + \|Fv(t)\| = 0.96|x| + 0.48$; $B'\nabla_x \nu(x) = 0.5x$, and for $\sigma = 0.1152$,

$$h(x) = \begin{cases} -x - 0.5 \text{ sgn } (x), & \text{if } \ |x| > 0.3 \\ \frac{-x(|x|+0.5)^2}{0.24}, & \text{if } \ |x| \le 0.3. \end{cases} \tag{27}$$

Thus the feedback control law $u\big(x(t)\big) = -10x(t) + h\big(x(t)\big)$ leads to uniformly ultimately bounded solutions of Eq. 24, as established by Thowsen.

However, unlike Thowsen's result which guarantees that $|x(t)| < 0.4$ for sufficiently large t, regardless of the delay size, we can use Eq. 23 to obtain the ultimate bound $\varepsilon^*(r) = \sqrt{0.0576 + 0.0864r}$. Thus for delay values $r < 1.18$ the present method gives tighter estimates on the region of ultimate confinement than that provided by Lyapunov function analysis, but for larger delays the situation is reversed. This is because the functional approach leads to an explicit delay-dependence in the estimate for the ultimate bound.

V. Conclusion

Presented in this paper is a Lyapunov functional approach to verifying the existence of controllers which lead to uniformly ultimately bounded solutions of uncertain systems governed by retarded functional differential equations with finite time-lag, subject to matching conditions. The proposed controller, which is the same as that developed in Thowsen [7] and depends on a positive switching parameter σ, generates a uniform ultimate bound which depends on the delay-size and σ. As shown in the example of Section IV this can lead to tighter estimates of the ultimate bound than afforded by analysis using Lyapunov functions.

ACKNOWLEDGEMENT. The author wishes to thank Professor George Leitmann for directing him to [7] and for several helpful comments.

References

[1] G. Leitmann, "Guaranteed asymptotic stability for some linear systems with bounded uncertainties," *J. Dynamic Syst. Meas. and Control 101*, 212–216 (1979).

[2] G. Leitmann, "On the efficacy of nonlinear control of uncertain linear systems," *J. Dynamic Syst. Meas. Control 102*, 95–102 (1981).

[3] G. Leitmann, "Guaranteed ultimate boundedness for a class of uncertain linear dynamical systems," *IEEE Trans. Auto. Control AC-26*, 1109–1110 (1978).

[4] M. Corless and G. Leitmann, "Continuous state feedback guaranteeing uniform ultimate boundedness for uncertain dynamic systems," *IEEE Trans. Auto. Control AC-26*, 1139–1144 (1978).

[5] M. Corless and J. Manela, "Control of uncertain discrete-time systems," *Proc. Amer. Control Conf.*, Seattle, WA, 515–520 (1986).

[6] M.E. Magaña and S.H. Zak, "Robust state feedback stabilization of discrete-time uncertain dynamical systems," *Proc. 26th IEEE Conf. on Decision and Control*, Los Angeles, CA, 1530–1534 (1987).

[7] A. Thowsen, "Uniform ultimate boundedness of the solutions of uncertain delay systems with state-dependent and memoryless feedback control," *Int. J. Control 37*, 1135–1143 (1983).

[8] J. Hale, *Theory of Functional Differential Equations*, Springer–Verlag, New York, 1977.

[9] T.A. Burton and S. Zhang, "Uniform boundedness, periodicity, and stability in ordinary and functional differential equations," *Annali di Matematica Pura ed Applicata 145*, 129–158 (1986).

[10] R.D. Driver, *Ordinary and Delay Differential Equations*, Springer–Verlag, New York, 1977.

STABILIZING DISCRETE CONTROL LAWS FOR HAMILTONIAN SYSTEMS

Kaveh Shamsa and Henryk Flashner

Department of Mechanical Engineering
University of Southern California
Los Angeles, CA 90089-1453

I. INTRODUCTION

In the past few years the area of automated mechanical systems has undergone tremendous development. Modern assembly lines are based on high precision performance of automated mechanical systems and manipulators. Future space missions require employment of space vehicles equipped with automatic control systems capable of performing the required maneuvers and suppressing the excited oscillations due to structural flexibility of these vehicles. The high performance requirements of the new generation of mechanical systems present challenging problems for the control system designer.

A difficulty in design of stable control laws for these systems is due to their nonlinear character. Dynamics of a system of interconnected rigid bodies are inherently nonlinear due to simultaneous high speed motions of many degrees of freedom and gravitational effects [21],[22]. Various approaches for control of interconnected rigid bodies which are based on linearized approximations (e. g. [22]) may result in inefficient or unstable designs. The approaches based on the theory of variable structures [23],[24] may result in highly oscillatory control forces, a phenomenon known as chattering.

Dynamics of a flexible mechanical system are in general nonlinear due to the couplings between the rigid body and elastic coordinates, and infinite dimensional due to the distributed parameter nature of these systems. A common approach for designing controllers for these systems is to use finite dimensional approximations for the dynamics of flexible systems, a methodology termed as modal truncation. Modal truncation may cause the well-known 'spillover' problems. Spillover refers to the phenomenon in which the control energy intended to go into modeled modes is exciting unmodeled modes that might cause instability (see [25]). However, it has been known that if co-located sensors and actuators are employed to control a flexible system, overall closed-loop stability is guaranteed for a broad class of continuous-time controllers (see for example [15],[16]).

The co-located sensors-actuators stability property disappears when in a discrete-time feedback system the analog controller is replaced by its digital equivalent. This is due to the sampling operation that introduces delay and destroys the passive structure of the controlled system. Analysis of the latter situation together with conditions for stability of the discrete- time system are studied in [26] and [27]. However, these studies are based on finite dimensional approximations, and and as a result the derived stability conditions depend highly on physical parameters of the plants and the number of modes employed to model their dynamics.

Despite the diverse nature of mechanical systems, they share a common energy conservative and/or dissipative property (Hamiltonian property) which may be employed to design stable control laws. In this paper the Hamiltonian property is used to develop energy dissipative and stable closed- loop systems for discrete-time control of mechanical systems. The proposed control laws are general in the sense that they are not restricted to a particular system, but span a very broad class of rigid and flexible mechanical systems. Use of the Hamiltonian property facilitates characterization of controllers which guarantee closed-loop stability irrespective of the physical parameters of the plants. Employment of the Hamiltonian property yields the conditions for stability of co-located discrete-time closed-loop control of flexible mechanical systems. As a result stability of the proposed feedback systems is sustained irrespective of the number of elastic modes used to model dynamics of the plants. This result is the extension of the well- known stabilizing properties of continuous-time co-located sensors-actuators controllers to discrete-time situations.

The paper has the following organization : In Section 2 the Hamiltonian property for mechanical systems is formulated. The general configuration for control of mechanical systems is posed in Section 3. Conditions for input-output and Lyapunov stability of the proposed feedback system are derived in Section 4. Implications of these results are discussed in Section 5. Validity of the results is verified in Section 6 by applying them to control of a two links rigid system, a linear structure modeled as a simply supported Euler-Bernouli beam, and a flexible system undergoing rigid body angular maneuvers.

II. DYNAMICS OF MECHANICAL SYSTEMS

The dynamics of mechanical systems can be characterized by an input-output relationship defined by the operator $\mathbf{G}$. This functional relates input forces and moments $\mathbf{Q} \in \mathbf{R}^n$ acting on the system and the output $\mathbf{y} \in \mathbf{R}^n$ that specifies the motion of the system. The output yis a function of generalized coordinates $\mathbf{q}$. Using the Lagrange formulation the equations of motion can be written as:

$$\frac{d}{dt}\frac{\partial L}{\partial \dot{q}_i} - \frac{\partial L}{\partial q_i} = R_i \quad i = 1,..,N \tag{1}$$

where the upperscript $\cdot$ denotes differentiation with respect to time, R_i are generalized forces, N is the number of degrees of freedom, and $L = L(q_1, q_2, \ldots; \dot{q}_1, \dot{q}_2, \ldots)$ is the Lagrangian of the system given by:

$$L = E_k - E_p \tag{2}$$

The function $E_k = E_k(q_1, q_2, \ldots; \dot{q}_1, \dot{q}_2, \ldots)$ defines the kinetic energy of the system and is a *quadratic positive definite form* of the velocities $\dot{q}_1, \dot{q}_2, \ldots$. The function $E_p = E_p(q_1, q_2, \ldots)$ is the potential of the system, and Q_i is the generalized force associated with the generalized coordinate q_i.

It should be noted that mechanical systems that contain flexible elements can be also modeled using Lagrange formulation. Assume that the motion of the system can be described by m rigid body degrees of freedom and additional vibrational degrees of freedom. Let the location of any point on a flexible element be denoted by ξ then assuming linear elasticity the flexible deflection with respect to the rigid configuration $\mathbf{w}\ (\xi, t)$ can be expressed as:

$$\mathbf{w}\ (\xi, t) = \sum_{i=m+1}^{M} \phi_i(\xi) q_i(t) \tag{3}$$

If the modal solution is known then ϕ_i represents the mode shapes for $i = 1, 2, \ldots$ and $M \to \infty$. If the modal solution is not known we use the method of assumed modes (see [18]) and then M is finite. In all flexible systems some energy dissipation occurs. Assuming linear viscous damping we can write the equations of motion of a general flexible mechanical system as:

$$\frac{d}{dt}\frac{\partial L}{\partial \dot{q}_i} - \frac{\partial L}{\partial q_i} = R_i \quad i = 1, 2, \ldots, m \tag{4}$$

$$\frac{d}{dt}\frac{\partial L}{\partial \dot{q}_i} - \frac{\partial L}{\partial q_i} = -c_i \dot{q}_i + R_i \quad i = m+1, m+2, \ldots \tag{5}$$

where m is the number of rigid degrees of freedom, and $c_i, i = m+1, \ldots$ are positive constants that model the structural damping of the system.

Multiplying both sides of equation (5) by $\dot{q}_i$ and summation yields:

$$\frac{dH}{dt} = \sum_{i=1} R_i \dot{q}_i - \sum_{i=m+1} c_i \dot{q}_i^2 \tag{6}$$

where H is the Hamiltonian of the system defined as:

$$H = E_k + E_p \tag{7}$$

Assuming that E_p possesses a finite lower bound, then without loss of generality H can be regarded as a *nonnegative* function. Moreover, it can be shown that for a flexible system, E_p is *positive definite* in terms of the elastic coordinates $q_{m+1}, q_{m+2}, \ldots$.

If the sensors and actuators of the system are collocated it can be shown that equation (6) can be written as:

$$\frac{dH}{dt} = \mathbf{Q}^T \dot{\mathbf{y}} - \sum_{i=m+1}^{M} c_i \dot{q}_i^2 \tag{8}$$

where $\mathbf{Q}$ is a vector of input forces acting on the system and $\mathbf{y}$ is a vector of sensed output at the location of the actuators.

III. STATEMENT OF THE PROBLEM

The objective in this paper is to study the discrete-time feedback system in Figure 1. The operator $\mathbf{G}$ denotes the dynamics of the plant as defined in Section 2 . Control

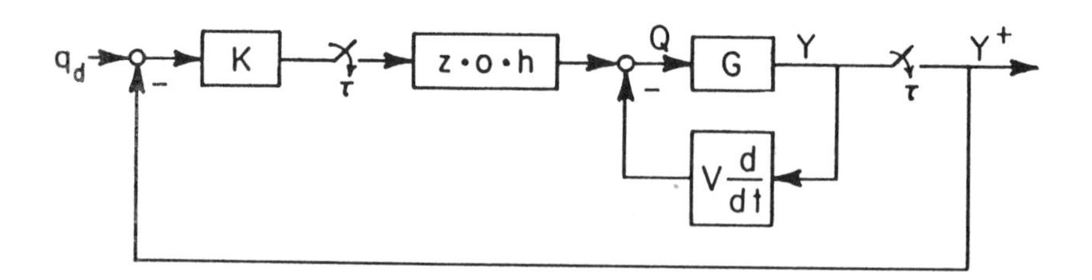

Figure 1: Basic feedback configuration

vector $\mathbf{Q} \in \mathbf{R}^n$, and output vector $y \in \mathbf{R}^n$ have the same physical meaning as defined in Section 2. The operator $\mathbf{K}$ denotes the dynamics of a digital controller which is assumed to be *linear time, invariant* and *causal*(see [4] for definition). The matrix $V \in \mathbf{R}^{n \times n}$ is a positive definite symmetric constant gain matrix that multiplies the analog velocity signal, τ is a sampling period, and z.o.h. stands for a *zero order hold* device [5]. Superscript $^+$ denotes a sequence of values of a time dependent quantity at the sampling instants, i.e.

$$\mathbf{y}^+ \equiv \mathbf{y}(l\tau)$$

where $l = 0, 1, 2, \ldots$ denotes sampling instants.

Using the fact that the operator $\mathbf{K}$ is linear by loop transformation we transform the system of Figure 1 to an equivalent system shown in Figure 2. The control input to the plant is therefore given by:

$$\mathbf{Q}\ (t) = \mathbf{q}_r(l) - \mathbf{u}(l) - \mathbf{V}\ \dot{\mathbf{y}}(t) \tag{9}$$

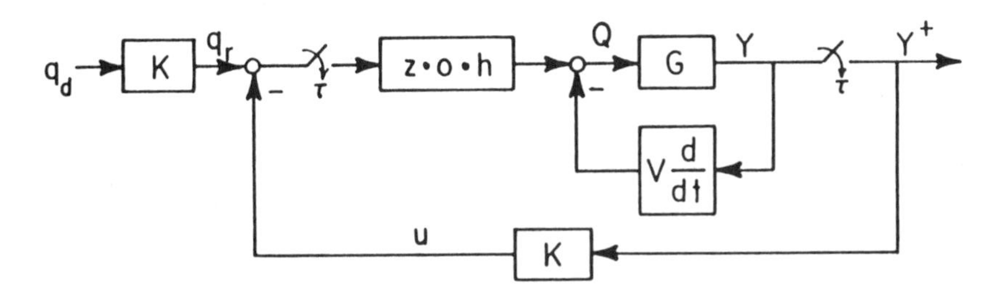

Figure 2: Equivalent feedback configuration

Note that the first two terms in the above equation are held constant over time intervals $l\tau \leq t < (l+1)\tau$, whereas the last term results from an analog velocity feedback implemented by, for example, tachometers and/or internal energy dissipation that is modeled as viscous damping. In the rest of the paper the feedback system of Figure 1 will be denoted by $\mathcal{F}$.

In the case where $\mathbf{G}$ expresses the dynamics of a flexible mechanical system, the control configuration depicted in Figure 2 is that of a co-located sensors and actuators control problem. This is due to the way that the inputs and outputs of the plant, $\mathbf{Q}$ and $\mathbf{y}$are defined. It should be also noted that the dynamics of the plant described by the operator $\mathbf{G}$ is in general nonlinear due to possible large angle spatial motion of many degrees of freedom.

In the following the stability of the feedback system $\mathcal{F}$ shown in Figure 2 is investigated using the notions of Lyapunov (see [6]) and input-output stability (see

[4]). Analysis is performed in inner product space of sequences (see [7]) defined below.

Given n-dimensional sequences $\mathbf{x}(l) \in \mathbf{R}^n$ and $\mathbf{y}(l) \in \mathbf{R}^n$, $l = 1, 2, \ldots$, an inner product is defined as :

$$< \mathbf{x}, \mathbf{y} > \equiv \sum_{l=0}^{\infty} \mathbf{x}^T(l)\mathbf{y}(l)$$

The norm of $\mathbf{x}$ induced by the above inner product is given by:

$$\|\mathbf{x}\| \equiv < \mathbf{x}, \mathbf{x} >^{\frac{1}{2}}$$

Truncation operation of the sequence $\mathbf{x}(l)$, $l = 1, 2, \ldots$ denoted by $\mathbf{x}_T(l)$ is defined by:

$$\mathbf{x}_T \equiv \mathbf{x}(l) \qquad for \ l \leq T \tag{10}$$

$$\mathbf{x}_T \equiv 0 \quad for \ l > T \quad T \in N \tag{11}$$

where N denotes the set of nonnegative integers. The truncated inner product of sequences is then defined as

$$< \mathbf{x}, \mathbf{y} >_T \equiv < \mathbf{x}_T, \mathbf{y}_T >$$

In the following discussion we denote the Euclidean norm by $| \cdot |$; $\lambda_{max}(\cdot)$ and $\lambda_{min}(\cdot)$ denote the minimum and maximum eigenvalues of a real symmetric matrix, respectively; the superscript * denotes a conjugate transpose operation; σ_{max} and σ_{min} denote the smallest and largest singular values of matrix $\mathbf{A}$, i.e.

$$\sigma_{max} = \lambda_{max}\{[\mathbf{A}^*\mathbf{A}]\}^{\frac{1}{2}} \quad , \sigma_{min} = \lambda_{min}\{[\mathbf{A}^*\mathbf{A}]\}^{\frac{1}{2}}$$

∇ denotes the backward difference operator; i.e. given a sequence $\mathbf{x} = \mathbf{x}(l)$, $\nabla\mathbf{x}$ is a sequence given by

$$\nabla\mathbf{x}(l) \equiv \mathbf{x}(l) - \mathbf{x}(l-1)$$

IV. MAIN RESULTS

Consider the feedback system $\mathcal{F}$ given in Figure 2 and denote the transfer function matrix of the linear operator $\mathbf{K}$ by $\mathbf{K}(z)$. We make the following assumptions:

(i) Let the state-space realization of $\frac{z}{z-1}\mathbf{K}(z)$ be given by:

$$\begin{aligned} \mathbf{x}(l+1) &= \mathbf{A}\mathbf{x}(l) + \mathbf{B}\mathbf{v}(l) \\ \mathbf{z}(l) &= \mathbf{C}\mathbf{x}(l) + \mathbf{D}\mathbf{v}(l) \end{aligned} \tag{12}$$

Assume that there exist positive definite symmetric matrices $\mathbf{P}, \mathbf{Q}, \mathbf{R}$ that satisfy satisfy the following equations:

$$\begin{aligned} \mathbf{A}^T\mathbf{P}\mathbf{A} - \mathbf{P} &= -\mathbf{Q} \\ \mathbf{B}^T\mathbf{P}\mathbf{A} &= \mathbf{C} \\ \mathbf{D}^T + \mathbf{D} - \mathbf{B}^T\mathbf{P}\mathbf{B} &= \mathbf{R} \end{aligned} \tag{13}$$

Satisfying the above equations implies that the transfer function matrices $\frac{z}{z-1}\mathbf{K}(z)$ and $\frac{z-1}{z}\mathbf{K}^{-1}(z)$ are positive real (see [10]). As a result $\mathbf{K}(z)$ is an exponentially stable transfer function matrix (see [4]).

(ii) Assume that $\det[\mathbf{K}(1)] \neq 0$, where $\det(\cdot)$ indicates determinant of a square matrix. Since $\frac{z-1}{z}\mathbf{K}^{-1}(z)$is a positive real transfer function matrix, then $\mathbf{K}^{-1}(z)$ is an exponentially stable transfer function matrix.

(iii) Assume that the sampling period τ satisfies the following inequality:

$$\lambda = \frac{\lambda_{min}(\mathbf{V})}{\tau} + \frac{\lambda_{min}(\mathbf{R})}{2} - \gamma_1 > 0 \tag{14}$$

where γ_1 is the gain of $\mathbf{K}(z)$ given by:

$$\gamma_1 = \sup_{z,|z|=1} \{\sigma_{max}[\mathbf{K}(z)]\} \tag{15}$$

Let the origin be the equilibrium state of the $\mathcal{F}$ then we have the following result:

Input-Output Stability

For any bounded input, the output of the feedback system $\mathcal{F}$ is bounded as follows:

$$\|\mathbf{y}^+\| \le \frac{1}{\gamma_2}\{\frac{\sigma_{max}\mathbf{C}}{2\lambda}[\gamma_1\|\mathbf{q}_d\| + \sqrt{\gamma_1^2\|\mathbf{q}_d\|^2 + 4\lambda c_0}] + \sigma_{max}(\mathbf{D})\sqrt{\frac{\gamma_1^2\|\mathbf{q}_d\|^2 + 2\lambda c_0}{\lambda \cdot \lambda_{min}(\mathbf{Q})}}\} \tag{16}$$

where

$$\begin{aligned} c_0 &= \frac{1}{2}\mathbf{x}^T(0)\mathbf{P}\mathbf{x}(0) + H(0) \\ \gamma_2 &= \inf_{z,\|z\|=1} \{\sigma_{min}[\mathbf{K}(z)]\} \end{aligned}$$

Proof:

Define a sequence $w(\cdot)$ as follows:

$$w(T) \equiv H(T\tau) + < \mathbf{u}, \nabla\mathbf{y}^+ >_T \quad T \in N \tag{17}$$

For the feedback system $\mathcal{F}$ of Figure 2 the signals $\mathbf{v}$,$\mathbf{z}$ defined in equation (12) correspond to $\nabla\mathbf{y}^+, \mathbf{u}$ respectively. Using the properties of positive dynamic system (see for example [9],[10]) we have the following relationship:

$$w(T) - w(0) = H(T\tau) - H(0) + \frac{1}{2}[\mathbf{x}^T(T+1)\mathbf{P}\mathbf{x}(T+1) - \mathbf{x}^T(0)\mathbf{P}\mathbf{x}(0)] +$$

$$+\frac{1}{2}\{\sum_{l=0}^{T}\{\mathbf{x}^T(l)\mathbf{Q}\ \mathbf{x}(l)+[\nabla\mathbf{y}^+(l)]^T\mathbf{R}\ [\nabla\mathbf{y}^+(l)]\}\} \tag{18}$$

From the definition of the sequence $w(\cdot)$ in equation (17) we have:

$$w(l+1)-w(l)=[\int_{l\tau}^{(l+1)\tau}\frac{dH}{dt}dt+[\mathbf{u}(l+1)]^T[\nabla\mathbf{y}^+(l+1)] \tag{19}$$

Using equations (8),(9) in equation (19):

$$w(l+1)-w(l)=[\int_{l\tau}^{(l+1)\tau}\{[\sum_{i,i>m}c_i\dot{q}_i^2]+\dot{y}^T\mathbf{V}\ \dot{y}\}dt+$$

$$\mathbf{q}_r^T(l)[\nabla\mathbf{y}^+(l+1)]+[\mathbf{u}(l+1)]^T[\nabla\mathbf{y}^+(l+1)] \tag{20}$$

Therefore we can deduce that

$$w(T)-w(0)=[\int_{0}^{T\tau}\{[\sum_{i,i>m}c_i\dot{q}_i^2]+\dot{y}^T\mathbf{V}\ \dot{y}\}dt+$$

$$+<\mathbf{q}_r,\nabla\mathbf{y}^+>_T+<\nabla\mathbf{u},\nabla\mathbf{y}^+>_T \tag{21}$$

From equations (18) and (21) we get:

$$H(T\tau)+\frac{1}{2}\mathbf{x}^T(T+1)\mathbf{P}\mathbf{x}(T+1)+\frac{1}{2}\{\sum_{l=0}^{T}\{\mathbf{x}^T(l)\mathbf{Q}\ \mathbf{x}(l)+$$

$$+[\nabla\mathbf{y}^+(l)]^T\mathbf{R}\ [\nabla\mathbf{y}^+(l)]\}\}\int_0^{T\tau}\{[\sum_{i,i>m}c_i\dot{q}_i^2]+\dot{y}^T\mathbf{V}\ \dot{y}\}dt+$$

$$-<\nabla\mathbf{u},\nabla\mathbf{y}^+>_T=<\mathbf{q}_r,\nabla\mathbf{y}^+>_T+H(0)+\frac{1}{2}\mathbf{x}^T(0)\mathbf{P}\mathbf{x}(0) \tag{22}$$

The linear operator $\mathbf{K}$ is described by an exponentially stable transfer function matrix and therefore it possesses a finite gain (see [4]) :

$$\|(\nabla\mathbf{u})_T\|\leq\gamma_1\|(\nabla\mathbf{y}^+)_T\|\quad\forall T\in N \tag{23}$$

$$\|(\mathbf{q}_r)_T\|\leq\gamma_1\|(\mathbf{q}_d)_T\|\quad\forall T\in N \tag{24}$$

where γ_1 is defined in equation (15).

Using Schwartz inequality it can be shown (see [13]) that the following inequality is satisfied:

$$\sum_{l=0}^{T-1}|\mathbf{y}((l+1)a)-\mathbf{y}(la)|^2\leq a\int_0^{T\tau}|\dot{(\mathbf{y})}|^2dt\quad\forall T\in N,\ T>0 \tag{25}$$

where a is an arbitrary constant.

Equation (22) together with inequalities (23) (24), and (25), the orthogonal equivalence property for quadratic positive definite forms [12], and Schwartz inequality for scalar products yield the following inequalities :

$$\frac{\lambda_{min}(\mathbf{Q})}{2}\|\mathbf{x}_T\|^2 + \{\lambda_{min}(\mathbf{V}) + \tau[\frac{\lambda_{min}(\mathbf{R})}{2} - \gamma_1]\}\int_0^{T\tau}\|\dot{\mathbf{y}}\|^2 dt +$$

$$+H(T\tau) + \sum_{i,i>m} c_i \int_0^{T\tau} \dot{q}_i^2 dt \leq \gamma_1\|(\mathbf{q}_d)_T\|\|(\nabla\mathbf{y}^+)_T\| +$$

$$+ H(0) + \frac{1}{2}\mathbf{x}^T(0)\mathbf{P}\mathbf{x}(0) \quad \forall T \in N \tag{26}$$

$$\frac{\lambda_{min}(\mathbf{Q})}{2}\|\mathbf{x}_T\|^2 + [\frac{\lambda_{min}}{\tau}(\mathbf{V}) + \frac{\lambda_{min}(\mathbf{R})}{2} - \gamma_1]\cdot\|(\nabla\mathbf{y}^+)_T\|^2 +$$

$$+H(T\tau) + \sum_{i,i>m} c_i \int_0^{T\tau} \dot{q}_i^2 dt \leq \gamma_1\|(\mathbf{q}_d)_T\|\|(\nabla\mathbf{y}^+)_T\|$$

$$+ H(0) + \frac{1}{2}\mathbf{x}^T(0)\mathbf{P}\mathbf{x}(0) \quad \forall T \in N \tag{27}$$

By letting $T \to \infty$ in inequality (27), using assumption (iii), and noting that H is nonnegative we have:

$$\|\nabla\mathbf{y}^+\| \leq \frac{1}{2\lambda}[\gamma_1\|\mathbf{q}_d\| + \sqrt{\gamma_1^2\|\mathbf{q}_d\|^2 + 4\lambda c_0}]\|\mathbf{x}\| \leq \sqrt{\frac{\gamma_1^2 + 2\lambda c_0}{\lambda\cdot\lambda_{min}(\mathbf{Q})}} \tag{28}$$

$$\|\mathbf{x}\| \leq \sqrt{\frac{\gamma_1^2\|\mathbf{q}_d\|^2 + 2\lambda c_0}{\lambda\cdot\lambda_{min}(\mathbf{Q})}} \tag{29}$$

where

$$c_0 = H(0) + \frac{1}{2}\mathbf{x}^T(0)\mathbf{P}\mathbf{x}(0)$$

Using equation (12), the triangular inequality for norms, and the orthogonal equivalence property for quadratic positive definite forms results in the following bound:

$$\|\mathbf{u}\| \leq \sigma_{max}(\mathbf{C})\cdot\|\mathbf{x}\| + \sigma_{max}(\mathbf{D})\cdot\|\nabla\mathbf{y}^+\| \tag{30}$$

Using the finite gain property of exponentially stable transfer function matrices we get a bound on the systems output:

$$\|\mathbf{y}^+\| \leq \{\sup_{z,|z|=1}\{\sigma_{max}[\mathbf{K}^{-1}(z)]\}\}\cdot\|\mathbf{u}\| = \frac{\|\mathbf{u}\|}{\inf_{z,|z|=1}\sigma_{min}[\mathbf{K}(z)]} \tag{31}$$

The inequalities given in (28)-(31) imply input-output stability of the feedback system and and the bound on the output given by in (16).

Lyapunov Stability

For the feedback system in $\mathcal{F}$ shown in Figure 2 define the state vector of the closed loop system as

$$\mathbf{x}_0 = [\mathbf{x}(l)\,\mathbf{q}_1(l\tau)\,\mathbf{q}_2(l\tau)\ldots\dot{\mathbf{q}}_1(l\tau)\dot{\mathbf{q}}_2(l\tau)\ldots]^T$$

where $\mathbf{x}$ is the state vector of the state-space realization of the transfer function matrix $\frac{(z-1)}{z}\mathbf{K}(z)$, as defined in equation (12). Then every free trajectory of the feedback system, i.e. with $\mathbf{q}_r = 0$, is bounded at all times and converges to an equilibrium state as time approaches infinity.
Note that we allow $\mathbf{x}_0$ to be infinite dimensional, therefore we use the notion of Lyapunov's stability for infinite dimensional systems, as given cited in [11].

Proof:
Inequality (28) and the fact that $\mathbf{q}_d = 0$ implies that:

$$\|\mathbf{x}\| < \infty \tag{32}$$

leads to the following bound of $\mathbf{x}$:

$$\sup_l(|\mathbf{x}(l)|) < \infty \tag{33}$$

Furthermore, inequality (32) and the fact that $\mathbf{x}$ is a monotonic bounded sequence yields:

$$\lim_{l\to\infty}\{\mathbf{x}(l)\} \tag{34}$$

By letting $T \to \infty$ in inequality (26), using assumption (iii) and the fact that H is nonnegative, we deduce the following:

$$\int_0^\infty |\dot{\mathbf{y}}|^2 dt < \infty \tag{35}$$

$$\int_0^\infty q_i^2 dt < \infty \quad i = m+1, m+2, \ldots \tag{36}$$

For $\mathbf{q}_d$ inequalities (28)-(31) yield:

$$\|\mathbf{u}\| < \infty \tag{37}$$

$$\|\mathbf{y}^+\| < \infty \tag{38}$$

Upon inserting equation (8) in equation (9) and setting $\mathbf{q}_r = 0$ we get:

$$H(s) \le H(0) - \int_0^s \nu^T \dot{\mathbf{y}}\, dt \tag{39}$$

where $\nu(t) \equiv \ell$ for $\ell \le t < (\ell+1)\tau$. Inequalities (35),(37),(39) together with the triangular inequality for norms and Schwartz's inequality for scalar products yields:

$$\sup_t\{H(t)\} < \infty \tag{40}$$

Since H is a positive definite function of $\dot{q}_1, \dot{q}_2, \ldots; q_{m+1}, q_{m+2}, \ldots$, then inequality (40) implies:

$$\sup_t(|\dot{q}_i|) < \infty \qquad i = 1, 2, \ldots \tag{41}$$

$$\sup_t(|q_i|) < \infty \qquad i = m+1, m+2, \ldots \tag{42}$$

Using the mean value theorem [8], for every t there exists an s such that:

$$q_i(t) = q_i(\ell\tau) + (t - \ell\tau)\dot{q}_i(s) \quad \ell\tau < t \leq (\ell+1)\tau \tag{43}$$

Inequalities (38),(41) together with equation (43) and the triangular inequality yield:

$$\sup_t(|q_i|) < \infty \qquad i = 1, 2, \ldots, m \tag{44}$$

Inequalities (35),(36),(42),(44) imply that all $q_i, i = 1, 2, \ldots$ have a finite limit as $t \to \infty$ (see [13] dor details)

V. DISCUSSION

(i) *Stability of multivariable discrete-time PD controller*:
Relaxing some of the conditions given in (i)-(iii) allows definition different classes of discrete time stabilizing control laws. Specifically, by assuming that the matrix **Q** in condition (iii) is only *positive semidefinite* and following similar proof steps, a large class of discrete controllers for Hamiltonian systems can be obtained, see [13] for details. This class includes the discrete-time multivariable PD controller. A theorem given in [13] specifies a bound on the sampling period that guaratees closed loop stability of a discrete PD controller and a Hamiltonian plant. In this case the input-output stability result applies only to finite dimensional systems. This result is an extension of the closed-loop stability properties proven for continuous- time PD controllers for Hamiltonian plants (see [14], [15], and [16]).

(ii) *Robustness with respect to parameter variations*:
The stabilizing control laws given in Section 4 are completely independent of the values of physical parameters of the plants and depend only on the Hamiltonian structure of the plant. Therefore, the stability results presented here are robust with respect to plant's parameter variations as long as the Hamiltonian structure of the plant is preserved.

(iii) *Robustness with respect to truncation*:
In the case that the operator G represents dynamics of a structurally flexible mechanical system, the proposed control laws guarantee stability of the feedback system in Figure 1 for any number of elastic coordinates used to model dynamics of the plant. This, again, is a consequence of the Hamiltonian structure of the plant and

its input-output relation given in equation (8). For linear the Hamiltonian plants, conditions (i)-(iii) guarantee that the poles of the feedback system in Figure 1 lie within the unit circle for any number of vibrational modes included in the model. In other words, the developed stability results are robust with respect to unmodeled dynamics and modal truncation.

VI. EXAMPLES

A. CONTROL OF A TWO LINKS RIGID MANIPULATOR

Consider a planar system consisting of two links of uniform density as shown in Figure 3. Link 1 is pivoted to the ground at point P_1 and link 2 is pivoted to link 1

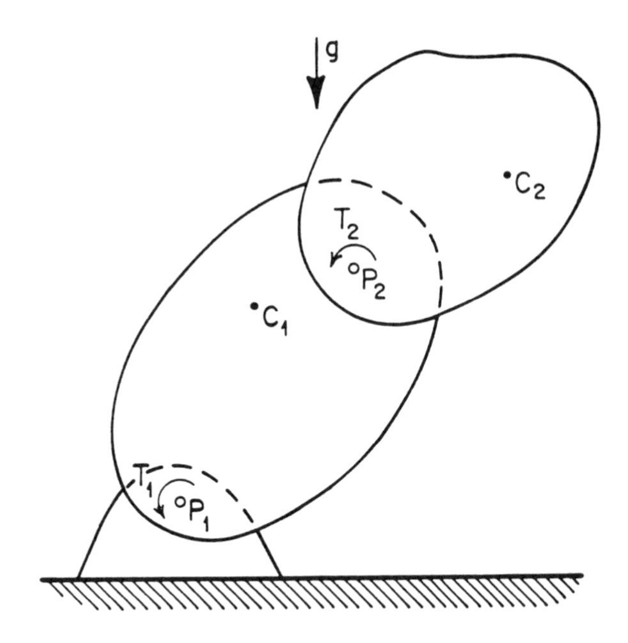

Figure 3: Model of a planar two link mechanical system

at point P_2. The point C_1 denotes the location of the center of mass of link 1 and C_2 is the center of mass of link 2. The motion of the system is described by absolute rotation of the line P_1P_2 denoted by θ_1, and the relative rotation of line P_2C_2 with respect to line P_1P_2 denoted by θ_2, as shown in the figure. The equations of motion of the system are given by:

$$(I_1 + m_2 d_1^2)\ddot{\theta}_1 + m_2 d_1 d_2 cos(\theta_1 - \theta_2)\ddot{\theta}_2 = -m_2 d_1 d_2 \dot{\theta}_2^2 sin(\theta_1 - \theta_2) \qquad (45)$$

$$- m_1 g d_3 sin\theta_1 - m_2 g d_1 sin\theta_1 - m_2 g d_2 sin(\theta_1 + \theta_2) + Q_1 \tag{46}$$

$$m_2 d_1 d_2 cos(\theta_1 - \theta_2)\ddot{\theta_1} + I_2 \ddot{\theta_2} = m_2 d_1 d_2 \dot{\theta}_1^2 sin(\theta_1 - \theta_2) \tag{47}$$

$$- m_2 g d_2 sin(\theta_1 + \theta_2) + Q_2 \tag{48}$$

Here I_1 is the moment of inertia of link 1 with respect to point P_1,I_2 is the moment of inertia of link 2 with respect to P_2, m_2 is the mass of link 2, d_1 is the distance P_1P_2 and d_2 is the distance P_2C_2, g is the acceleration of gravity, m_1 is the mass of link 1, d_3 is the distance P_1C_1,Q_1 and Q_2 are generalized forces given by

$$Q_1 = T_1 - T_2, \quad Q_2 = T_2$$

where T_1 is the torque applied to link 1 at point P_1 and T_2 is the torque acting on link 2 at point P_2.

Consider the two-link manipulator described above to be a plant controlled by the feedback configuration shown in Figure 1. We assume the following numerical values:

$$I_1 = I_2 = 2\ kg \cdot m^2,\ m_1 = m_2 = 1\ kg, d_1 = d_2 = 1\ m,\ d_3 = 0.5m,\ g = 10\frac{m}{sec^2}.$$

$\mathbf{Q} = [Q_1\ Q_2]^T$ is the input to the plant and $\mathbf{y} = [\theta_1\ \theta_2]^T$ is the output of the plant. The controller $\mathbf{K}(z)$ is assumed to be:

$$\mathbf{K}(z) = k \begin{bmatrix} 1 + 0.5\frac{z-1}{z} & 0 \\ 0 & 1 + 0.5\frac{z-1}{z} \end{bmatrix} \tag{49}$$

where k is a positive scalar. We choose the sampling period $\tau = 0.05sec$, and the gain matrix of the analog velocity feedback $\mathbf{V}$ as:

$$\mathbf{V} = \begin{bmatrix} 10 & 0 \\ 0 & 10 \end{bmatrix} \tag{50}$$

Using equation (14) the feedback system consisting of the two link manipulator and the controller described above is input-output stable for $k < 400$.

A simulation study was carried for the above system for $k = 30, 40, 50, 65, 80, 125$. In the simulations we assumed zero initial conditions for the state:

$$\theta_1 = \theta_2 = \dot{\theta}_1 = \dot{\theta}_2 = 0 \tag{51}$$

The input signal is given by the sequence $\mathbf{q}(\ell)_d$ as follows:

$$\mathbf{q}(\ell)_d = \begin{bmatrix} 2\sin(l+1)\tau & \sin(l+1)\tau \end{bmatrix}^T \quad l < \frac{\pi}{\tau} \tag{52}$$

$$\mathbf{q}_d = \begin{bmatrix} 0 & 0 \end{bmatrix} \quad otherwise \tag{53}$$

where $l = 0, 1, 2, \ldots$ is the indicator of sampling instants as defined previously. The response of the feedback system for $k = 40$ and $k = 125$ is shown in Figures 4 and 5, and the Euclidean norm of the error signal $\mathbf{e}(l) = \mathbf{q}_d(\ell) - \mathbf{y}^+(l+1)$ is given in the Table 1.

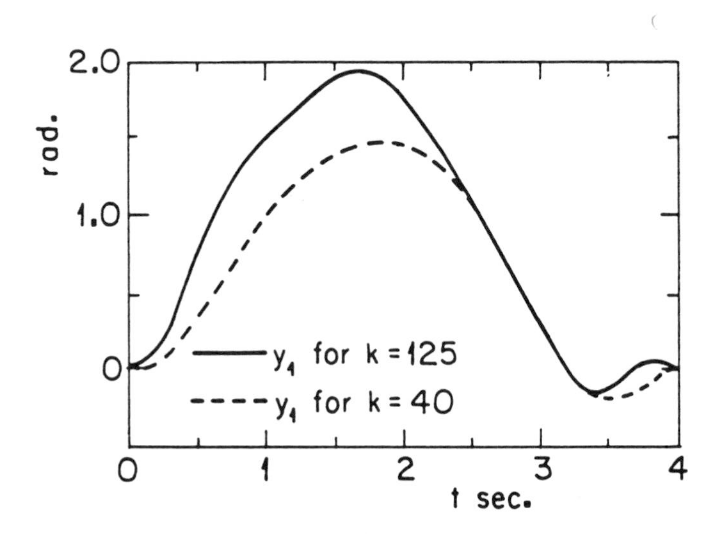

Figure 4: Response of link 1 of the planar manipulator

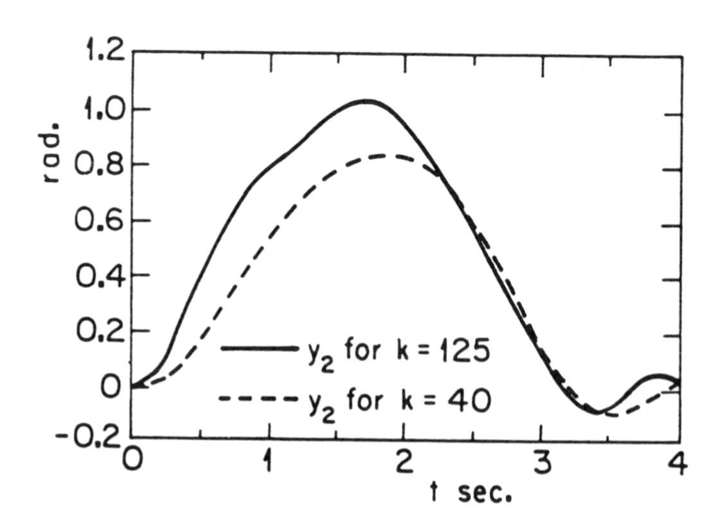

Figure 5: Response of link 2 of the planar manipulator

k	$\|\|e\|\|$
30	5.223670
40	4.061023
50	3.271763
65	2.508333
80	2.030934
125	1.326872

Table 1 demonstrates the input-output stability of the system and shows that $\|\mathbf{e}\|$ decreases when k increases. As shown before there is an upper bound on k, and therefore, a lower bound on $\|\mathbf{e}\|$. In other words there exists an upper limit on the performance of the feedback system.

B. STRUCTURAL CONTROL

For the feedback system of Figure 1 assume a plant that can be modeled by a uniform Euler-Bernouli simply supported beam, as shown in Figure 6. The length

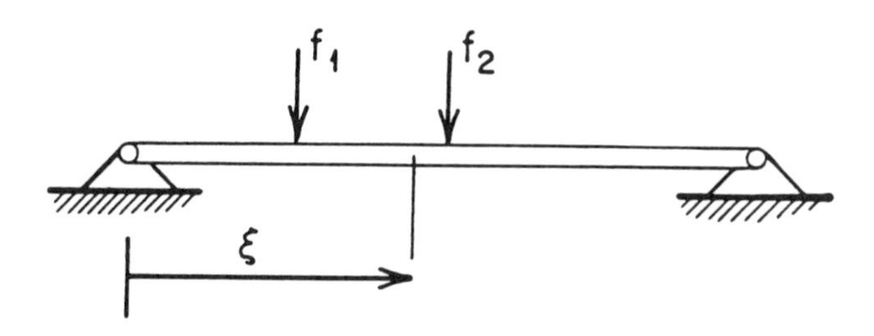

Figure 6: Controlled simply supported beam

of the beam is l and the mass per unit length is m. The beam is acted upon by two transverse forces f_1 and f_2 at $\xi = \frac{l}{\sqrt{15}}$ and $\frac{l}{2}$,respectively. The variable ξ denotes the location along the beam. Using modal analysis the deflection of the beam may expanded in terms of eigenfunctions:

$$w(\xi, t) = \sum_{i=1}^{M} \phi_i(\xi) q_i(t) \tag{54}$$

$$\phi_i = \sqrt{\frac{2}{ml}} \sin \frac{i\pi\xi}{l} \tag{55}$$

The dynamic equations of the plant are given by (see [18]):

$$\ddot{q}_i + 2\zeta\omega_i q_i + \omega_i^2 q_i = \sqrt{\frac{2}{ml}}[f_1 \sin\frac{i\pi}{\sqrt{15}} + f_2 \sin\frac{i\pi}{2}] \quad i = 1,2,\ldots,M \tag{56}$$

where

$$\omega_i = (i\pi)^2\sqrt{\frac{EI}{ml^4}} \quad i = 1,2,\ldots,M \tag{57}$$

In the above equations E is the Young's modulus of elasticity, I is the cross-section moment of inertia of the beam about the neutral axis and ζ is the structural damping coefficient.

Note that the complete solution is for $M \to \infty$, however for control design purposes we analyze only a finite number of modes.

Let $\mathbf{Q} = [f_1 f_2]^T$ be the input of the plant and let $\mathbf{y} = [w(\frac{l}{\sqrt{15}}, t), w(\frac{l}{2}, t)]^T$ be its output. Choose the transfer function matrix of the controller as :

$$\mathbf{K}(z) = g\begin{bmatrix} 1 + 0.5\frac{z-1}{z} & 0 \\ 0 & 1 + 0.5\frac{z-1}{z} \end{bmatrix} \tag{58}$$

where g is a positive scalar. Choose the sampling period $\tau = 0.05sec$, and the matrix of velocity feedback gain $\mathbf{V}$:

$$\mathbf{V} = 0.051g\begin{bmatrix} 1 & 0 \\ 0 & 1 \end{bmatrix} \tag{59}$$

According to the analysis of Section 4, the feedback system is stable for any value of the gain g and any number of modes M. To demonstrate the validity of the analysis Figure 7 presents locus of the poles of the feedback system for $0 < g < 20000$. The following numerical values for the physical parameters were assumed:

$$M = 6,\ ml = 2\,kg,\ \omega_i = (i\pi)^2\,\frac{1}{sec},\ \zeta = .01$$

As can be observed the poles of the closed loop system stay within the unit circle, thus confirming exponential stability. Similar results were obtained for $M = 10$ and $M = 20$.

C. LARGE ANGLE MOTION OF A FLEXIBLE SYSTEM

Consider a planar system consisting of a rigid circular hub with a flexible appendage modeled as a uniform clamped-free Euler-Bernouli beam. The motion of the system is described in terms of the generalized coordinates q_1 which denotes rotation of the hub about its center, and $w(\xi, t)$ that denotes the flexible deflection of the appendage with respect to the rigid body configuration at location ξ along the beam, as shown in Figure 8. This configuration models a flexible spacecraft in rotational maneuvers (see [19] and [20]). The system's motion is controlled by an external torque T

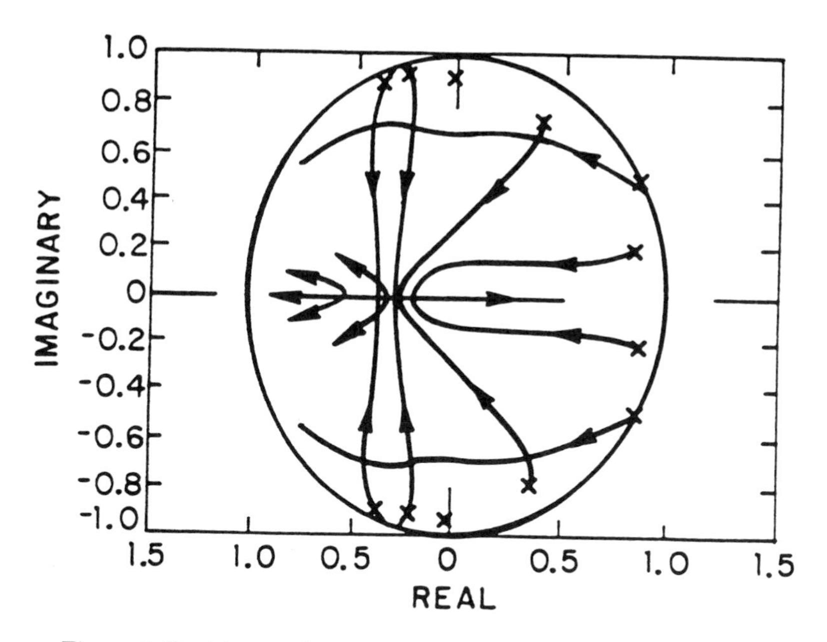

Figure 7: Root-locus of a controlled simply supported beam

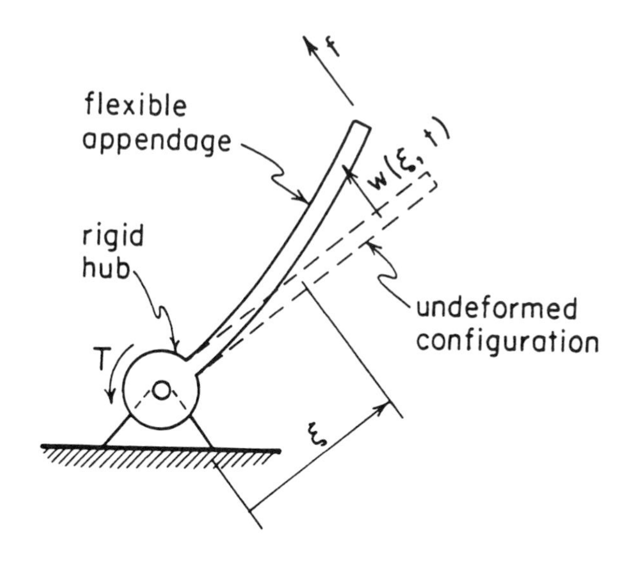

Figure 8: Model of a flexible appendage

acting on the hub and an inertial force f acting at the tip of the beam in a direction perpendicular to the beam. Neglecting the effect of longitudinal motion during bending and using the method of assumed modes the two mode approximation of the flexible motion is given by:

$$w(\xi,t) = \phi_2(\xi,t)q_2(t) + \phi_3(\xi,t)q_3(t) \tag{60}$$

$$\phi_2(\xi) = (\frac{5}{mL})^{\frac{1}{2}} \cdot (\frac{\xi - r}{L})^2 \tag{61}$$

$$\phi_3(\xi) = (\frac{105}{mL})^{\frac{1}{2}} \cdot [(\frac{\xi - r}{L})^2 - \frac{6}{5}(\frac{\xi - r}{L})^3] \tag{62}$$

The equations of motion are given by:

$$[J + m(\frac{(r+L)^3 - r^3}{3}) + q_2^2 + q_3^2]\ddot{q}_1 + (5mL)^{\frac{1}{2}}(\frac{L}{4} + \frac{r}{3})\ddot{q}_2 + \tag{63}$$

$$(105mL)^{\frac{1}{2}} \cdot (\frac{L}{100} + \frac{r}{30})\ddot{q}_3 + \tag{64}$$

$$2\dot{q}_1(\dot{q}_2 q_2 + \dot{q}_3 q_3) = Q_1 \tag{65}$$

$$(5mL)^{\frac{1}{2}} \cdot (\frac{L}{4} + \frac{r}{3})\ddot{q}_1 + \ddot{q}_2 + c_2\dot{q}_2 + \frac{EI}{mL^4}(20q_2 - 16\sqrt{21}q_3) \tag{66}$$

$$- q_2\dot{q}_1^2 = (\frac{5}{mL})^{\frac{1}{2}} f \tag{67}$$

$$(105mL)^{\frac{1}{2}} \cdot (\frac{L}{100} + \frac{r}{30})\ddot{q}_1 + \ddot{q}_3 + c_3\dot{q}_3 + \frac{EI}{mL^4}(\frac{3612}{5}q_3 - 16\sqrt{21}q_2) \tag{68}$$

$$- q_3\dot{q}_1^2 = -(\frac{105}{mL})^{\frac{1}{2}} f \tag{69}$$

Here r is the radius of the hub, J is the moment of inertia of the hub about its central axis perpendicular to plane of motion, L is the length of the beam, m is the beam's mass per unit length, E is the Young's modulus of elasticity, and I is the moment of cross-sectional area of the appendage with respect to the neutral axis in bending. The generalized force Q_1 is given by

$$Q_1 = T + f(r + l)$$

The coefficients c_2 and c_3 denote the structural damping coefficients of the two modeled modes.

Consider the aforementioned system to be the plant of the feedback system of Figure 1, with $Q = [Q_1\ f]^T$ and $y = [q_1\ w(L,t)]^T$ being the input and output of the plant, respectively. Suppose that the reference trajectory $\mathbf{q}_d = 0$ and it is desired to choose a feedback controller which brings the state of the system to the origin, for any initial value of the state. Using the results of Section 4 we select the control laws

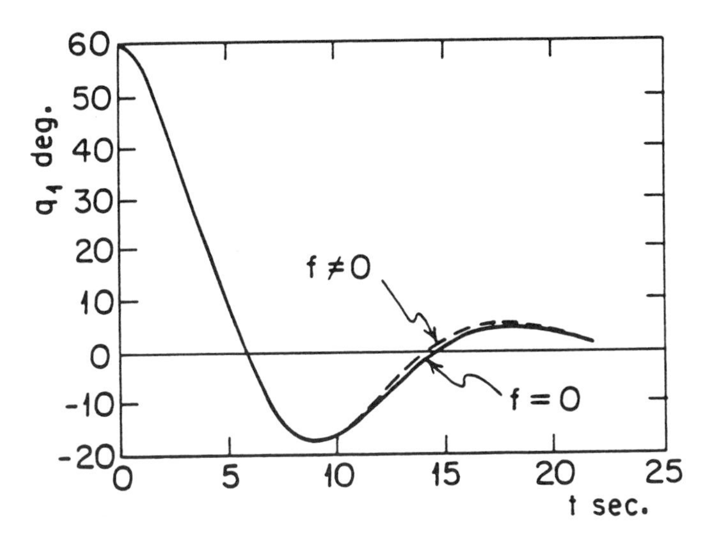

Figure 9: Rigid body motion of flexible appendage

for two cases. In the first case only the hub torque is employed with the matrices that define the feedback control given by:

$$\mathbf{K}(z) = 9\begin{bmatrix} 1+9\frac{z-1}{z} & 0 \\ 0 & 0 \end{bmatrix} \quad \mathbf{V} = \begin{bmatrix} 10 & 0 \\ 0 & 0 \end{bmatrix} \tag{70}$$

In the second case we employ both the hub torque and the control force at the end of the flexible beam. For this the control law is defined by the following matrices:

$$\mathbf{K}(z) = 9\begin{bmatrix} 1+9\frac{z-1}{z} & 0 \\ 0 & 1+9\frac{z-1}{z} \end{bmatrix} \quad \mathbf{V} = \begin{bmatrix} 10 & 0 \\ 0 & 10 \end{bmatrix} \tag{71}$$

In both case the sampling time is chosen to be $\tau = 0.1sec.$

The behavior of the system have been simulated for both control configurations and the results are shown in Figures 9 and 10, respectively. In the simulation studies the following it was assumed that a large angle maneuver of 60^0 of the system is required. Hence the initial conditions were:

$$q_1(0) = \frac{\pi}{6}, \; q_i(0) = 0, \;\; i = 2,3. \;\; \dot{q}_i(0) = 0, \quad i = 1,2,3.$$

The physical parameters of the plant are as follows:

$$L = 4\,m, \; r = 0.5\,m, \; EI = 128\,N\cdot\, m^2, m = 0.5\,\frac{kg}{m}, J = 50\,kg\cdot m^2 \tag{72}$$

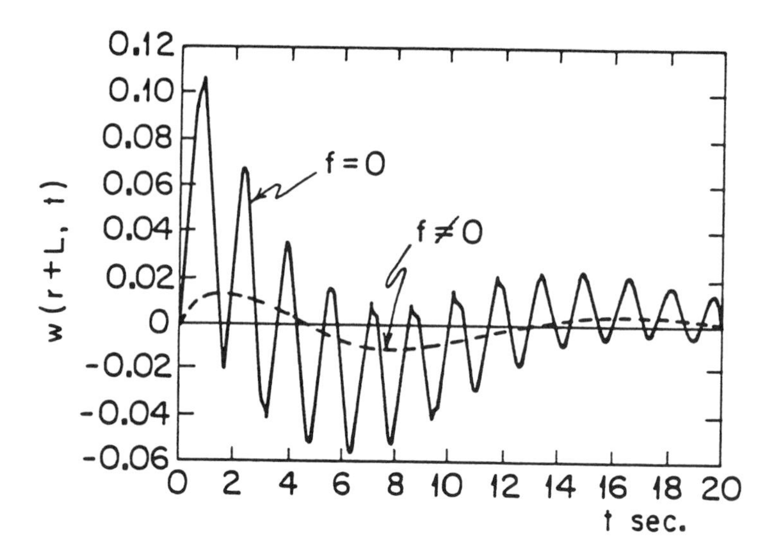

Figure 10: Flexible motion of flexible appendage

$$c_2 = 0.08\,N\,\frac{sec}{m}, c_3 = 0.6\,N\frac{sec}{m} \tag{73}$$

As can be seen from the figures the control law results in a completed maneuver. However, as expected flexible deflection of the appendage is much smaller when the end force is employed.

References

[1] Boresi A. P., Lynn P. P., "Elasticity in Engineering Mechanics", Prentice-Hall, Englewood Cliffs, New Jersey, 1974.

[2] Fung Y. C., "Foundations of Solid Mechanics", Prentice- Hall, Englewood Cliffs, New Jersey, 1965.

[3] Goldstein H., "Classical Mechanics", Addison-Wesley, London, 1960.

[4] Desoer C. A., Vidyasagar M., "Feedback Systems, Input-Output Properties", Academic Press, New York, 1975.

[5] Astrom K., Wittenmark B., "Computer Controlled Systems", Prentice-Hall, Englewood Cliffs, New Jersey, 1984.

[6] Kalman R. E., Bertram J. E., "Control System Analysis and Design via the 'Second Method' of Lyapunov", ASME Trans. Basic Engineering, Vol. 82, June 1960, pp. 371-400.

[7] Kantorovitch L. V., Akilov G. P., "Functional Analysis in Normed Spaces", (English Translation), MacMillan, New York, 1964.

[8] Rudin W., "Principles of Mathematical Analysis", McGraw- Hill, New York, 1976.

[9] Popov V. M., "Hyperstability of Automatic Control Systems", Springer-Verlag, New York, 1973.

[10] Landau Y. D., "Adaptive Control", Marcel Dekker, New York, 1979.

[11] Baker R. A., Bergen A. R., "Lyapunov Stability and Lyapunov Functions of Infinite Dimensional Systems", IEEE Trans. Automatic Control, Vol. AC-14, August 1969, PP. 325-334.

[12] Moore J. T., "Elements of Linear Algebra and Matrix Theory", McGraw-Hill, New York, 1968.

[13] Shamsa K., "Stability Results for Control of Mechanical Systems using Energy Concepts", Dissertation in Partial Fulfillment of Doctor of Philosophy Requirements, University of Southern California, 1988.

[14] Takegaki M., Suguru A., "A New Feedback Method for Dynamic Control of Manipulators", ASME Trans. Dynamic Systems, Measurement and Control, Vol. 102, June 1981, pp. 119-125.

[15] Gevarter W. B., "Basic Relations for Control of Flexible Vehicles", AIAA Journal, Vol. 8, April 1970, pp. 666-672.

[16] Benhabib R. J., Iwens R. P., Jackson R. L., "Stability of Distributed Control for Large Flexible Structures using Positivity Concepts", AIAA Journal of Guidance and Control, Vol. 5, September-October 1981, pp. 487-494.

[17] Kailath T., "Linear Systems", Prentice-Hall, Englewood Cliffs, New Jersey, 1977.

[18] Meirovitch L., "Analytical Methods in Vibrations", MacMillan, New York, 1967.

[19] Breakwell J. A., "Optimal Feedback Slewing of Flexible Spacecraft", AIAA Journal of Guidance and Control, Vol. 4, September-October 1981.

[20] Turner J. D., Chun H. M., "Optimal Distributed Control of a Flexible Spacecraft During a Large Angle Rotational Maneuver", The Charles Stark Draper Laboratory, Cambridge, Massachusetts.

[21] Luh J. Y. S., "Anatomy of Industrial Robots and their Controls", IEEE Trans. Automatic Control, Vol. AC-28, February 1983, pp. 133-153.

[22] Paul R. P., "Robot Manipulators : Mathematics, Programming and Control", MIT Press, Cambridge, Massachusetts, 1982.

[23] Young K. D., "Controller Design for a Manipulator using Theory of Variable Structures", IEEE Trans. Systems, Man and Cybernetics, Vol. SMC-8, February 1978, pp. 101-109.

[24] Vukobratovic K. M., Stokic M. D., "Control of Manipulation Robots : Theory and Applications", Springer-Verlag, New York, 1982.

[25] Flashner H., Benhabib R. J., "Issues in the Design of Large Space Structures Control Systems", AIAA Dynamics Specialists Conference, Palm Springs, California, 1984.

[26] Balas M., "Discrete Time Stability of Continuous Time Controllers for Large Space Structures", AIAA Journal of Guidance and Control, Vol. 5, September-October 1982, pp. 181-190.

[27] McClamroch N. H., "Sampled Data Control of Flexible Structures using Constant Gain Velocity Feedback", AIAA Journal of Guidance and Control, Vol. 7, November-December 1984, pp. 747-750.

CONTROL AND DYNAMIC SYSTEMS, VOL. 35

COMPONENT MODEL REDUCTION IN CANONICAL CORRELATION COORDINATES

Robert E. Skelton, Jae H. Kim and Dong Da

School of Aeronautics and Astronautics

Purdue University

West Lafayette, Indiana 47907

I. INTRODUCTION

Large scale systems usually are composed of several interconnected dynamic components. Each component is often manufactured by a different company, using different types of models for each component and it is often desired to reduce the model of each component separately as opposed to a centralized approach that ignores the identity of each component. For example it is useful to know what level of detail is required in the actuator dynamics or the solar panel dynamics, or the antenna dynamics of a space system. This paper is concerned with the simplification of each dynamic component in a large scale system of N components. Our aim is to reduce the order of each component while keeping the output performance of the entire system of N components as close as possible. This paper combines two methods, canonical correlation analysis [1,5] and component cost analysis [2-4].

The dynamics of the ith body in an N-body system is described as

$$M_i\ddot{q}_i + D_i\dot{q}_i + K_iq_i = B_iw + Z_i^*f \ , \ i=1, \ \cdots \ , N \tag{1.1}$$

where w is an n_u-vector of white noise disturbance in the actuators. For a tree-topological structure f is a vector made up of (N-1) non-working constraint forces and torques, described by the constraints

$$\sum_{i=1}^{N} Z_i\dot{q}_i = Z\dot{q} = 0 \, . \tag{1.2}$$

We shall restrict our attention to constraints of the form (1.2) with constant Z. Generally, M_i, D_i and K_i are functions of q, $\dot{q}$ and linearization about a specific $\bar{q}(t)$ is required. We assume that such linearization has already taken place and that M_i, K_i, D_i, B_i, Z_i are all constant matrices.

Most of the available model reduction theory presents a centralized approach for the overall system and does not address the question of component model reduction. Some exceptions are the component cost analysis method [2-4], the canonical correlation analysis method [1,5] and the work of Craig [6,7]. This paper combines the first two of these methods.

The idea of the canonical correlation analysis is to place the component with state vector x_k and the rest of the system with state x_{N-k} in such coordinates that

$$E\begin{bmatrix} x_{N-k} \\ x_k \end{bmatrix}(x_{N-k}^* \; x_k^*) = X = \begin{bmatrix} I_{N-k} & \Delta \\ \Delta^* & I_k \end{bmatrix} \quad , \quad x_k = T_k \begin{bmatrix} q_k \\ \dot{q}_k \end{bmatrix} \quad , \tag{1.3}$$

where E is an expectation operator and I_k is an identity matrix of the size of x_k. The upper left square corner of Δ is diagonal with entries δ_α. If dim $x_k <$ dim x_{N-k} then Δ has the form

$$\Delta = \begin{bmatrix} \text{diag}(\delta_1 \, , \, \delta_2 \, , \; \cdots) \\ 0 \end{bmatrix} , \quad 0 \le \delta_\alpha \le 1 \; , \quad \alpha = 1, \; \cdots \; N_k \; .$$

The δ_α's are called the component canonical correlation coefficients, and $\begin{bmatrix} x_{N-k} \\ x_k \end{bmatrix}$ the canonical correlation coordinates. $\delta_\alpha = E[(x_k)_\alpha (x_{N-k})_\alpha]$ represents the correlation of the component variable in the state of the component to be reduced with the overall system. If $\delta_\alpha << 1$, the αth state in the component is a candidate for deletion since this state variable is correlated neither with any other state variable in the body being reduced, nor with any state variable in the rest of the system. The coordinate transformation T_k that takes the system to canonical correlation coordinates will be described in the next section.

The second method of component reduction is component cost analysis. The idea of the component cost analysis is based on the decomposition of the norm of the impulse response (in the deterministic problem) or the norm of the response to white noise (in the stochastic problem), which we state here,

$$V \triangleq \lim_{t \to \infty} E(y^* Q_y y + f^* Q_f f) = V^{(k)} + V^{(N-k)} \quad , \quad V^{(k)} = \sum_{\alpha=1}^{N_k} V_\alpha^{(k)} \tag{1.4}$$

where $V^{(k)}$ is the cost of state variables in x_k and $V_\alpha^{(k)}$ is the contribution of each state variable in the component x_k we seek to reduce. We have chosen the performance measure V to include both the vector y (the overall system output) and the vector f of constraint forces. The motivation for including f is to keep the nonworking constraint forces near their original mean squared values. Previous component cost applications [2, 3] have not included such terms. We expect these terms to be important in N body problems, since errors in the forces at the hinge points may be greatly magnified at the end of a "chain" of bodies. From this motivation we suggest that the sequence in which the N bodies are reduced is proportional to the mass or inertia of each body.

The definition of the component cost $V_\alpha^{(k)}$ is

$$V_\alpha^{(k)} = \lim_{t\to\infty} E\left[\frac{\partial(y^* Q_y y + f^* Q_f f)}{\partial x_\alpha} x_\alpha\right] \tag{1.5}$$

where x_α is the αth state variable in vector x_k in the state form of (1.1), $\dot{x} = Ax + B(u + w)$, $\begin{bmatrix} y \\ f \end{bmatrix} = Cx$. $V_\alpha^{(k)}$ is finite if there are no unstable modes observable in either y or x_α. It has been shown [2] that $V_\alpha^{(k)}$ has the value given by the calculation

$$V_\alpha^{(k)} = [XC^* QC]_{n+\alpha,n+\alpha}, \quad n \triangleq \sum_{j=1}^{N-1} N_j \tag{1.6}$$

where X is the state covariance satisfying

$$0 = XA^* + AX + BWB^* \ , \ E w(t) w^T(\tau) = W\delta(t-\tau) .$$

The state $x_\alpha^{(k)}$ which has the smallest $V_\alpha^{(k)}$ is a candidate for deletion.

We shall show that the canonical correlation analysis method compromises performance in terms of y(t) and f(t), but tends to preserve stability properties very nicely. We shall show that the component cost method sometimes compromises stability properties, but tends to preserve performance of y(t) and f(t) when system is stable. We shall combine these methods to take advantage of both desirable properties.

II. SYSTEM DESCRIPTION

The multi-component mechanical system (1.1) is described in matrix form by

$$\begin{bmatrix} M_1 & & & \\ & M_2 & & \\ & & \ddots & \\ & & & M_N \end{bmatrix} \begin{bmatrix} \ddot{q}_1 \\ \vdots \\ \ddot{q}_N \end{bmatrix} + \begin{bmatrix} D_1 & & & \\ & D_2 & & \\ & & \ddots & \\ & & & D_N \end{bmatrix} \begin{bmatrix} \dot{q}_1 \\ \vdots \\ \dot{q}_N \end{bmatrix} + \begin{bmatrix} K_1 & & & \\ & K_2 & & \\ & & \ddots & \\ & & & K_N \end{bmatrix} \begin{bmatrix} q_1 \\ \vdots \\ q_N \end{bmatrix}$$

$$= \begin{bmatrix} B_1 \\ \vdots \\ B_N \end{bmatrix} w + \begin{bmatrix} Z_1^* \\ \vdots \\ Z_N^* \end{bmatrix} f \tag{2.1}$$

$$y = [P_1 \ \dots \ P_N] \begin{bmatrix} q_1 \\ \vdots \\ q_N \end{bmatrix} + [R_1 \ \dots \ R_N] \begin{bmatrix} \dot{q}_1 \\ \vdots \\ \dot{q}_N \end{bmatrix} = Pq + R\dot{q} \ , \ y \in \mathbb{R}^{n_y} . \tag{2.2}$$

Also there exist the constraints

$$[Z_1 \ \dots \ Z_N] \begin{bmatrix} \dot{q}_1 \\ \vdots \\ \dot{q}_N \end{bmatrix} = Z\dot{q} = 0 . \tag{2.3}$$

where N is the number of the components of the system. The matrix dimensions are

$q_i \in \mathbb{R}^{N_i}$	configuration coordinate of component i
$w \in \mathbb{R}^{n_u}$	input vector
$f \in \mathbb{R}^{r}$	non-working constraint force vector
$M_i^* = M_i > 0$	$M_i \in \mathbb{R}^{N_i \times N_i}$

$$K_i^* = K_i \geq 0 \qquad K_i \in \mathbb{R}^{N_i \times N_i}$$
$$B_i \in \mathbb{R}^{N_i \times n_u} \qquad Z_i \in \mathbb{R}^{r \times N_i}$$
$$P_i \in \mathbb{R}^{n_y \times N_i} \qquad R_i \in \mathbb{R}^{n_y \times N_i}$$

We shall assume independent constraints so that Z has linearly independent rows. In matrix form our system is described by

$$\begin{cases} M\ddot{q} + D\dot{q} + Kq = Bw + Z^* f \\ Z\dot{q} = 0 \\ y = Pq + R\dot{q} \, . \end{cases} \tag{2.4}$$

The nonworking constraint force f can be determined from (2.4). Since Z is constant, differentiation of the constrtaint equation with respect to time leads to

$$Z\ddot{q} = 0 \, . \tag{2.5a}$$

Also we have

$$\ddot{q} = -M^{-1}D\dot{q} - M^{-1}Kq + M^{-1}Bw + M^{-1}Z^* f \, . \tag{2.5b}$$

Since Z has linearly independent rows, $ZM^{-1}Z^*$ is invertible. Hence, the two equations (2.5a) and (2.5b) yield

$$f = [ZM^{-1}Z^*]^{-1} ZM^{-1} [D\dot{q} + Kq - Bw] \, . \tag{2.6}$$

System (2.4) is not minimal. That is, one can reduce the dimension of q without changing the transfer function between input w and output y. One way to accomplish this reduction without changing coordinates is the following; Rearrange the order of the elements of the vector q in (2.3) (and hence the columns of Z) until Z has the structure $Z = [Z_a \;\; Z_b]$ where Z_a is square and nonsingular. The q_a is redundant and can be eliminated by substitution of

$$q = \begin{bmatrix} q_a(t) \\ q_b(t) \end{bmatrix} = \begin{bmatrix} -Z_a^{-1}Z_b \\ I \end{bmatrix} q_b \triangleq Z_o q_b \tag{2.7}$$

Since $Z_o^* Z^* = 0$, we get the equation of motion in the smaller vector q_b by premultiplying the first equation of (2.4) by matrix Z_o^*,

$$\mathcal{M}\ddot{q} + \mathcal{D}\dot{q} + \mathcal{K}q = \mathcal{B}w \tag{2.8}$$

where

$$\mathcal{M} \triangleq Z_o^* M Z_o \ , \ \mathcal{D} \triangleq Z_o^* D Z_o \ , \ \mathcal{K} \triangleq Z_o^* K Z_o \ ,$$

$$\mathcal{B} \triangleq Z_o^* B \ , \ \text{and} \ q \triangleq q_b \ .$$

Now augment (2.6) to y in the third equation of (2.4) to get the output equations

$$y \triangleq \begin{bmatrix} y \\ f \end{bmatrix} = \mathcal{P}q + \mathcal{R}\dot{q} + Ew \tag{2.9}$$

where

$$\mathcal{P} = \begin{bmatrix} P \\ (ZM^{-1}Z^*)^{-1}ZM^{-1}K \end{bmatrix} Z_o$$

$$\mathcal{R} = \begin{bmatrix} R \\ (ZM^{-1}Z^*)^{-1}ZM^{-1}D \end{bmatrix} Z_o$$

$$E = \begin{bmatrix} 0 \\ -(ZM^{-1}Z^*)^{-1}ZM^{-1}B \end{bmatrix} .$$

Define the state of the ith component

$$x_i = \begin{bmatrix} q_i \\ \dot{q}_i \end{bmatrix} \ , \quad x_i \in \mathbb{R}^{2n_i} \ ,$$

Notice that n_i is the dimension of q_i (after elimination of redundancy) and N_i is the dimension of q_i (before elimination of redundancy).

Then defining the system state vector

$$x \triangleq [x_1^* \; x_2^* \dots x_N^*]^* \quad , \quad x \in \mathbb{R}^n \quad , \quad n \triangleq 2\sum_{i=1}^{N} n_i$$

yields the state equation of the system

$$\begin{cases} \dot{x} = Ax + Bw \\ y = Cx + Ew \end{cases} \tag{2.10}$$

where

$$A = S^* \begin{bmatrix} 0 & I \\ -M^{-1}K & -M^{-1}D \end{bmatrix} S$$

$$B = S^* \begin{bmatrix} 0 \\ M^{-1}B \end{bmatrix}$$

$$C = [\, P \quad R \,]\, S$$

and S is the unitary matrix such that

$$\begin{bmatrix} q \\ \dot{q} \end{bmatrix} = S\,x\,.$$

III. CANONICAL CORRELATION ANALYSIS

In order to obtain the canonical coordinate for the kth component of the system, we should first put x_k at the bottom of the state vector

$$x^k \triangleq [x_1^* \; x_2^* \dots x_{k-1}^* \; x_{k+1}^* \dots x_N^* \;\; x_k^*]^* \triangleq [x_{N-k}^* \;\; x_k^*]^* \tag{3.1}$$

This arrangement can be obtained by the transformation

$$x = T_k x^k \tag{3.2a}$$

where T_k has the form

$$T_k^{-1} = \begin{bmatrix} I_1 & 0 & 0 \\ 0 & 0 & I_2 \\ 0 & I_3 & 0 \end{bmatrix} \tag{3.2b}$$

where I_i are identity matrix and thier dimensions are

$$I_1 \in \mathbb{R}^{m \times m} \qquad m \triangleq 2\sum_{i=1}^{k-1} n_i$$

$$I_2 \in \mathbb{R}^{s \times s} \qquad s \triangleq 2\sum_{i=k+1}^{N} n_i$$

$$I_3 \in \mathbb{R}^{(2n_k) \times (2n_k)}$$

Remark: We define I_i to be an 0×0 matrix (it does not exist) if $I_i \in \mathbb{R}^{m \times m}$ and $m = 0$.

Hence for reduction of the kth body (component) we use the model

$$\begin{cases} \dot{x}^k = A_k x^k + B_k w \\ y = C_k x^k + Ew \end{cases} \tag{3.3}$$

where

$$A_k = T_k^{-1} A T_k$$
$$B_k = T_k^{-1} B$$
$$C_k = C T_k \ .$$

Let $E w(t) = 0$ and $E w(t) w^*(\tau) = W\delta(t-\tau)$. Because term Ew is not changed by coordinate transformation, we can put it aside, and add it back after model reduction. Assume that (A_k, B_k) is controllable, and A_k is stable. Then $X > 0$ satisfies

$$0 = XA_k^* + A_k X + B_k W B_k^*$$

Let $X = \begin{bmatrix} X_{11} & X_{12} \\ X_{12}^* & X_{22} \end{bmatrix}$

where

$$X_{11} \in \mathbb{R}^{(n-2n_k)\times(n-2n_k)}$$
$$X_{22} \in \mathbb{R}^{(2n_k)\times(2n_k)}$$
$$X_{12} \in \mathbb{R}^{(n-2n_k)\times(2n_k)}$$
$$X_{11} > 0$$
$$X_{22} > 0$$

The coordinate transform T that puts vector x^k in C^3(canonical correlation coordinates) [see (1.3)] is computed as follows [5],

$$x^k = T\mathbf{x}^k$$

where

$$T = \begin{bmatrix} T_1 & 0 \\ 0 & T_2 \end{bmatrix} \qquad \begin{matrix} T_1 \in \mathbb{R}^{(n-2n_k)\times(n-2n_k)} \\ T_2 \in \mathbb{R}^{(2n_k)\times(2n_k)} \end{matrix}$$

where T_1 and T_2 are computed from

$$X_{11} = T_a T_a^*$$
$$X_{22} = T_b T_b^*$$
$$T_1 = T_a U$$
$$T_2 = T_b V$$

and the singular value decomposition of $T_a^{-1} X_{12} T_b^{-*} = U\Delta V^*$,

$$\Delta = \begin{bmatrix} \text{diag}\,(\delta_1, \ldots, \delta_{2n_k}) \\ 0 \end{bmatrix} \quad \text{if } (n-2n_k) > 2n_k . \tag{3.4}$$

Then we obtain the C^3

$$\begin{cases} \dot{\mathbf{x}}^k = A_c \mathbf{x}^k + B_c w \\ y = C_c \mathbf{x}^k \end{cases} \tag{3.5}$$

where

$$A_c = T^{-1}A_kT = T^{-1}T_k^{-1}AT_kT$$
$$B_c = T^{-1}B_k = T^{-1}T_k^{-1}B$$
$$C_c = C_kT = CT_kT\,.$$

In these coordinates we have the property:

$$0 = \mathbf{X}^k A_c^T + A_c\mathbf{X}^k + B_cWB_c^*$$

$$\lim_{t\to\infty} E[\mathbf{x}^k\mathbf{x}^{k^*}] \triangleq \mathbf{X}^k = \begin{bmatrix} I_1 & \Delta \\ \Delta^* & I_2 \end{bmatrix}$$

$I_1 \in \mathbb{R}^{(n-2n_k)\times(n-2n_k)}$ identity matrix

$I_2 \in \mathbb{R}^{(2n_k)\times(2n_k)}$ identity matrix .

The state variables corresponding to small $\delta_\alpha << 1$ are to be deleted.

To reduce the model of the $(k+1)^{th}$ body (component), simply set k = k + 1 and return (3.1) $\rightarrow$ (3.5), continue until all bodies are reduced.

IV. COMPONENT COST ANALYSIS

The designer is always interested in the output performance. It is reasonable to delete the state which has very small effect on the output performance. The Component Cost Analysis(CCA) deletes such states.

The basic idea of the CCA has been stated in (1.4) thru (1.6) in the Introduction. That idea is extended to pq-CCA by including more terms in (1.6). Let p = nonnegative integer, q = positive integer, and ρ_i be weighting factors. For the system $\dot{x} = Ax + Bw$ and $y = \mathcal{O}_{pq}x$, define $\mathcal{O}_{pq}$ and $\mathcal{Q}_{pq}$ by

$$\mathcal{O}_{pq} = \begin{bmatrix} CA^{-p} \\ \vdots \\ CA^{-1} \\ C \\ CA \\ \vdots \\ CA^{q-1} \end{bmatrix}, \quad \mathcal{Q}_{pq} = \text{block diag} [\ \cdots \ \rho_i Q \ \cdots \ ] . \tag{4.1}$$

Then the cost value of interest is

$$V = tr[X\mathcal{O}^*_{pq}\mathcal{Q}_{pq}\mathcal{O}_{pq}] , \tag{4.2a}$$

which may also be written

$$V = tr[X \sum_{i=-p}^{q-1} (CA^i)^* (\rho_i Q)(CA^i)] . \tag{4.2b}$$

The CA^{-1} term in $\mathcal{O}_{pq}$ helps preserve steady- state step responses and the C term helps preserve the mean-squared value of the output vector y . The terms CA, ..., CA^{q-1} help preserve the first q time rates of changes in y(t). The overall effect, of course, is a combination of each of these terms. The cost (4.2) reduces to other published results [2-4] for special choices of p, q and weight $\mathcal{Q}_{pq}$. In other words (4.2) is a very general expression of the cost for pq-CCA, and the cost decomposition is

$$V = tr[X\mathcal{O}^*_{pq}\mathcal{Q}_{pq}\mathcal{O}_{pq}] = \sum_{i=1}^{n} V_i , \tag{4.3a}$$

where

$$V_i = [X\mathcal{O}^*_{pq}\mathcal{Q}_{pq}\mathcal{O}_{pq}]_{ii} . \tag{4.3b}$$

Note that (1.6) is a special case of (4.3b) with $p = 0$, $q = 1$ and $\rho_i = 1$.

For the given system one should *properly* select p, q and $\mathcal{Q}_{pq}$ in (4.3). Although the work by Villemagne and Skelton [6] is for system model reduction, it is closely related to the (p,q) selection required above. We cite their result to further motivate our strategy (4.3a) and (4.3b).

Theorem [6, Theorem 5.2]

Suppose the realization (A, B, C) *of a linear system were driven by a zero mean unit intensity white noise process, w(t), yielding the state covariance matrix* X *satisfying* $0 = XA^* + AX + BB^*$. *Define the matrix* O_{pq} *and its singular value decomposition* :

$$O_{pq} \triangleq \begin{bmatrix} CA^{-p} \\ \vdots \\ CA^{q-1} \end{bmatrix} = [U_1 \ U_2] \begin{bmatrix} \Sigma & 0 \\ 0 & 0 \end{bmatrix} \begin{bmatrix} V_1^* \\ V_2^* \end{bmatrix} . \tag{4.4}$$

The reduced order system (A_R, B_R, C_R) *defined by the projection*

$$\begin{aligned} A_R \triangleq LAR \ , \ B_R \triangleq LB \ , \ C_R \triangleq CR \\ L \triangleq V_1^* \ , \ R \triangleq XL^*(LXL^*)^{-1} \end{aligned} \tag{4.5}$$

matches these 2(p + q) *matrix parameters of the original system* (A, B, C):

$$\begin{aligned} CA^{-i}XC^*, \ CA^{-i}B, \ i = 1, 2, ..., p \\ CA^{j}XC^*, \ CA^{j}B, \ j = 0, 1, ..., q-1 \ . \end{aligned} \tag{4.6}$$

These four types of parameters govern the behavior of the first p derivatives of both the transfer function (matrix) $C(j\omega I - A)^{-1}B$ and the power spectral density at $\omega = 0$ and also the first q derivatives of both the transfer function and the power spectral density at $\omega = \infty$. Hence, matching such parameters guarantee the correct steady state response to step inputs (if $q \geq 1$) and the correct impulse response $Ce^{At}B$ and autocorrelation $Ce^{At}XC^*$ in the neighborhood of t = 0 (this neighborhood is broadened by an increase in q).

For our problem, (in C^3, see (3.5)), let Q_i,i=1,2,3 be partitioned parts of

$$O_{pq}^* Q_{pq} O_{pq} = \sum_{i=-p}^{q-1} (CA^i)^* (\rho_i Q)(CA^i) \triangleq \begin{bmatrix} Q_1 & Q_2 \\ Q_2^* & Q_3 \end{bmatrix} \tag{4.7}$$

where

$$Q_1 \in \mathbb{R}^{(n-2n_k)\times(n-2n_k)}, \; Q_3 \in \mathbb{R}^{(2n_k)\times(2n_k)}, \; Q_2 \in \mathbb{R}^{(n-2n_k)\times(2n_k)}.$$

Then V_i becomes (in C^3)

$$V = \sum_{i=1}^{n-2n_k} [Q_1 + \Delta Q_2^*]_{ii} + \sum_{i=1}^{2n_k} [Q_3 + \Delta^* Q_2]_{ii} \tag{4.8}$$

$$= \sum_{i=1}^{n-2n_k} V_i^{(N-k)} + \sum_{i=1}^{2n_k} V_i^{(k)}$$

where $V_i^{(k)}$ is the component cost of the ith state of the kth component (which we intend to reduce). $V_i^{(N-k)}$ is the component cost of the ith state of other components.

A combination of the canonical correlation analysis and the component cost analysis may be expressed as

$$\gamma_i \triangleq \left[\frac{\delta_i}{\sum_{j=1}^{2n_k} \delta_j}\right]\left[\frac{V_i^{(k)}}{\sum_{j=1}^{2n_k} V_j^{(k)}}\right]. \tag{4.9}$$

Then γ_i represents the relative importance of the ith state in the component k with respect to both correlation and cost of the state. Hence the states with the smallest γ_i are considered for deletion.

V. EXAMPLE: DOUBLE PINNED EULER-BERNOULLI BEAMS

Consider a 2-body system which is composed of two uniform Euler-Bernoulli beams as shown in Figure 1. The first beam is pinned to the inertial frame with stiffness k_1 and viscous damping coefficient c_1, and the second beam is connected to the first beam via another pin with k_2 and c_2. ω_0 is the minimum undamped natural frequency of the beam. The geometrical dimension of the aluminum beam used in this example is 1" x 0.125" x 100" . Each beam is identical and modeled with ten elastic free-free modes. We call this the "full-order model". Rayleigh damping is

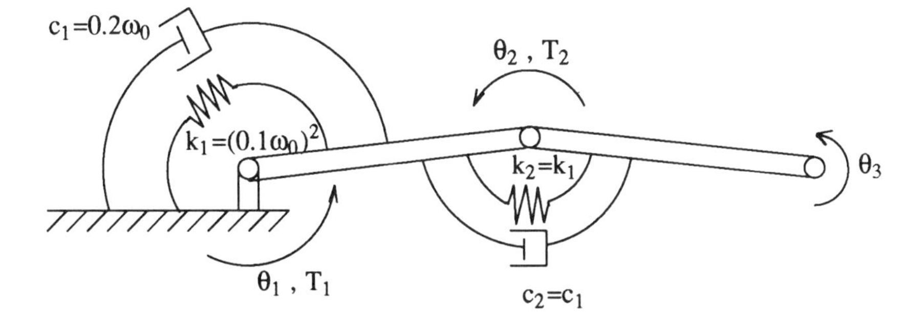

Figure 1. Two Beam System

assumed ; $2\xi_i\omega_i = \alpha + \beta\omega_i^2$, $i = 1, \cdots, n$. α and β are obtained by presetting $\xi_{min} = 0.005$ and $\xi_{max} = 0.1$.

For simulation purposes, impulsive torques of magnitudes 10^{-2} in-lb are applied at pins 1 and 2. For model reduction purpose, actuator noise is assumed to be white noise with intensity of 10^{-4}. Selected as outputs are $y_1 = \theta_1$ = the inertial rotations at pin 1, $y_2 = \theta_2$ = the relative rotation at pin 2, and $y_3 = \theta_3$ = the inertial rotations at the tip of beam 2 (see Figure 1). Nonworking constraint forces at pins 1 and 2 are also included in the output equations since we wish to preserve these forces in the reduced-order model. The output weighting matrix is given as

$$Q = \text{diag}(\cdots [CXC^*]_{ii}^{-1} \cdots) .$$

The reason for choosing Q in this way is that we wish to have consistent units in Q. Based on the Theorem stated in Section IV and some numerical experiences we selected $p = 1$ and $q = 1$ for pq-CCA.

The dynamical equations for each beam are formulated by using Kane's method [9], linearized about zero nominal motion and then assembled into the system equations of motion. After transforming the system equations to C^3, we reduced the full order model in two ways : 1) by canonical correlation analysis (C^3 method), 2) by component cost analysis (pq-CCA method in C^3). For this 2-body system, C^3 method deletes the states corresponding to the smallest δ_i. For the pq-CCA method in C^3 we delete two states at a time , one for each beam, which has the smallest index γ_i in (4.9).

In order to evaluate the reduced-order model, we computed

1. Output Variance ; $Y_i = [Eyy^*]_{ii}$, i = 1,...,5
2. Steady-State Step Response ; $CA^{-1}B$
3. Impulse Response due to T_1 and T_2 .

The output variances of the full-order model are given in Table 1.

Output	y_1	y_2	y_3	f_1	f_2
Variance	0.20392	0.20161	0.46633	4.9854×10^{-4}	1.1784×10^{-4}
Unit	$(\text{degree})^2$			$(\text{lbs})^2$	

Table 1. Output Variances of Full-Order Model

The output variance errors versus the number of deleted states are plotted in Figure 2 for both reduction methods, where Y_i and Y_i^R represent the ith output variances of the full-order model and the reduced-order model, respectively. Given in Table 2 are the steady-state step responses for the full-order and reduced-order models when torques of magnitude 10^{-2} in-lb are simultaneously applied at both pins. Note that in the reduction process at each iteration we deleted one state for each component.

These tables and figures together with the impulse responses in Figures 3 and 4 show that for this example, the pq-CCA method in C^3 is superior to the C^3 method alone in the output performance. More precisely, pq-CCA method in C^3 gives better transient characteristics (see Figures 3 and 4) and better norm of the response(see Figure 2). This is as expected since pq-CCA tends to preserve the norm of the response. Table

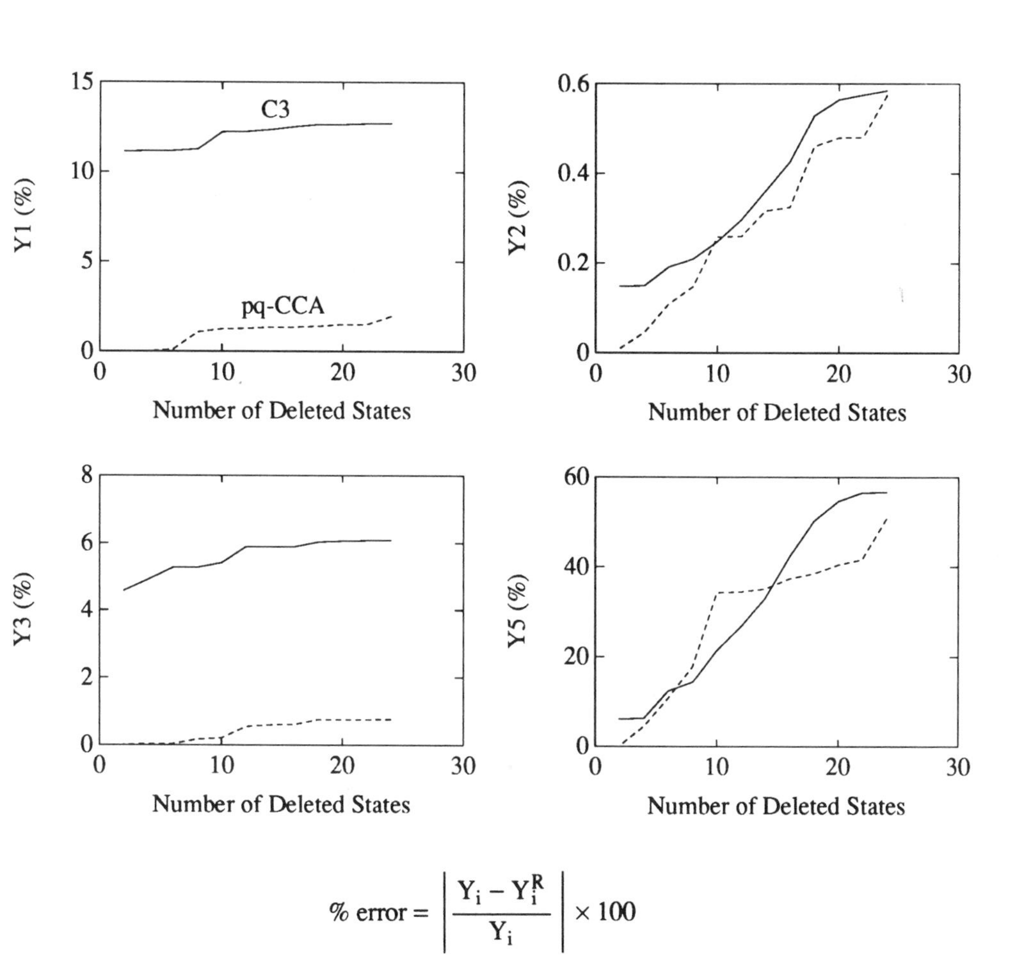

$$\% \text{ error} = \left| \frac{Y_i - Y_i^R}{Y_i} \right| \times 100$$

Figure 2. Output Variance Errors

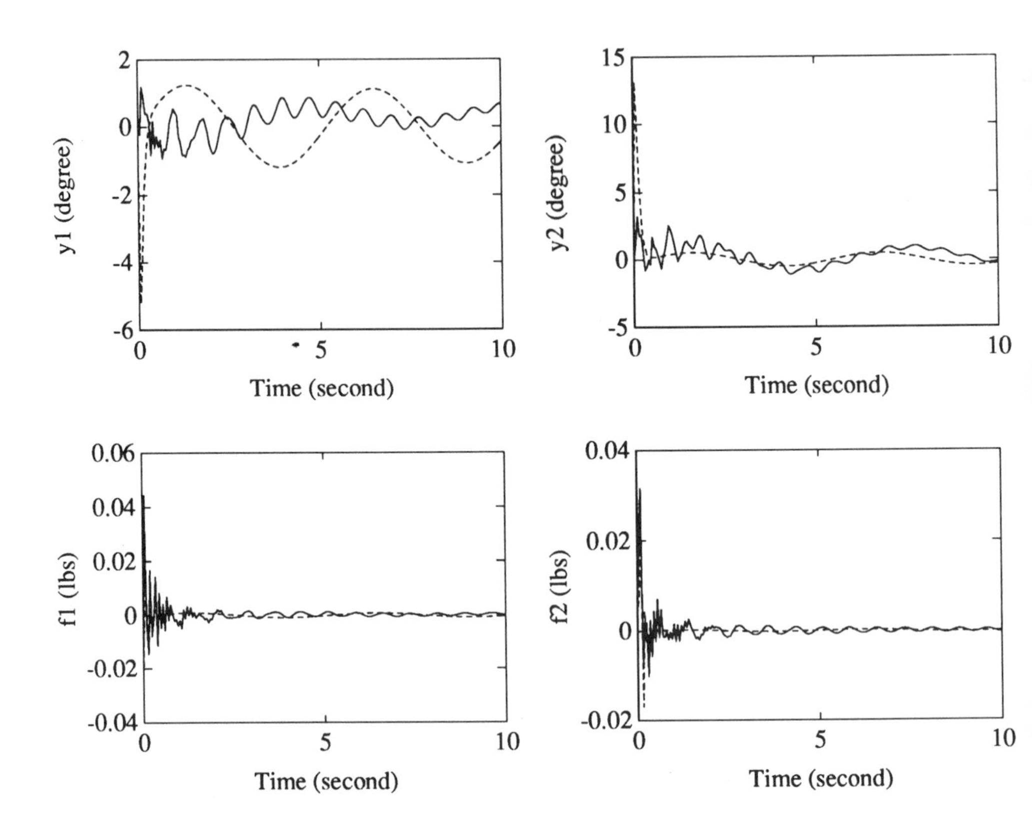

Figure 3. Impulse Response ; C^3 Method

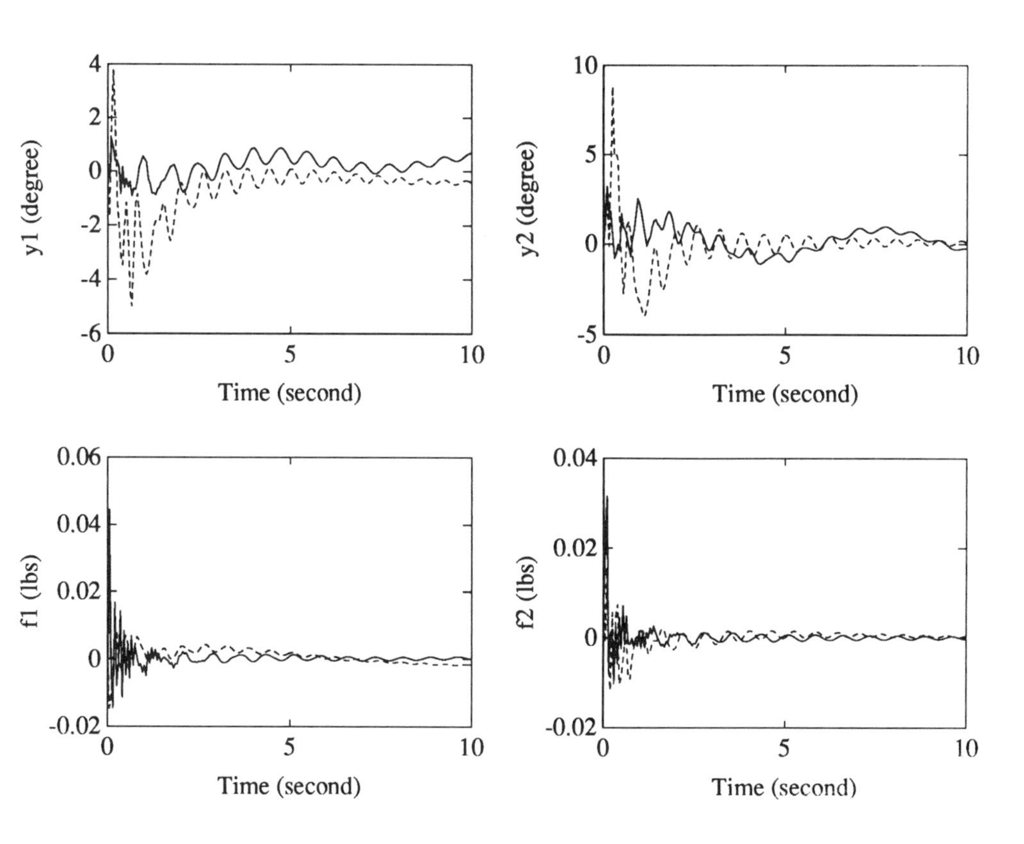

Figure 4. Impulse Response ; pq-CCA Method in C^3

2 shows that both methods have poor steady-state step response, and this is an appropriate topic for further research.

	Full-Order Model (n_x=44)	Reduced-Order Model	
		C^3 (n_x=28)	pq-CCA in C^3 (n_x=28)
y_1 (degree)	-0.22750	-0.03078(86.5)*	0.68349(400.4)
y_2 (degree)	-0.22750	-0.23823(-4.7)	0.30497(234.1)
y_3 (degree)	-0.45501	-0.27267(40.1)	1.37370(401.9)

* number in ( ) indicates percent error.

$$y_i \triangleq [CA^{-1}B]_{i1} + [CA^{-1}B]_{i2}$$

Table 2. Steady-State Step Response

VI. CONCLUSION

Component model reduction approaches are developed in the canonical correlation coordinates. Truncation in these coordinates is performed by two different criteria. The first is the canonical correlation analysis and the second is the generalized component cost analysis(pq-CCA). The component cost method is superior when accuracy in the RMS outputs is desired (because component costs include output information). Truncation using the canonical correlation analysis alone is very effective when it is desired to preserve similar stability margins(for further discussion of this matter, see [5]). The pq-component cost analysis method in the canonical correlation coordinates takes advantages of both analyses. An algorithm is developed to successively truncate the model of each component in an N-component system. The double Euler-Bernoulli beam example shows that the component approach for model reduction can be effective on a multibody system.

REFERENCES

[1] Akaike, A., "Markovian Representations of Stochastic Processes by Canonical Variables," *SIAM J. Contr.*, Vol.13, pp. 162-173, 1975.

[2] Skelton, R.E. and Yousuff, A., "Component Cost Analysis of Large Scale System", *Int. J. Control,* Vol.37, No. 2, pp. 285-304, 1983.

[3] Skelton, R.E. and Singh, R., "Component Model Reduction by Component Cost Analysis", *AIAA Guid. Control Conf.,* Minn., 1988.

[4] Skelton, R.E., *Dynamic Systems Control*; *Linear Systems Analysis and Synthesis,* John Wiley & Sons, New York, 1988.

[5] Villemagne, C. and Skelton, R.E., "Controller Reduction Using Canonical Interactions", *IEEE,Trans. Auto. Contr.,* Vol.33, No.8, 1988.

[6] Villemagne, C. and Skelton, R.E., "Model Reduction Using a Projection Formulation", *Int. J. Control,* Vol.46, No. 6, pp. 2141-2169, 1987.

[7] Craig, R.R., *Structural Dynamics* ; *An Introduction to Computer Methods,* John Wiley & Sons, Inc., New York, 1981.

[8] Craig, R.R., "A Review of Time-Domain and Frequency-Domain Component-Mode Synthesis Methods", *Int. J. of Analytical and Experimental Modal Analysis,* Vol. 2, No. 2, pp. 59-72, April 1987.

[9] Kane, T.R., Likins, P.W. and Levinson, D.A., *Spacecraft Dynamics,* McGraw-Hill Book Company, New York, 1983.

CONTROLLING A FLEXIBLE PLATE TO MIMIC A RIGID ONE*

THOMAS L. VINCENT
YEONG CHING LIN
SHIV P. JOSHI

Aerospace and Mechanical Engineering
University of Arizona
Tucson, Arizona

I. INTRODUCTION

Positioning a flexible structure, such as a flat plate, will generally excite unwanted flexible modes in the structure. In this paper, we examine the problem of designing a control system for a flexible flat plate, using force actuators, which will position the plate with minimal flexible mode excitation. The problem as formulated here is non-traditional since the location of the actuators is considered to be part of the total control design.

We have previously shown [1-3] that traditional state variable feedback methods will generally not yield a satisfactory design for flexible structures. Rather, an open-loop design based on a modal analysis is more effective for positioning. A modal analysis is used to determine both actuator location and

*Research Supported by Honeywell, Sperry Space Systems Division

the open-loop control force to minimize the amount of energy which goes into the flexible modes. Indeed, if it is possible to position the plate so that no energy goes into the flexible modes, the flexible plate will behave as a rigid one. Since this cannot be accomplished realistically with a finite number of actuators in a finite amount of time, our objective is more modest. We say that a flexible plate is an m% mimic of a rigid one if m% of the energy used to position the plate goes into rigid body motion. Our objective is to make m as large as possible for a given number of actuators and a given positioning requirement.

Since a flexible structure is a continuous system, it will have an infinite number of flexible modes. By modeling the plate in terms of a finite number of differential equations, we will be approximating the continuous system in terms of a consistent system with a finite number of modes [4-6]. The degree of accuracy depends on the order of the consistent system [7,8]. Suppose, for example, the consistent system has r rigid body modes and f flexible body modes. Then, in order to arbitrarily position the plate, at least r actuators are needed [9]. If only r actuators used to position the body, there will be some spillover into the flexible energy in the system. In order to avoid this spillover with an $r + f$ degree of freedom consistent system, one could use $r + f$ actuators and move the system model as a rigid body with no energy imparted into flexible modes. Since the system model becomes more accurate as $f \to \infty$, this would be an impractical way of removing all of the flexible energy. However, this procedure does suggest a practical way of eliminating a large portion of the imparted flexible energy associated with positioning the plate. In particular, we will demonstrate here that by using $r + q$ actuators ($1 \leq q \leq f$), it is possible to either suppress q flexible body modes or to suppress a combination of modes so as to minimize total flexible energy input. We develop a procedure which can be used to select actuator location and force input so that either selective modes are suppressed or the amount of flexible energy is minimized.

II. THEORETICAL CONSIDERATIONS

According to theory, the continuous dynamical system representation of a plate is given by [10,11]

$$m(p, q)\ \frac{\partial^2 y(p, q, t)}{\partial t^2} + \frac{Eh^3}{12(1 - \nu)}\ \nabla^4 y(p, q, t) = f(p, q, t)\ , \tag{1}$$

where y is displacement perpendicular to the coordinate axes p and q. The p and q axes are in the plane of the undeflected plate with p in the direction of the width, and q in the direction of the height. Time is denoted by t, h is plate thickness, E is the elastic modulus, ν is Poisson's ratio, and ∇^4 is the differential operator defined by

$$\nabla^4 = \frac{\partial^4}{\partial^4 p} + 2\,\frac{\partial^4}{\partial^2 p \partial^2 q} + \frac{\partial^4}{\partial^4 q}\ .$$

The first term represents inertial forces, the second term represents internal resultant forces due to flexural stiffness, and the term on the right-hand side represents external forces, including control actions. In general, the closed-form solutions to continuous formulation are not available for complex loading and boundary conditions.

A continuous structure with infinite degrees of freedom is represented by finite degrees of freedom in various ways. When the motion of the plate is defined in terms of finite elements in the structure, the formulation is known as a finite element method. The finite element method discretizes a plate into a finite number of smaller plates and expresses the displacement at any point of the continuous structure in terms of a finite number of displacements at the boundaries of the elements,

$$w(p, q, t) = \sum_{i=1}^{3n} N_i(p, q)\ v_i(t)\ , \tag{2}$$

where w(p, q, t) is the displacement at a point in an element, $v_i(t)$ are the generalized displacements at the nodes, and $N_i(p, q)$ are the shape functions. Substituting Eq. (2) into the kinetic and potential energy expressions, we obtain the following forms:

$$T(t) = \frac{1}{2} \sum_{i=1}^{3n} \sum_{j=1}^{3n} m_{ij} \dot{v}_i(t)\ \dot{v}_j(t) \quad \text{and} \quad V(t) = \sum_{i=1}^{3n} \sum_{j=1}^{3n} k_{ij} v_i(t)\ v_j(t)\ . \tag{3,4}$$

where n is the number of nodes. Substituting Eqs. (3) and (4) into Lagrange's equations of motion, we obtain

$$M\ddot{Y} + KY = F , \tag{5}$$

where M and K are mass and stiffness matrices, respectively, F is a vector of generalized nodal forces, and

$$Y = \begin{bmatrix} y_1 \\ \vdots \\ y_n \\ \theta_1 \\ \vdots \\ \theta_n \\ \phi_1 \\ \vdots \\ \phi_n \end{bmatrix} \tag{6}$$

where y_i is the displacement at each node point i, θ_i is the rotation about the p axis at each node point i, and ϕ_i is the rotation about the q axis at each node point i.

The shape functions for the plate element of width a and height b are given as follows [4, 12] (transverse deflection and rotation are the degrees of freedom at each node):

$$\begin{aligned}
N_1 &= (1+2p)(1-p)^2(1+2q)(1-q)^2 & N_7 &= -(3-2p)p^2(1-q)q^2 b \\
N_2 &= (3-2p)p^2(1+2q)(1-q)^2 & N_8 &= -(1+2p)(1-p)^2(1-q)q^2 b \\
N_3 &= (3-2p)p^2(3-2q)q^2 & N_9 &= -p(1-p)^2(1+2q)(1-q)^2 a \\
N_4 &= (1+2p)(1-p)^2(3-2q)q^2 & N_{10} &= (1-p)p^2(1+2q)(1-q)^2 a \\
N_5 &= (1+2p)(1-p)^2 q(1-q)^2 b & N_{11} &= (1-p)p^2(3-2q)q^2 a \\
N_6 &= (3-2p)p^2 q(1-q)^2 b & N_{12} &= -p(1-p)^2(3-2q)q^2 a
\end{aligned} \tag{7}$$

The stiffness and mass matrices are given in Tables 1 and 2.

Equation (5) is a set of second-order differential equations. Control input to the system will be by means of force and/or torque actuators located on the various nodes. A diagonal actuator placement matrix B,

Table 1. Stiffness elements matrix for the free-free plate ($k = Eh^2/[12(1-\nu^2)ab]$).

$\frac{156}{35}(\beta^2+\beta^{-2})+\frac{72}{25}$											
$-\frac{156}{35}\beta^2+\frac{54}{35}\beta^{-2}-\frac{72}{25}$	$\frac{156}{35}(\beta^2+\beta^{-2})+\frac{72}{25}$										
$-\frac{54}{35}(\beta^2+\beta^{-2})+\frac{72}{25}$	$\frac{54}{35}\beta^2-\frac{156}{35}\beta^{-2}-\frac{72}{25}$	$\frac{156}{35}(\beta^2+\beta^{-2})+\frac{72}{25}$									
$\frac{54}{35}\beta^2-\frac{156}{35}\beta^{-2}-\frac{72}{25}$	$-\frac{54}{35}(\beta^2+\beta^{-2})+\frac{72}{25}$	$-\frac{156}{35}\beta^2+\frac{54}{35}\beta^{-2}-\frac{72}{25}$	$\frac{156}{35}(\beta^2+\beta^{-2})+\frac{72}{25}$								
$\left[\frac{22}{35}\beta^2+\frac{78}{35}\beta^{-2}+\frac{6}{25}(1+5\nu)\right]b$	$\left[-\frac{22}{35}\beta^2+\frac{27}{35}\beta^{-2}-\frac{6}{25}(1+5\nu)\right]b$	$\left(-\frac{13}{35}\beta^2-\frac{27}{35}\beta^{-2}+\frac{6}{25}\right)b$	$\left(\frac{13}{35}\beta^2-\frac{78}{35}\beta^{-2}-\frac{6}{25}\right)b$	$\left(\frac{4}{35}\beta^2+\frac{52}{35}\beta^{-2}+\frac{8}{25}\right)b^2$							
$\left[-\frac{22}{35}\beta^2+\frac{27}{35}\beta^{-2}-\frac{6}{25}(1+5\nu)\right]b$	$\left[\frac{22}{35}\beta^2+\frac{78}{35}\beta^{-2}+\frac{6}{25}(1+5\nu)\right]b$	$\left(\frac{13}{35}\beta^2-\frac{78}{35}\beta^{-2}-\frac{6}{25}\right)b$	$\left(-\frac{13}{35}\beta^2-\frac{27}{35}\beta^{-2}+\frac{6}{25}\right)b$	$\left(-\frac{4}{35}\beta^2+\frac{18}{35}\beta^{-2}-\frac{8}{25}\right)b^2$	$\left(\frac{4}{35}\beta^2+\frac{52}{35}\beta^{-2}+\frac{8}{25}\right)b^2$						
$\left(\frac{13}{35}\beta^2+\frac{27}{35}\beta^{-2}-\frac{6}{25}\right)b$	$\left(-\frac{13}{35}\beta^2+\frac{78}{35}\beta^{-2}+\frac{6}{25}\right)b$	$-\left[\frac{22}{35}\beta^2+\frac{78}{35}\beta^{-2}+\frac{6}{25}(1+5\nu)\right]b$	$\left[\frac{22}{35}\beta^2-\frac{27}{35}\beta^{-2}+\frac{6}{25}(1+5\nu)\right]b$	$\left(\frac{3}{35}\beta^2+\frac{9}{35}\beta^{-2}+\frac{2}{25}\right)b^2$	$\left(-\frac{3}{35}\beta^2+\frac{26}{35}\beta^{-2}-\frac{2}{25}\right)b^2$	$\left(\frac{4}{35}\beta^2+\frac{52}{35}\beta^{-2}+\frac{8}{25}\right)b^2$					
$\left(-\frac{13}{35}\beta^2+\frac{78}{35}\beta^{-2}+\frac{6}{25}\right)b$	$\left(\frac{13}{35}\beta^2+\frac{27}{35}\beta^{-2}-\frac{6}{25}\right)b$	$\left[\frac{22}{35}\beta^2-\frac{27}{35}\beta^{-2}+\frac{6}{25}(1+5\nu)\right]b$	$-\left[\frac{22}{35}\beta^2-\frac{78}{35}\beta^{-2}+\frac{6}{25}(1+5\nu)\right]b$	$\left(-\frac{3}{35}\beta^2+\frac{26}{35}\beta^{-2}-\frac{2}{25}\right)b^2$	$\left(\frac{3}{35}\beta^2+\frac{9}{35}\beta^{-2}+\frac{2}{25}\right)b^2$	$\left(-\frac{4}{35}\beta^2+\frac{18}{35}\beta^{-2}-\frac{8}{25}\right)b^2$	$\left(\frac{4}{35}\beta^2+\frac{52}{35}\beta^{-2}+\frac{8}{25}\right)b^2$				
$-\left[\frac{78}{35}\beta^2+\frac{22}{35}\beta^{-2}+\frac{6}{25}(1+5\nu)\right]a$	$\left(\frac{78}{35}\beta^2-\frac{13}{35}\beta^{-2}+\frac{6}{25}\right)a$	$\left(\frac{27}{35}\beta^2+\frac{13}{35}\beta^{-2}-\frac{6}{25}\right)a$	$\left[-\frac{27}{35}\beta^2+\frac{22}{35}\beta^{-2}+\frac{6}{25}(1+5\nu)\right]a$	$-\left[\frac{11}{35}(\beta^2+\beta^{-2})+\frac{1}{50}(1+60\nu)\right]ab$	$\left[\frac{11}{35}\beta^2-\frac{13}{70}\beta^{-2}+\frac{1}{50}(1+5\nu)\right]ab$	$\left[-\frac{13}{70}(\beta^2+\beta^{-2})+\frac{1}{50}\right]ab$	$\left[\frac{13}{70}\beta^2-\frac{11}{35}\beta^{-2}-\frac{1}{50}(1+5\nu)\right]ab$	$\left(\frac{52}{35}\beta^2+\frac{4}{35}\beta^{-2}+\frac{8}{25}\right)a^2$			
$\left(-\frac{78}{35}\beta^2+\frac{13}{35}\beta^{-2}-\frac{6}{25}\right)a$	$\left[\frac{78}{35}\beta^2+\frac{22}{35}\beta^{-2}+\frac{6}{25}(1+5\nu)\right]a$	$\left[\frac{27}{35}\beta^2-\frac{22}{35}\beta^{-2}-\frac{6}{25}(1+5\nu)\right]a$	$\left(-\frac{27}{35}\beta^2-\frac{13}{35}\beta^{-2}+\frac{6}{25}\right)a$	$\left[-\frac{11}{35}\beta^2+\frac{13}{70}\beta^{-2}-\frac{1}{50}(1+5\nu)\right]ab$	$\left[\frac{11}{35}(\beta^2+\beta^{-2})+\frac{1}{50}(1+60\nu)\right]ab$	$\left[-\frac{13}{70}\beta^2+\frac{11}{35}\beta^{-2}+\frac{1}{50}(1+5\nu)\right]ab$	$\left[\frac{13}{70}(\beta^2+\beta^{-2})-\frac{1}{50}\right]ab$	$\left(\frac{26}{35}\beta^2-\frac{3}{35}\beta^{-2}-\frac{2}{25}\right)a^2$	$\left(\frac{52}{35}\beta^2+\frac{4}{35}\beta^{-2}+\frac{8}{25}\right)a^2$		
$\left(-\frac{27}{35}\beta^2-\frac{13}{35}\beta^{-2}+\frac{6}{25}\right)a$	$\left[\frac{27}{35}\beta^2-\frac{22}{35}\beta^{-2}-\frac{6}{25}(1+5\nu)\right]a$	$\left[\frac{78}{35}\beta^2+\frac{22}{35}\beta^{-2}+\frac{6}{25}(1+5\nu)\right]a$	$\left(-\frac{78}{35}\beta^2+\frac{13}{35}\beta^{-2}-\frac{6}{25}\right)a$	$\left[-\frac{13}{70}(\beta^2+\beta^{-2})+\frac{1}{50}\right]ab$	$\left[\frac{13}{70}\beta^2-\frac{11}{35}\beta^{-2}-\frac{1}{50}(1+5\nu)\right]ab$	$-\left[\frac{11}{35}(\beta^2+\beta^{-2})+\frac{1}{50}(1+60\nu)\right]ab$	$\left[\frac{11}{35}\beta^2-\frac{13}{70}\beta^{-2}+\frac{1}{50}(1+5\nu)\right]ab$	$\left(\frac{9}{35}\beta^2+\frac{3}{35}\beta^{-2}+\frac{2}{25}\right)a^2$	$\left(\frac{18}{35}\beta^2-\frac{4}{35}\beta^{-2}-\frac{8}{25}\right)a^2$	$\left(\frac{52}{35}\beta^2+\frac{4}{35}\beta^{-2}+\frac{8}{25}\right)a^2$	
$\left[-\frac{27}{35}\beta^2+\frac{22}{35}\beta^{-2}+\frac{6}{25}(1+5\nu)\right]a$	$\left(\frac{27}{35}\beta^2+\frac{13}{35}\beta^{-2}-\frac{6}{25}\right)a$	$\left(\frac{78}{35}\beta^2-\frac{13}{35}\beta^{-2}+\frac{6}{25}\right)a$	$-\left[\frac{78}{35}\beta^2+\frac{22}{35}\beta^{-2}+\frac{6}{25}(1+5\nu)\right]a$	$\left[-\frac{13}{70}\beta^2+\frac{11}{35}\beta^{-2}+\frac{1}{50}(1+5\nu)\right]ab$	$\left[\frac{13}{70}(\beta^2+\beta^{-2})-\frac{1}{50}\right]ab$	$\left[-\frac{11}{35}\beta^2+\frac{13}{70}\beta^{-2}-\frac{1}{50}(1+5\nu)\right]ab$	$\left[\frac{11}{35}(\beta^2+\beta^{-2})+\frac{1}{50}(1+60\nu)\right]ab$	$\left(\frac{18}{35}\beta^2-\frac{4}{35}\beta^{-2}-\frac{8}{25}\right)a^2$	$\left(\frac{9}{35}\beta^2+\frac{3}{35}\beta^{-2}+\frac{2}{25}\right)a^2$	$\left(\frac{26}{35}\beta^2-\frac{3}{35}\beta^{-2}-\frac{2}{25}\right)a^2$	$\left(\frac{52}{35}\beta^2+\frac{4}{35}\beta^{-2}+\frac{8}{25}\right)a^2$

$\beta = \frac{b}{a}$

Table 2. Consistent mass matrix for the free-free plate with $m = \rho hab/176{,}400$.

1	24336											
2	8424	24336										
3	2916	8424	24336									
4	8424	2916	8424	24336								
5	3432b	1188b	702b	2028b	$624b^2$							
6	1188b	3432b	2028b	702b	$216b^2$	$624b^2$						
7	-702b	-2028b	-3432b	-1188b	$-162b^2$	$-468b^2$	$624b^2$					
8	-2028b	- 702b	-1188b	-3432b	$-468b^2$	$-162b^2$	$216b^2$	$624b^2$				
9	-3432a	-2028a	- 702a	-1188a	-484ab	-286ab	169ab	286ab	$624a^2$			
10	2028a	3432a	1188a	702a	286ab	484ab	-286ab	-169ab	$-468a^2$	$624a^2$		
11	702a	1188a	3432a	2028a	169ab	286ab	-484ab	-286ab	$-162a^2$	$216a^2$	$624a^2$	
12	-1188a	- 702a	-2028a	-3432a	-286ab	-169ab	286ab	484ab	$216a^2$	$-162a^2$	$-468a^2$	$624a^2$
	1	2	3	4	5	6	7	8	9	10	11	12

$$B = \begin{bmatrix} b_1 & & \cdots & 0 \\ \vdots & b_2 & \ddots & \vdots \\ 0 & & \cdots & b_{3n} \end{bmatrix} \tag{8}$$

is used to specify whether force or torque actuators are located on a particular node or not. In particular, b_i will have the value zero or one. If $b_i = 0$,

for any $i = 1, ..., n$ $\rightarrow$ no force actuator at node i

for any $i = n + 1, ..., 2n$ $\rightarrow$ no θ torque actuator at node $i - n$

for any $i = 2n + 1, ..., 3n$ $\rightarrow$ no ϕ torque actuator at node $i - 2n$.

Similarly, $b_i = 1$ implies the existence of an actuator on a node. The control design process which follows could be used to determine values for the b_i's. The actuators are assumed to produce generalized forces subject to control bounds of the form

$$|u_i| \le U_i \qquad i = 1, ..., 3n , \tag{9}$$

where U_i, $i = 1, ..., 3n$, are fixed constants. It follows that

$$F = Bu . \tag{10}$$

Equation (5) may be reduced to a space representation for computer simulation purposes by choosing $X_1 = Y$ and $X_2 = \dot{Y}$. We then obtain, from (5) and (10),

$$\dot{X}_1 = \dot{Y} = X_2 \tag{11}$$

$$\dot{X}_2 = \ddot{Y} = - M^{-1}KX_1 + M^{-1}Bu . \tag{12}$$

Equations (11) and (12) are rewritten as

$$\dot{X} = \begin{Bmatrix} \dot{X}_1 \\ \dot{X}_2 \end{Bmatrix} = \begin{bmatrix} 0 & I \\ -M^{-1}K & 0 \end{bmatrix} \begin{Bmatrix} X_1 \\ X_2 \end{Bmatrix} + \begin{Bmatrix} 0 \\ M^{-1}B \end{Bmatrix} u . \tag{13}$$

Equation (13) is in state space form, where u is the vector of external forces and X is the vector of state variables.

Another way of choosing state variables is to let

$$M\dot{Y} = P . \tag{14}$$

Then,

$$\dot{P} = M\ddot{Y} = - KY + Bu \tag{15}$$

and

$$\dot{Y} = M^{-1}P . \tag{16}$$

Hence, we have

$$\dot{X} = \begin{Bmatrix} \dot{Y} \\ \dot{P} \end{Bmatrix} = \begin{bmatrix} 0 & M^{-1} \\ -K & 0 \end{bmatrix} \begin{Bmatrix} Y \\ P \end{Bmatrix} + \begin{Bmatrix} 0 \\ B \end{Bmatrix} u . \tag{17}$$

Consider now the frequencies associated with the solution to the equations

$$|M\lambda^2 - K| = 0 . \tag{18}$$

These frequencies will be real [4,7]. Any zero frequency corresponds to a rigid body mode and any non-zero frequency corresponds to a flexible body mode [4,6].

Corresponding to these frequencies, we obtain modal vectors from the equations,

$$[M\lambda_i^2 - K]\xi_i = 0 \qquad i = 1, ..., 3n , \tag{19}$$

which may be used to form the transformation matrix

$$\Phi = [\xi_1, ..., \xi_{3n}] . \tag{20}$$

Now consider the transformation

$$Y = \Phi z , \tag{21}$$

where z represents a new set of generalized coordinates. Transforming (5) [with F replaced by Bu] by (21), we obtain

$$M\Phi\ddot{z} + K\Phi z = Bu . \tag{22}$$

Multiplying by Φ^T yields

$$\Phi^T M \Phi \ddot{z} + \Phi^T K \Phi z = \Phi^T B u . \tag{23}$$

The transformation Φ has the property that

$$\Phi^T M \Phi \triangleq \overline{M} \tag{24}$$

and

$$\Phi^T K \Phi \triangleq \overline{K} \tag{25}$$

are diagonal matrices [4,6,7]. Hence, (23) is a decoupled system and z represents decoupled generalized coordinates. If we also define

$$\Phi^T B = \overline{B} , \tag{26}$$

then (23) may be written more compactly as

$$\overline{M}\ddot{z} + \overline{K}z = \overline{B}u . \tag{27}$$

III. SUPPRESSION METHODS

Equation (27), written out for the first few terms, will be of the form:

$$\begin{aligned}
\overline{m}_1 \ddot{z}_1 &= \overline{b}_{11} u_1 + \overline{b}_{12} u_2 + \ldots + \overline{b}_{1,3n} u_{3n} \\
\overline{m}_2 \ddot{z}_2 &= \overline{b}_{21} u_1 + \overline{b}_{22} u_2 + \ldots + \overline{b}_{2,3n} u_{3n} \\
\overline{m}_3 \ddot{z}_3 &= \overline{b}_{31} u_1 + \overline{b}_{32} u_2 + \ldots + \overline{b}_{3,3n} u_{3n} \\
\overline{m}_4 \ddot{z}_4 + \overline{k}_4 z_4 &= \overline{b}_{41} u_1 + \overline{b}_{42} u_2 + \ldots + \overline{b}_{4,3n} u_{3n} \\
\overline{m}_5 \ddot{z}_5 + \overline{k}_5 z_5 &= \overline{b}_{51} u_1 + \overline{b}_{52} u_2 + \ldots + \overline{b}_{5,3n} u_{3n} \\
&\vdots \\
\overline{m}_{3n} \ddot{z}_{3n} + \overline{k}_{3n} z_{3n} &= \overline{b}_{3n,1} u_1 + \overline{b}_{3n,2} u_2 + \ldots + \overline{b}_{3n,3n} u_{3n}
\end{aligned} \tag{28}$$

where n is the number of nodes. The first three equations correspond to the three rigid body modes and the remaining equations describe the elastic body modes. The number of equations depends on the number of nodes used in the finite element analysis.

Consider now a rigid plate of the same configuration with the same dimensions and mass as the flexible one. Displacement and angular rotation of this plate may be modeled by

$$m\ddot{y} = F_y$$
$$I_{pp}\ddot{\theta} = M_{pp} \tag{29}$$
$$I_{qq}\ddot{\phi} = M_{qq}$$

where m is the total mass of the plate and I_{pp} and I_{qq} are the principal moments of inertia of the plate about the horizontal and vertical axes through the center of mass of the plate. The generalized displacements, forces, and masses are related to (29) as follows:

$$m = \bar{m}_1 \qquad y = z_1$$
$$I_{pp} = \bar{m}_2 \qquad \theta = z_2 \tag{30}$$
$$I_{qq} = \bar{m}_3 \qquad \phi = z_3$$

and

$$F_y = \bar{b}_{11} u_1 + \dots + \bar{b}_{1,3n} u_{3n}$$
$$M_{pp} = \bar{b}_{21} u_1 + \dots + \bar{b}_{2,3n} u_{3n} \tag{31}$$
$$M_{qq} = \bar{b}_{31} u_1 + \dots + \bar{b}_{3,3n} u_{3n} \ .$$

Since the flexible plate is to mimic the rigid one, we use (29) to determine $F_y(t)$, $M_{pp}(t)$, and $M_{qq}(t)$ to produce the desired positioning. The corresponding control inputs to the actuators must satisfy (31). Clearly, we must have three or more actuators for arbitrary positioning. For example, suppose that n = 25, as illustrated in Figure 1, and that we had just three force actuators located at nodes 7, 13, and 19. Then, we have $b_7 = b_{13} = b_{19} = 1$ with $b_i = 0$, i = 1, ..., 75, i ≠ 7, 13, 19. Thus, from (31)

$$F_y = \bar{b}_{1,7} u_7 + \bar{b}_{1,13} u_{13} + \bar{b}_{1,19} u_{19}$$
$$M_{pp} = \bar{b}_{2,7} u_7 + \bar{b}_{2,13} u_{13} + \bar{b}_{2,19} u_{19} \tag{32}$$
$$M_{qq} = \bar{b}_{3,7} u_7 + \bar{b}_{3,13} u_{13} + \bar{b}_{3,19} u_{19} \ .$$

Provided an inverse exists for the matrix

$$\begin{bmatrix} \bar{b}_{1,7} & \bar{b}_{1,13} & \bar{b}_{1,19} \\ \bar{b}_{2,7} & \bar{b}_{2,13} & \bar{b}_{2,19} \\ \bar{b}_{3,7} & \bar{b}_{3,13} & \bar{b}_{3,19} \end{bmatrix}, \tag{33}$$

we are then able to solve for $u_7(t)$, $u_{13}(t)$, and $u_{19}(t)$ to produce a motion in the plate. How well this motion mimics the desired rigid plate motion depends on the degree of flexibility of the actual plate. Using only three actuators on a flexible plate will produce an m% mimic, with m < 100. The value for m will depend on the location and type of actuators used. Clearly, using only three actuators, m could be maximized by considering all possible locations and actuator types. However, even under optimal actuator placement, additional and usually significant improvement can be obtained by increasing the number of actuators used. With additional actuators, we can either seek to suppress some flexible body modes or seek to minimize the amount of energy which goes into the flexible body modes.

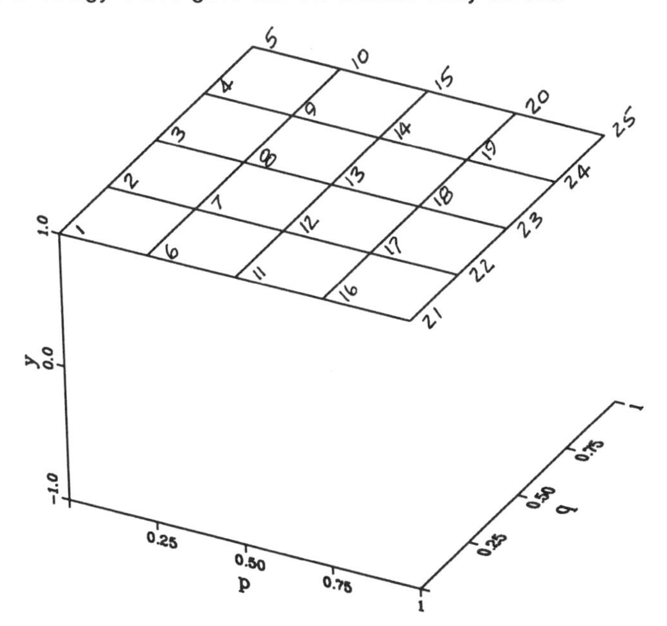

Fig.1. First rigid body mode (displacement).

A. Mode Suppression

Suppose for example that five force actuators are available for positioning the square plate illustrated in Figure 1. Suppose that they are located at nodes 7, 9, 13, 17, and 19. We then have $b_7 = b_9 = b_{13} = b_{17} = b_{19} = 1$ with $b_i = 0$, $i = 1, 75$; $i \neq 7, 9, 13, 17, 19$. We now have from (31)

$$\begin{aligned} F_y &= \bar{b}_{1,7}u_7 + \bar{b}_{1,9}u_9 + \bar{b}_{1,13}u_{13} + \bar{b}_{1,17}u_{17} + \bar{b}_{1,19}u_{19} \\ M_{pp} &= \bar{b}_{2,7}u_7 + \bar{b}_{2,9}u_9 + \bar{b}_{2,13}u_{13} + \bar{b}_{2,17}u_{17} + \bar{b}_{2,19}u_{19} \\ M_{qq} &= \bar{b}_{3,7}u_7 + \bar{b}_{3,9}u_9 + \bar{b}_{3,13}u_{13} + \bar{b}_{3,17}u_{17} + \bar{b}_{3,19}u_{19} \,. \end{aligned} \tag{34}$$

Since there are more unknowns (the u_i's) in (34) than equations, we may now impose additional requirements. In particular, if we wish to suppress all motion in the first two flexible modes, we have from (28) the conditions that

$$\begin{aligned} 0 &= \bar{b}_{4,7}u_7 + \bar{b}_{4,9}u_9 + \bar{b}_{4,13}u_{13} + \bar{b}_{4,17}u_{17} + \bar{b}_{4,19}u_{19} \\ 0 &= \bar{b}_{5,7}u_7 + \bar{b}_{5,9}u_9 + \bar{b}_{5,13}u_{13} + \bar{b}_{5,17}u_{17} + \bar{b}_{5,19}u_{19} \,. \end{aligned} \tag{35}$$

Provided that an inverse exists for the matrix

$$\begin{bmatrix} \bar{b}_{1,7} & \bar{b}_{1,9} & \bar{b}_{1,13} & \bar{b}_{1,17} & \bar{b}_{1,19} \\ \bar{b}_{2,7} & \bar{b}_{2,9} & \bar{b}_{2,13} & \bar{b}_{2,17} & \bar{b}_{2,19} \\ \bar{b}_{3,7} & \bar{b}_{3,9} & \bar{b}_{3,13} & \bar{b}_{3,17} & \bar{b}_{3,19} \\ \bar{b}_{4,7} & \bar{b}_{4,9} & \bar{b}_{4,13} & \bar{b}_{4,17} & \bar{b}_{4,19} \\ \bar{b}_{5,7} & \bar{b}_{5,9} & \bar{b}_{5,13} & \bar{b}_{5,17} & \bar{b}_{5,19} \end{bmatrix}, \tag{36}$$

we will be able to solve for $u_7(t)$, $u_9(t)$, $u_{13}(t)$, $u_{17}(t)$, and $u_{19}(t)$ to not only produce the "rigid body" motion in the plate but to suppress the first two flexible modes as well.

A similar procedure may be used for a different number of actuators, actuator locations, and actuator types. For a fixed number of actuators and for the selected suppressed modes, the actuator types and locations can be optimized in order to maximize m.

Depending on design requirements, mode suppression may be a desirable feature. However, this mode suppression technique does not guarantee that m will be maximized subject to a given number of actuators. This is simply because a given mode suppression configuration may end up dumping energy into unsuppressed modes.

B. Flexible Energy Suppression

If we replace the right-hand side of the fourth and remaining equations of (28) with inputs f_4, f_5, ..., we obtain

$$\begin{aligned} \bar{m}_4 \ddot{z}_4 + \bar{k}_4 z_4 &= f_4 \\ \bar{m}_5 \ddot{z}_5 + \bar{k}_5 z_5 &= f_5 \\ \vdots \qquad & \quad \vdots \end{aligned} \tag{37}$$

or, more generally,

$$\ddot{z}_i + \omega_i^2 z_i = \frac{f_i}{\bar{m}_i} \qquad i = 4, 5, ..., 3n \tag{38}$$

where $\omega_i^2 = \bar{k}_i / \bar{m}_i$. With all initial conditions zero, the solution for each mode is given by

$$z_i = \frac{1}{\omega_i \bar{m}_i} \int_0^t f_i(\tau) \sin \omega_i (t - \tau) d\tau \, . \tag{39}$$

Thus,

$$\dot{z}_i = \frac{1}{\bar{m}_i} \int_0^t f_i(\tau) \cos \omega_i (t - \tau) d\tau \, . \tag{40}$$

The total energy in each mode is given by

$$E_i = \frac{1}{2} \bar{m}_i \dot{z}_i^2 + \frac{1}{2} \bar{k}_i z_i^2 \, , \tag{41}$$

which may be written as

$$2 E_i \bar{m}_i = (\bar{m}_i \dot{z}_i)^2 + (\omega_i \bar{m}_i z_i)^2 \, . \tag{42}$$

Thus,

$$2\,E_i\,\bar{m}_i = \left[\int_0^t f_i(\tau)\cos\omega_i(t-\tau)d\tau\right]^2 + \left[\int_0^t f_i(\tau)\sin\omega_i(t-\tau)d\tau\right]^2, \tag{43}$$

which reduces to

$$2\,E_i\,\bar{m}_i = \left[\int_0^t f_i(\tau)\cos\omega_i\,\tau\,d\tau\right]^2 + \left[\int_0^t f_i(\tau)\sin\omega_i\,\tau\,d\tau\right]^2. \tag{44}$$

From the definition of the Fourier transform of $f_i(t)$,

$$F_i(\omega) = \int_{-\infty}^{\infty} f_i(\tau)\cos\omega\,\tau\,d\tau - i\int_{-\infty}^{\infty} f_i(\tau)\sin\omega\,\tau\,d\tau, \tag{45}$$

we obtain the well-known result that if $f_i(t)$ is zero everywhere except on the interval (0, t) [13],

$$2\,E_i\,\bar{m}_i = |F_i(\omega_i)|^2. \tag{46}$$

We will now examine the energy under three common inputs.

Under a constant input f_i, given by

$$f_i = A_i, \tag{47}$$

the energy in the flexible modes as given by (44) reduces to

$$2\,E_i\,\bar{m}_i = \frac{A_i^2}{\omega_i^2}(2 - 2\cos\omega_i\,t) \qquad i = 4, 5, \ldots, 3n. \tag{48}$$

Over a sufficiently long time period, this yields an average energy of

$$E_i = \frac{A_i^2}{\bar{k}_i}. \tag{49}$$

With the input given by (each actuator input is a sine function at frequency ω)

$$f_i = A_i \sin \omega t , \tag{50}$$

the energy as given by (44) reduces to

$$2 E_i \bar{m}_i = \frac{A_i^2}{\omega_i^2 \left[1 - \left(\frac{\omega}{\omega_i}\right)^2\right]^2} \times \left[\left[\sin \omega t - \frac{\omega}{\omega_i} \sin \omega_i t\right]^2 + \left(\frac{\omega}{\omega_i}\right)^2 (\cos \omega t - \sin \omega_i t)^2\right] . \tag{51}$$

Over a sufficiently long time period, this yields an average energy of

$$E_i = \frac{1}{4} \frac{\left[1 + 3 \left(\frac{\omega}{\omega_i}\right)^2\right]}{\left[1 - \left(\frac{\omega}{\omega_i}\right)^2\right]^2} \frac{A_i^2}{\bar{k}_i} . \tag{52}$$

Note that for $\omega/\omega_i << 1$, this reduces to

$$E_i = \frac{A_i^2}{4\bar{k}_i} , \tag{53}$$

and for $\omega/\omega_i >> 1$, we obtain

$$E_i = \frac{3A_i^2}{4\bar{k}_i} \left[\frac{1}{\left(\frac{\omega}{\omega_i}\right)^2}\right] . \tag{54}$$

It is of interest to compare the previous case with an input given by (each actuator input is a cosine function at ω)

$$f_i = A_i \cos \omega t . \tag{55}$$

In this case, the energy is given by

$$2\,E_i\,\bar{m}_i = \frac{A_i^2}{\omega_i^2\left[1-\left(\frac{\omega}{\omega_i}\right)^2\right]^2} \times \left[\left(\frac{\omega}{\omega_n}\sin\omega\,t - \sin\omega_n\,t\right)^2 + (\cos\omega_n\,t - \cos\omega\,t)^2\right]. \tag{56}$$

Over a sufficiently long time period, this yields an average energy of

$$E_i = \frac{1}{4}\,\frac{\left[3+\left(\frac{\omega}{\omega_i}\right)^2\right]}{\left[1-\left(\frac{\omega}{\omega_i}\right)^2\right]^2}\,\frac{A_i^2}{\bar{k}_i}, \tag{57}$$

which differs from our previous result. If $\omega/\omega_i << 1$, this reduces to

$$E_i = \frac{3}{4}\,\frac{A_i^2}{\bar{k}_i}. \tag{58}$$

If $\omega/\omega_i >> 1$, we obtain

$$E_i = \frac{1}{4}\,\frac{A_i^2}{\bar{k}_i}\left[\frac{1}{\left(\frac{\omega}{\omega_i}\right)^2}\right]. \tag{59}$$

We will confine our analysis to these three inputs. Under these inputs, it follows that the average energy present in a given mode will be proportional to the square of the maximum amplitude squared, A_i^2, divided by an appropriate "spring constant" $\tilde{k}_i$, where for a constant input

$$\tilde{k}_i = \bar{k}_i\,, \tag{60}$$

for a sine input,

$$\tilde{k}_i = 4\bar{k}_i \frac{\left[1 - \left(\frac{\omega}{\omega_i}\right)^2\right]^2}{\left[1 + 3\left(\frac{\omega}{\omega_i}\right)^2\right]}, \tag{61}$$

and for a cosine input,

$$\tilde{k}_i = 4\bar{k}_i \frac{\left[1 - \left(\frac{\omega}{\omega_i}\right)^2\right]^2}{\left[3 + \left(\frac{\omega}{\omega_i}\right)^2\right]}. \tag{62}$$

This observation gives us a way of choosing the control for any extra actuators. For example, consider again the situation with five force actuators discussed above and suppose we wish to minimize the energy going into the first three flexible modes, which is given by

$$G = \frac{A_4^2}{\tilde{k}_4} + \frac{A_5^2}{\tilde{k}_5} + \frac{A_6^2}{\tilde{k}_6}. \tag{63}$$

If we use a "bar" to denote the maximum amplitude of a constant, sine, or cosine varying function (e.g., $F_y = \bar{F}_y \sin\omega t$, $u_7 = \bar{u}_7 \sin\omega t$), we may then write the constraints for minimizing (63) by first noting from our definition of f_i [see (28) and (37)] that we have

$$\begin{aligned}
A_4 &= \bar{b}_{4,7}\bar{u}_7 + \bar{b}_{4,9}\bar{u}_9 + \bar{b}_{4,13}\bar{u}_{13} + \bar{b}_{4,17}\bar{u}_{17} + \bar{b}_{4,19}\bar{u}_{19} \\
A_5 &= \bar{b}_{5,7}\bar{u}_7 + \bar{b}_{5,9}\bar{u}_9 + \bar{b}_{5,13}\bar{u}_{13} + \bar{b}_{5,17}\bar{u}_{17} + \bar{b}_{5,19}\bar{u}_{19} \\
A_6 &= \bar{b}_{6,7}\bar{u}_7 + \bar{b}_{6,9}\bar{u}_9 + \bar{b}_{6,13}\bar{u}_{13} + \bar{b}_{6,17}\bar{u}_{17} + \bar{b}_{6,19}\bar{u}_{19}
\end{aligned} \tag{64}$$

and from Eq. (34),

$$\begin{aligned}
\bar{F}_y &= \bar{b}_{1,7}\bar{u}_7 + \bar{b}_{1,9}\bar{u}_9 + \bar{b}_{1,13}\bar{u}_{13} + \bar{b}_{1,17}\bar{u}_{17} + \bar{b}_{1,19}\bar{u}_{19} \\
\bar{M}_{pp} &= \bar{b}_{2,7}\bar{u}_7 + \bar{b}_{2,9}\bar{u}_9 + \bar{b}_{2,13}\bar{u}_{13} + \bar{b}_{2,17}\bar{u}_{17} + \bar{b}_{2,19}\bar{u}_{19} \\
\bar{M}_{qq} &= \bar{b}_{3,7}\bar{u}_7 + \bar{b}_{3,9}\bar{u}_9 + \bar{b}_{3,13}\bar{u}_{13} + \bar{b}_{3,17}\bar{u}_{17} + \bar{b}_{3,19}\bar{u}_{19}
\end{aligned} \tag{65}$$

For a solution to exist, the first three rows of (36) must form a matrix of maximum rank.

C. Computational Considerations

If one chooses the actuator types, number, and locations ahead of time, then the control inputs for both mode suppression and energy suppression may be easily calculated as suggested above. The equations for mode suppression are linear, and we have used LFTDS and LFSDS from IMSL to make these calculations. Energy suppression requires solving a quadratic cost function subject to linear constraints. The program QPROG from IMSL was found to be useful in this regard.

If the actuator types and locations are not given, then a selection process must be implemented. Under either mode or energy suppression, a reasonable objective is to minimize the amount of energy which goes into the flexible modes. We will assume the number of actuators is fixed since the energy in the flexible modes could always be made arbitrarily small by making the number of actuators arbitrarily large. The following procedures could be used:

1. Mode Suppression

For the given number of actuators, one must select all possible actuator locations as represented by matrices of the form of (36). [In the five-actuator example, force actuators were placed at nodes 7, 9, 13, 17, and 19.] For the plate of Figure 1 with five actuators, this corresponds to 17,259,390 different matrices. Most of these matrices will not have an inverse and may be discarded. For those that do have an inverse, one may then calculate the energy which goes into some of the adjacent modes by means of an equation of the form of (63). One may then select which of the configurations minimizes this energy.

2. Energy Suppression

For the given number of actuators, one must again select all possible actuator locations and then examine matrices of the form of the first three rows of (36). For the plate of Figure 1 with five actuators, this again corresponds to 17,259,390 different matrices. For those matrices of maximum rank, the optimization problem described above may be solved. The final actuator locations and types are determined by the minimum of all the minimizations.

Clearly, such a brute force approach leaves much to be desired in terms of computer time. Even if we restrict the class of actuators to just force actuators, 53,130 different matrices still need to be examined. Algorithms which will reduce this computational task have been developed and will be discussed elsewhere.

IV. FREE PLATE EXAMPLE

Consider the uniform flexible plate illustrated in Figure 1. It is assumed that there are no forces acting on the plate other than those that will be imposed by means of actuators. The plate is aluminum, 1 meter square, with a thickness of 0.635×10^{-3} m, and a density distribution of $\rho = 2637.9$ kg/m^3 with Young's modulus of 6.82×10^{10} N/m^2 and Poisson's ratio $\nu = 0.33$. Henceforth, this particular plate will be referred to as the "UA plate."

Our objective is to design a controller for the flexible plate so that it will mimic a rigid one. To make the example specific, we want to have the UA plate oscillate at a frequency of 4.6 Hz in a direction perpendicular to the plane of the plate using only force actuators. The amplitude of oscillation is to be 0.140 mm.

Using a finite element analysis, one may obtain M, K, and B matrices so that the system may be modeled as equations of the form of (5). For the 25-node plate illustrated in Figure 1, we obtain the K and M matrices as given in Tables 1 and 2. The first 10 modal frequencies in hertz according to the finite element analysis are given by

$$\begin{aligned} \lambda_1 &= 0.002 & \lambda_6 &= 3.845 \\ \lambda_2 &= 0.002 & \lambda_7 &= 5.631 \\ \lambda_3 &= 0.002 & \lambda_8 &= 5.631 \\ \lambda_4 &= 2.201 & \lambda_9 &= 9.657 \\ \lambda_5 &= 3.030 & \lambda_{10} &= 9.657 \end{aligned} \tag{66}$$

These compare favorably with experimental results for frequencies in hertz [11,14,15]:

$$\begin{array}{ll} \lambda_1 = 0.000 & \lambda_6 = 3.842 \\ \lambda_2 = 0.000 & \lambda_7 = 5.507 \\ \lambda_3 = 0.000 & \lambda_8 = 5.507 \\ \lambda_4 = 2.122 & \lambda_9 = 9.676 \\ \lambda_5 = 3.112 & \lambda_{10} = 9.676 \end{array} \tag{67}$$

Using (66) along with the K and M matrices as given in Tables 1 and 2, we obtain the modal vectors from (19). The 75 $\times$ 75 transformation matrix Φ as defined by (20), as well as its transpose, then follows. Due to the size of this matrix, we do not reproduce it here. Indeed, it is not necessary as it is far more useful to examine a number of 5 $\times$ 5 matrices formed from Φ^T. The first three rows of Φ^T relate the actuator inputs to the first three rigid body modes. The remaining 72 rows of Φ^T relate the actuator inputs to the flexible body modes [e.g., see (28)]. Since our example involves only force inputs, we are concerned with only the first 25 elements of each row. The 1st element is associated with the force actuator at the 1st node, the 2nd element is associated with the force actuator at the 2nd node, etc. Thus, if we form a 5 $\times$ 5 force effectiveness matrix of the form

$$F_j = \begin{bmatrix} \phi^T_{j,1} & \phi^T_{j,2} & \phi^T_{j,3} & \phi^T_{j,4} & \phi^T_{j,5} \\ \phi^T_{j,6} & \cdot & & & \\ \phi^T_{j,11} & & \cdot & & \\ \phi^T_{j,16} & & & \cdot & \\ \phi^T_{j,21} & & & & \phi^T_{j,25} \end{bmatrix} \tag{68}$$

where $\phi^T_{j,k}$ is the jth row and kth column element of the Φ^T matrix, we have a convenient representation of how a force actuator at a particular node interacts with that mode. For example, a force actuator at node 3 (see Figure 1) will enter the jth mode through $\phi_{j,3}^T$. Table 3 lists F_j for j = 1, ..., 10.

A. Mode Shapes

Each of the modes listed in Table 3 will have a characteristic mode shape. These are illustrated in Figures 1-11. A 64-node analysis was used to generate Figures 4-11 in order to more accurately portray the shapes obtained. Figures 1-3 illustrate the rigid body modes and Figures 4-11 illustrate the first eight flexible body modes.

The zero displacement curves for the first eight flexible mode shapes are denoted by the heavy solid lines of Figures 12-19. Figures 12-14 are in agreement with the experimental results [11,14].

B. Plate Dynamics

In order to illustrate the techniques of mode and energy suppression, we will examine positioning of the plate with one, two, and three force actuators at specified node points. Since our objective is to only displace the plate normal to itself, only a single actuator is needed for this rigid body requirement. We will examine the single-actuator case first. This will allow us to illustrate some of the computations used for the other cases, as well as providing a standard for comparison.

If we place a single actuator at node 13, we see from Table 3 that this actuator will affect only the 1st and 6th modes (of the 11 listed). The equivalent mass of the UA plate is 1.675 kg. The design requirement is that the displacement of the UA plate oscillate according to

$$y = (0.140 \text{ mm})[1 - 2\pi(4.6)t] . \tag{69}$$

From the first equation of (29), we obtain

$$F_y = (0.196 \text{ N}) \cos 2\pi(4.6)t . \tag{70}$$

Thus, from conditions of the form of (34), we find $u_{13}(t)$. In particular, we have

$$F_y = u_{13} . \tag{71}$$

Table 3. The F_j matrices for the first 10 modes.

1st Mode (Rigid Body in y)					6th Mode (3rd Flexible)				
1.0000	1.0000	1.0000	1.0000	1.0000	0.4294	0.1839	0.0616	0.1839	0.429
1.0000	1.0000	1.0000	1.0000	1.0000	0.1839	-0.0376	-0.1477	-0.0376	0.1839
1.0000	1.0000	1.0000	1.0000	1.0000	0.0616	-0.1477	-0.2555	-0.1477	0.0616
1.0000	1.0000	1.0000	1.0000	1.0000	0.1839	-0.0376	-0.1477	-0.0376	0.1839
1.0000	1.0000	1.0000	1.0000	1.0000	0.4294	0.1839	0.0616	0.1839	0.4294
2nd Mode (Rigid Body in θ)					7th Mode (4th Flexible)				
-0.5984	-0.3484	-0.0984	0.1516	0.4016	0.1859	-0.0550	-0.1693	-0.0198	0.2357
-0.5492	-0.2992	-0.0492	0.2008	0.4508	0.1538	-0.0031	-0.0774	-0.0039	0.1450
-0.5000	-0.2500	0.0000	0.2500	0.5000	0.0200	0.0091	0.0000	-0.0091	-0.0200
-0.4508	-0.2008	0.0492	0.2992	0.5492	-0.1450	0.0039	0.0774	0.0031	-0.1538
-0.4016	-0.1516	0.0984	0.3484	0.5984	-0.2357	0.0198	0.1693	0.0550	-0.1859
3rd Mode (Rigid Body in θ)					8th Mode (5th Flexible)				
0.4016	0.4508	0.5000	0.5492	0.5984	0.2357	0.1450	-0.0200	-0.1538	-0.1859
0.1516	0.2008	0.2500	0.2992	0.3484	-0.0198	-0.0039	-0.0091	0.0031	0.0550
-0.0984	-0.0492	0.0000	0.0492	0.0984	-0.1693	-0.0774	0.0000	0.0774	0.1693
-0.3484	-0.2992	-0.2500	-0.2008	-0.1516	-0.0550	-0.0031	0.0091	0.0039	0.0198
-0.5984	-0.5492	-0.5000	-0.4508	-0.4016	0.1859	0.1538	0.0200	-0.1450	-0.2357
4th Mode (1st Flexible)					9th Mode (6th Flexible)				
-0.4437	-0.2495	0.0000	0.2495	0.4437	-0.1434	0.0642	-0.0063	-0.0800	0.1253
-0.2495	-0.1333	0.0000	0.1333	0.2495	-0.1132	0.0702	0.0051	-0.0613	0.1229
0.0000	0.0000	0.0000	0.0000	0.0000	-0.0940	0.0763	0.0000	-0.0763	0.0940
0.2495	0.1333	0.0000	-0.1333	-0.2495	-0.1229	0.0613	-0.0051	-0.0702	0.1132
0.4437	0.2495	0.0000	-0.2495	-0.4437	-0.1253	0.0800	0.0063	-0.0642	0.1434
5th Mode (2nd Flexible)					10th Mode (7th Flexible)				
0.0000	0.2086	0.3021	0.2086	0.0000	0.1253	0.1229	0.0940	0.1132	0.1434
-0.2086	0.0000	0.0889	0.0000	-0.2086	-0.0800	-0.0613	-0.0763	-0.0702	-0.0642
-0.3021	-0.0889	0.0000	-0.0889	-0.3021	-0.0063	0.0051	0.0000	-0.0051	0.0063
-0.2086	0.0000	0.0889	0.0000	-0.2086	0.0642	0.0702	0.0763	0.0613	0.0800
0.0000	0.2086	0.3021	0.2086	0.0000	-0.1434	-0.1132	-0.0940	-0.1229	-0.1253

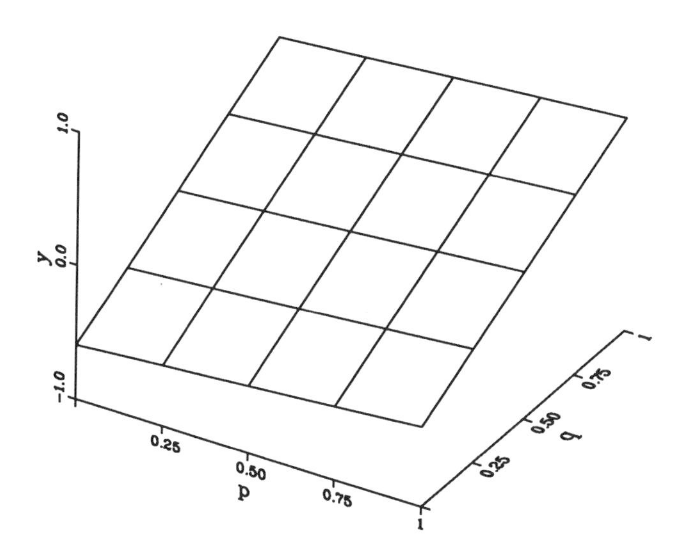

Fig. 2. Second rigid body mode (rotation about p).

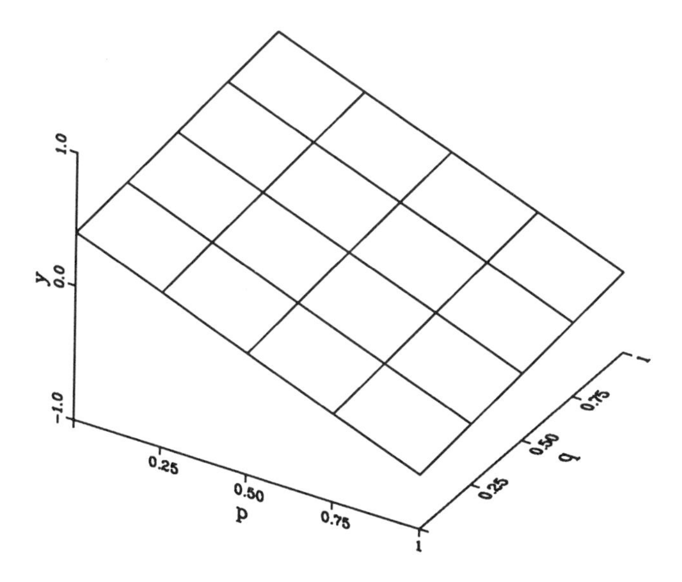

Fig. 3. Third rigid body mode (rotation about q).

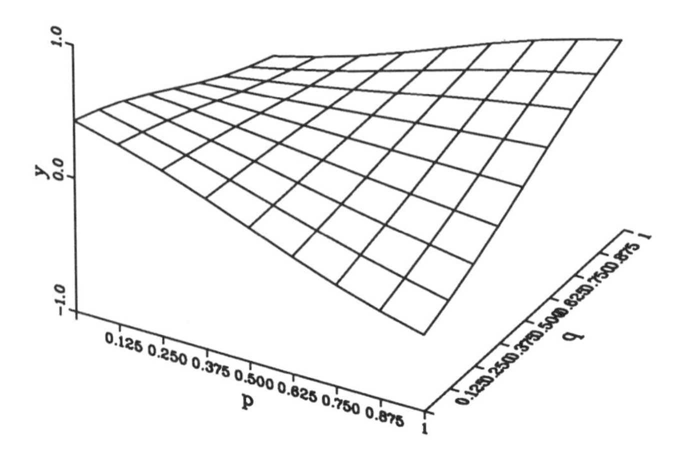

Fig. 4. First flexible body mode.

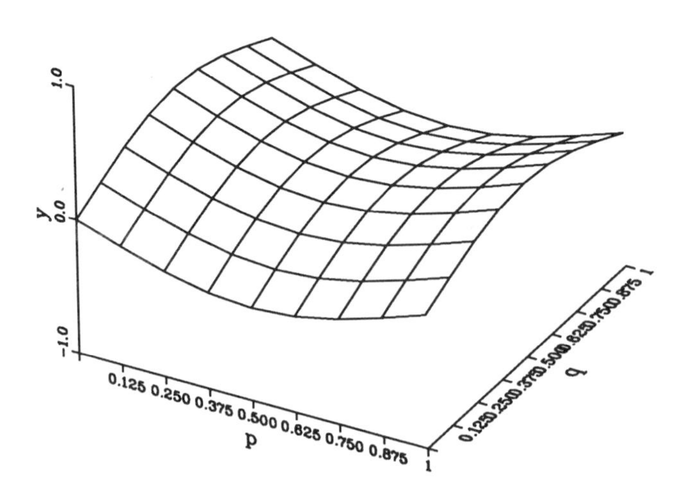

Fig. 5. Second flexible body mode.

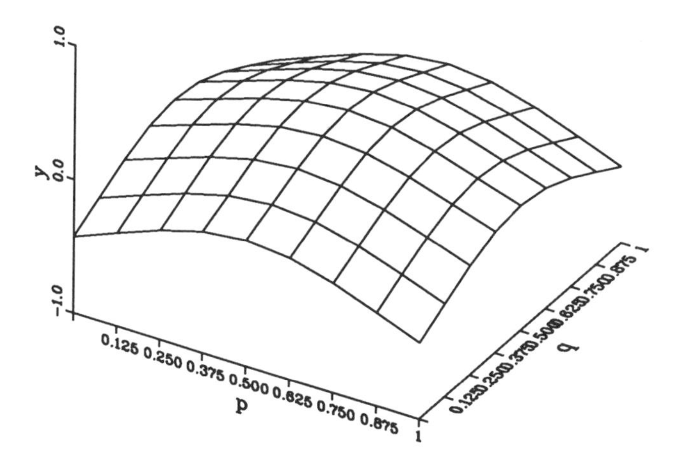

Fig. 6. Third flexible body mode.

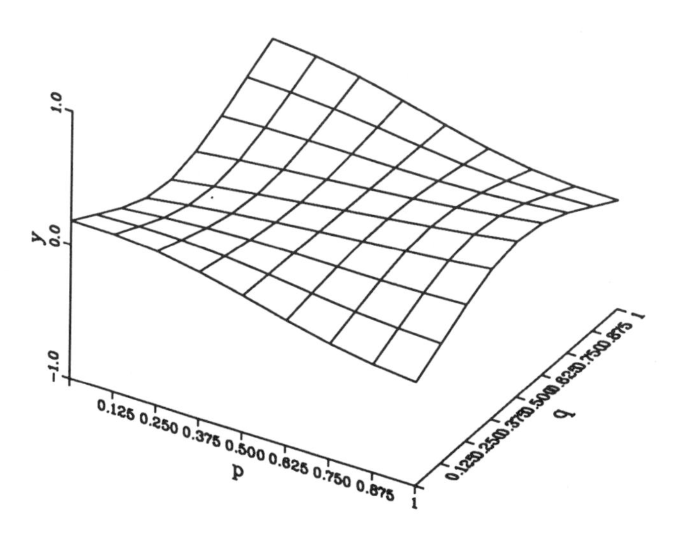

Fig. 7. Fourth flexible body mode.

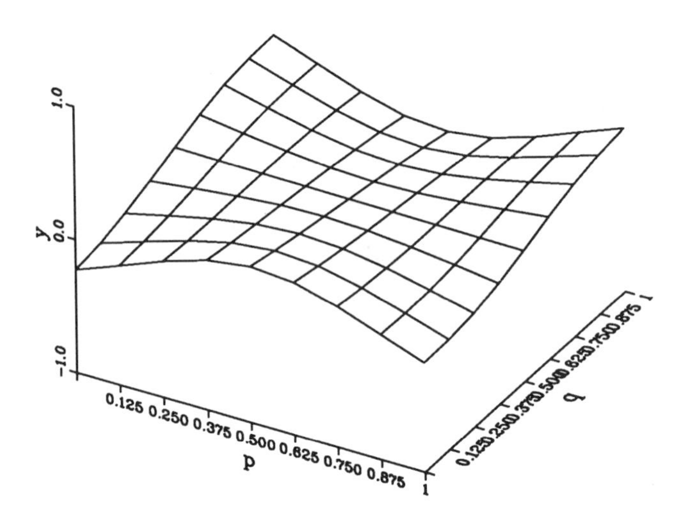

Fig. 8. Fifth flexible body mode.

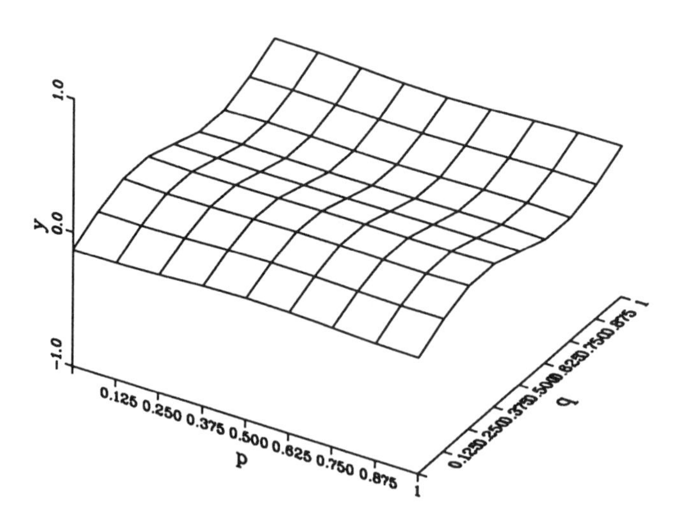

Fig. 9. Sixth flexible body mode.

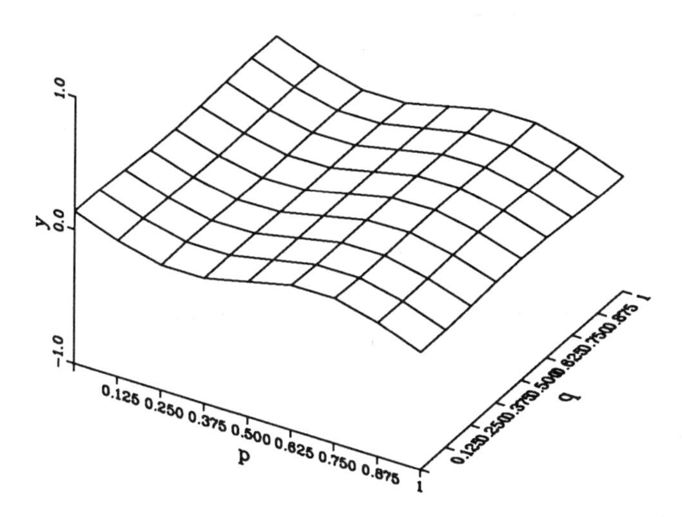

Fig. 10. Seventh flexible body mode.

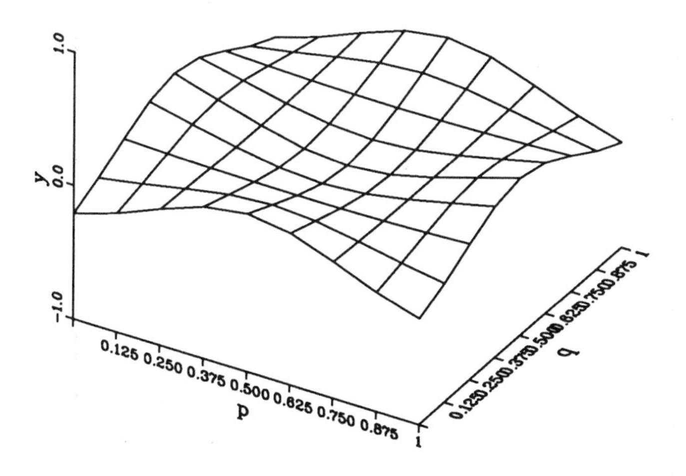

Fig. 11. Eighth flexible body mode.

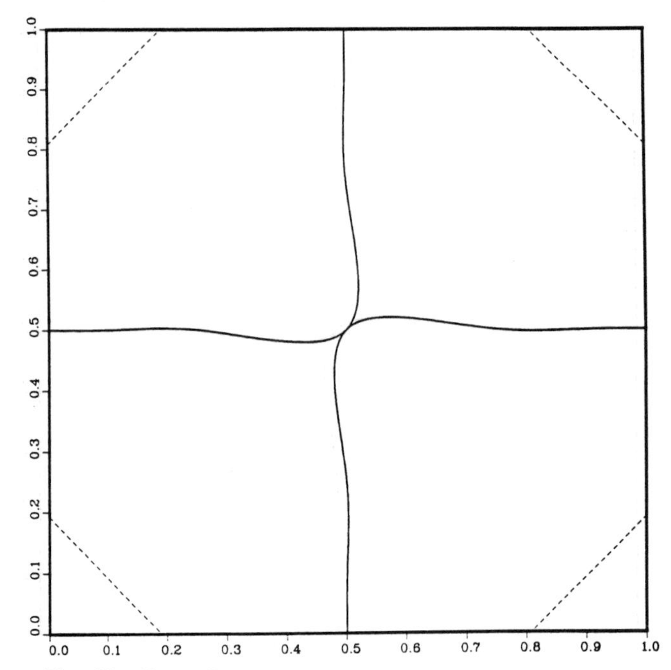

Fig. 12. Zero displacement curve for first flexible mode.

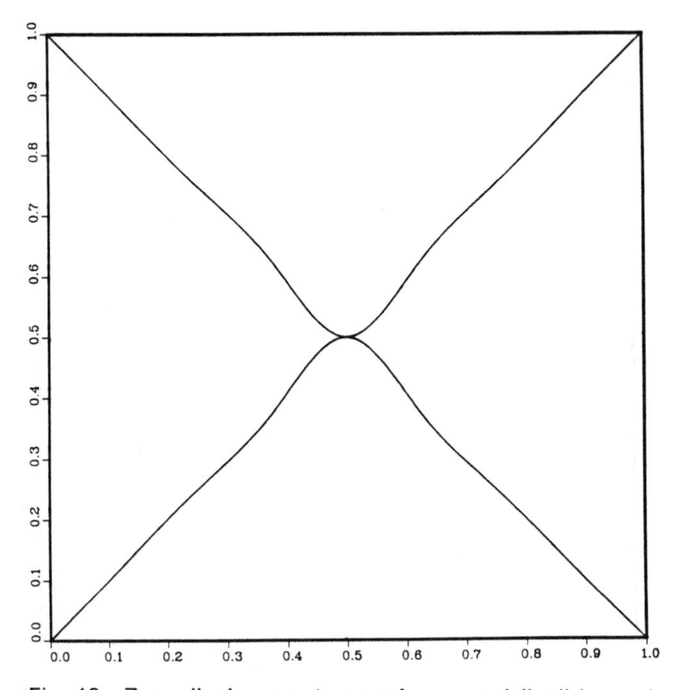

Fig. 13. Zero displacement curve for second flexible mode.

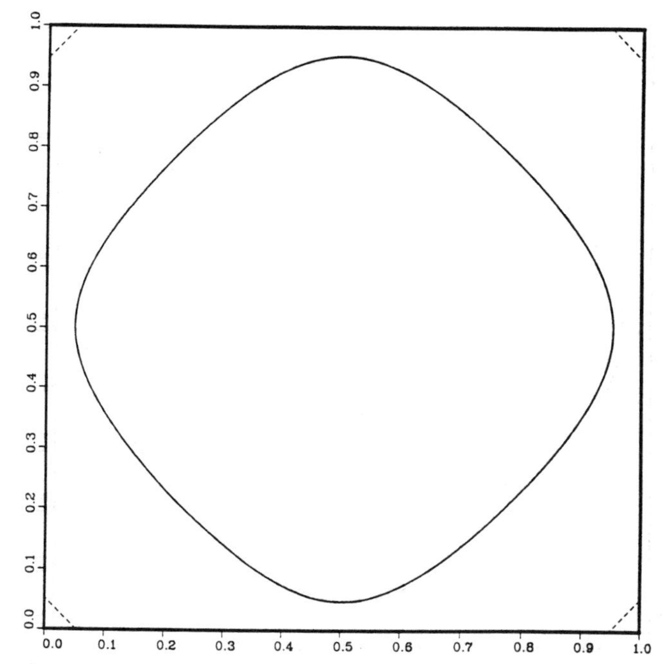

Fig. 14. Zero displacement curve for third flexible mode.

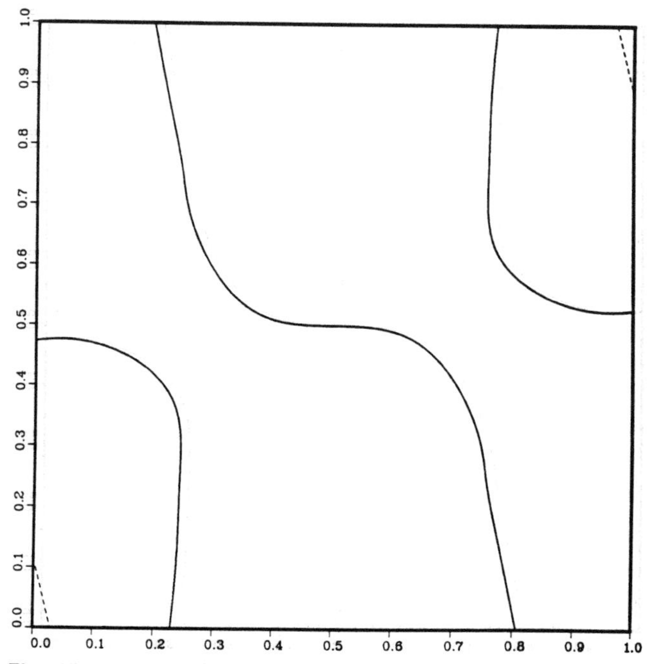

Fig. 15. Zero displacement curve for fourth flexible mode.

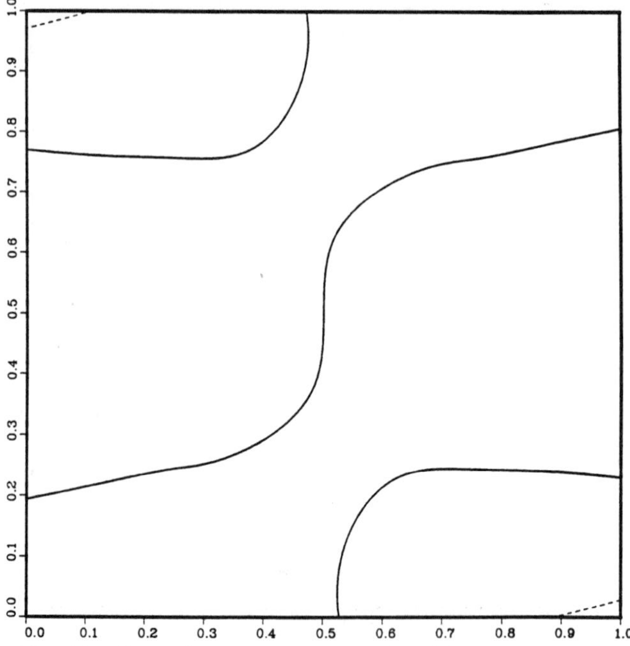

Fig. 16. Zero displacement curve for fifth flexible mode.

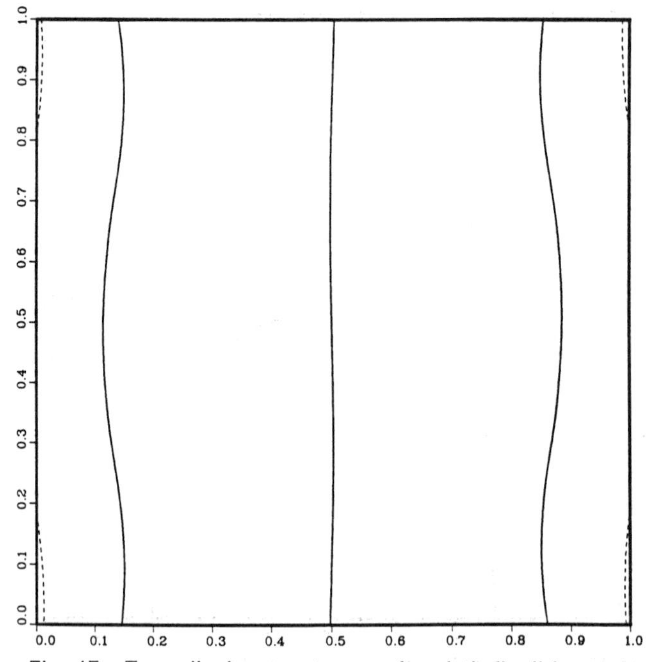

Fig. 17. Zero displacement curve for sixth flexible mode.

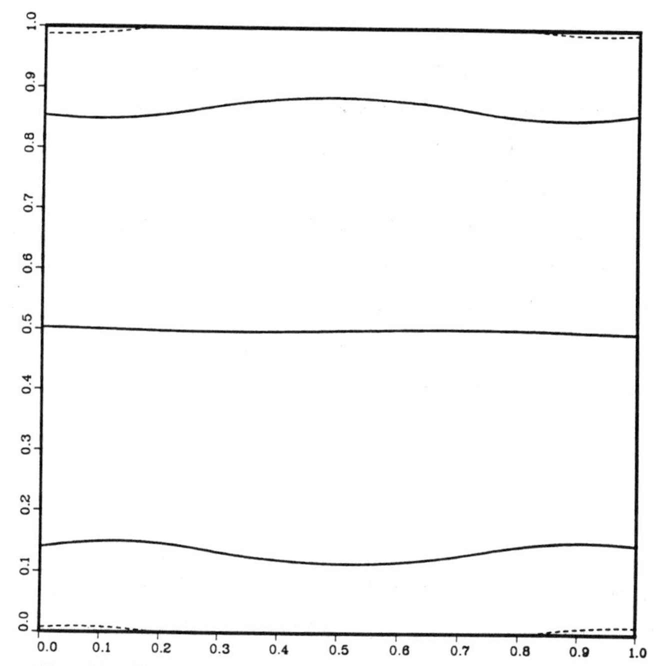

Fig. 18. Zero displacement curve for seventh flexible mode.

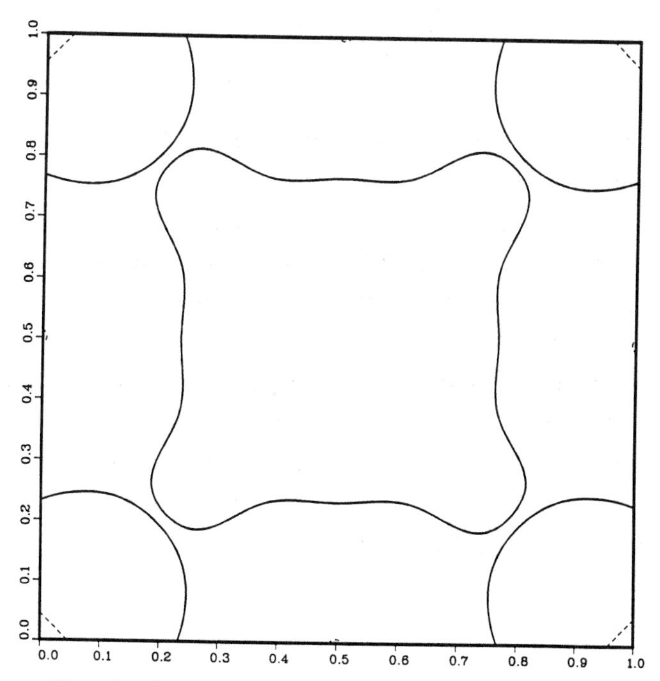

Fig. 19. Zero displacement curve for eighth flexible mode.

Applying this control to the flexible plate as modeled by (5), we obtain its dynamical response. We will characterize this response here in terms of the FFT spectrum of the displacement obtained at nodes 7, 11, 13, 21, and 23, as illustrated in Figures 20a-20e. As predicted, the 1st mode at 0 Hz and the 6th mode at 3.845 Hz, along with the driving frequency at 4.6 Hz, dominate the dynamics. It is of interest to compare the average energy in the plate with the average energy in each of the excited modes. This may be easily done during the computer simulation run, provided the run is long enough to provide a good average. The total average energy in the plate is given by

$$E_T = \frac{1}{T}\int_0^T \frac{1}{2}(\dot{Y}M\dot{Y} + YKY)dt\ , \tag{72}$$

where Y as defined by (6) is the vector of generalized displacements and M and K are the mass and stiffness matrices previously defined. The total average energy in each mode is given by

$$E_i = \frac{1}{T}\int_0^T \frac{1}{2}(\bar{m}_i\dot{z}_i^2 + \bar{k}_i z_i^2)dt\ , \tag{73}$$

where z_i are the decoupled generalized displacements defined by (21) and $\bar{m}_i$ and $\bar{k}_i$ are the mass and stiffness associated with each mode.

For this one-actuator case, we obtain the energies as summarized in Table 4. Clearly, most of the elastic energy is in the 6th mode. We see that 99.1% of the total energy has gone into the elastic modes, so we have only a 0.9% mimic with the single-actuator case.

In an experimental setting, the energy in a given elastic mode would have to be determined from measurements. For example, if displacement sensors were placed at nodes 7, 11, 13, 21, and 23, then an FFT of these data would produce results similar to Figures 20a-20e. The displacement at a given node j will be a sum of all the modal contributions,

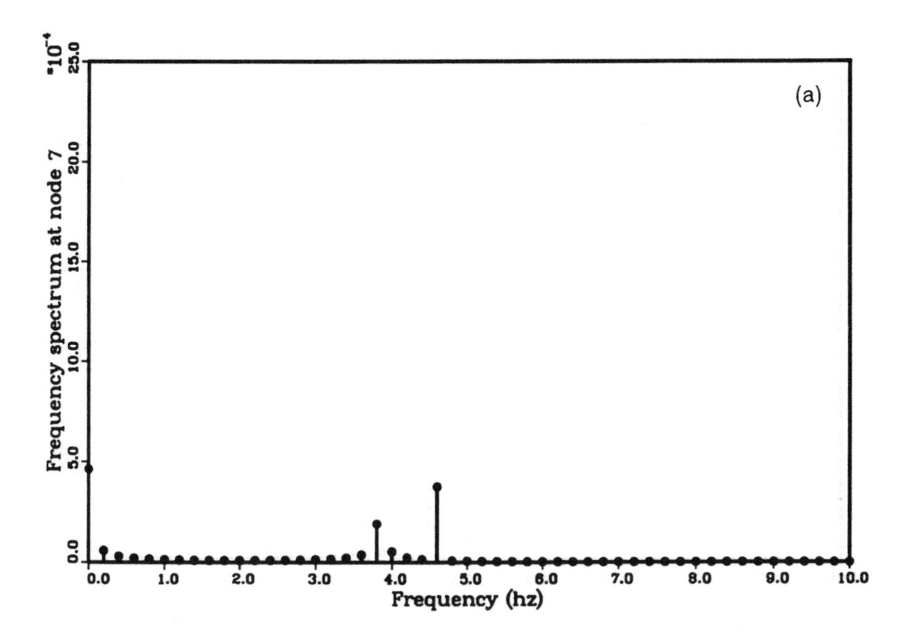

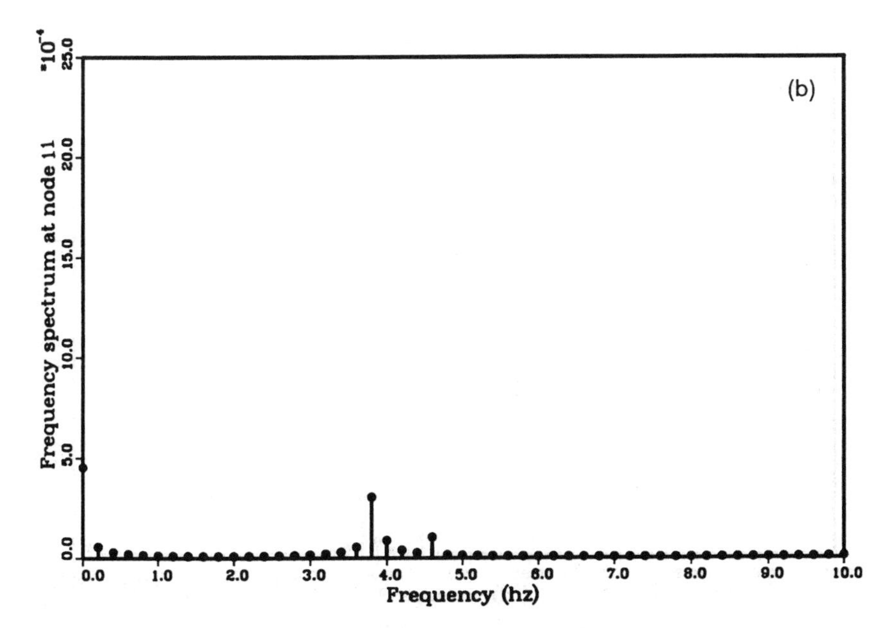

Fig. 20. FFT obtained using a single actuator.

Fig. 20.--Continued.

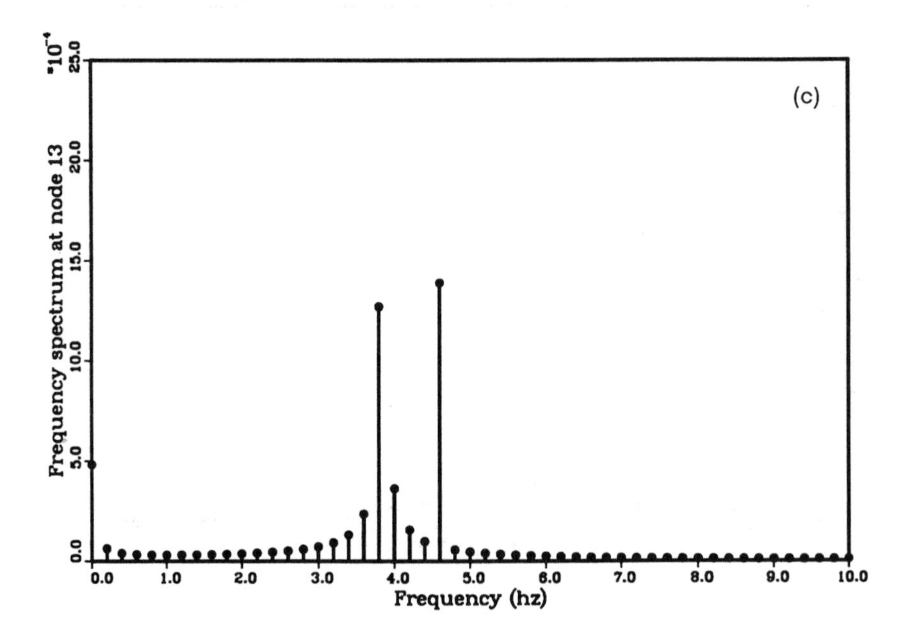

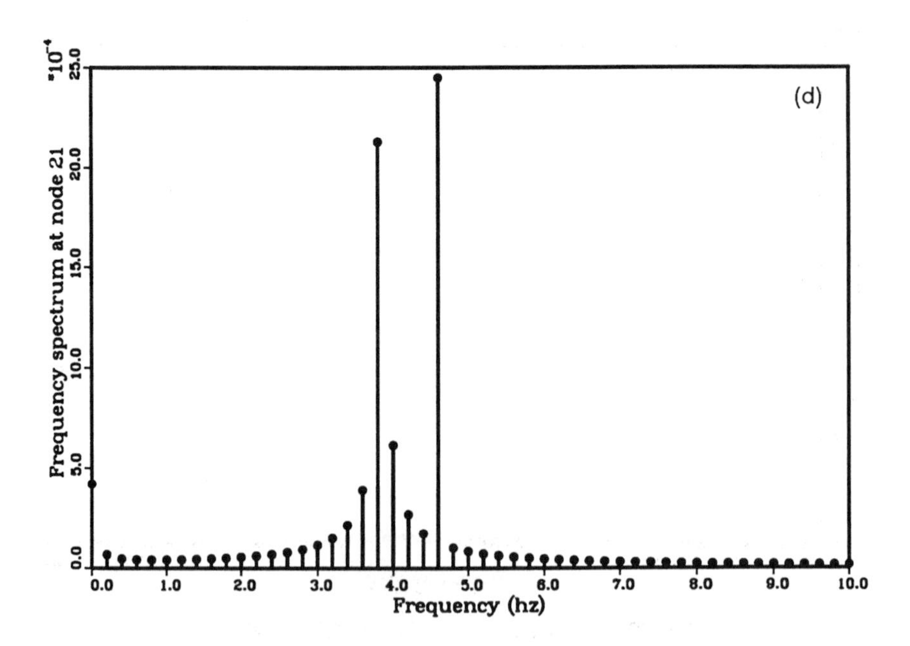

Fig. 20.--Continued.

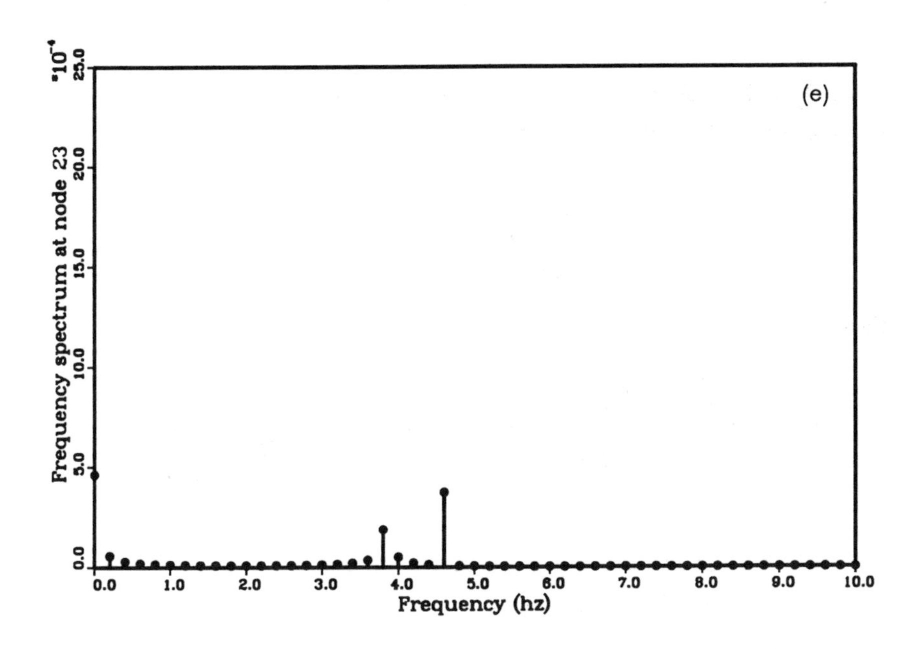

Table 4. Average energy (Newton-meters) components for the four example cases.

Case	E_T Total Average Energy	Modes				
		E_1 1st Rigid	E_4 1st Elastic	E_5 2nd Elastic	E_6 3rd Elastic	$\sum_{i=4}^{75} E_i$ All Elastic
1 actuator	7.5995×10^{-4}	0.0687×10^{-4}	0.0	0.0	7.2198×10^{-4}	7.5308×10^{-4}
2 actuators	3.8925×10^{-4}	0.0687×10^{-4}	0.0	2.8752×10^{-4}	0.4815×10^{-4}	3.8238×10^{-4}
3 actuators (mode suppression)	2.2281×10^{-4}	0.0687×10^{-4}	0.0	1.8684×10^{-4}	0.0	2.1595×10^{-4}
3 actuators (energy suppression)	1.7393×10^{-4}	0.0687×10^{-4}	0.0	1.2765×10^{-4}	0.2482×10^{-4}	1.6706×10^{-4}

$$y_j = R_j + \sum_{i=4}^{75} \phi_{ji} z_i , \tag{74}$$

where R_j is the rigid body contribution which will be at the driving frequency. The solution to (37) with $f_i = A_i \cos\omega t$ is given by

$$z_i = \frac{A_i/\bar{k}_i}{\left[1 - \left[\frac{\omega}{\omega_i}\right]^2\right]} (\cos \omega t - \cos \omega_i t) . \tag{75}$$

Thus,

$$y_j = R_j + \sum_{i=4}^{75} \frac{\phi_{ji} A_i/\bar{k}_i}{\left[1 - \left[\frac{\omega}{\omega_i}\right]^2\right]} \cos \omega t - \sum_{i=4}^{75} \frac{\phi_{ji} A_i/\bar{k}_i}{\left[1 - \left[\frac{\omega}{\omega_i}\right]^2\right]} \cos \omega_i t . \tag{76}$$

This may be compared with the representation obtained from the FFT (no damping in the actual system is assumed),

$$y_j = D_j \sin(\omega t + \alpha) - \sum_{i=4}^{75} D_{ji} \cos \omega_i t , \tag{77}$$

where D_j is the j-node Fourier coefficient corresponding to the driving frequency and D_{ji} is the jth-node Fourier coefficient corresponding to the i mode. Comparing (76) and (77), it follows that

$$D_{ji} = \frac{\phi_{ji} A_i/\bar{k}_i}{\left[1 - \left[\frac{\omega}{\omega_i}\right]^2\right]} . \tag{78}$$

Thus,

$$\frac{A_i^2/\bar{k}_i}{\left[1 - \left[\frac{\omega}{\omega_i}\right]^2\right]^2} = \left[\frac{D_{ji}}{\phi_{ji}}\right]^2 \bar{k}_i . \tag{79}$$

It then follows from (57) that

$$E_i = \frac{\bar{k}_i}{4}\left[3 + \left(\frac{\omega}{\omega_i}\right)^2\right]\left(\frac{D_{ji}}{\phi_{ji}}\right)^2 . \tag{80}$$

Note that the elastic energy in any given mode may be calculated at any node for which that mode is significantly represented. For example, the 6th mode is well represented at node 21 and node 13. From these FFT spectrums, we obtain

$$D_{21,6} = 0.2129 \times 10^{-2} \qquad D_{13,6} = 0.1270 \times 10^{-2} . \tag{81}$$

Correspondingly, we obtain from Table 3

$$\phi_{21,6} = \phi^T_{6,21} = 0.42944 \qquad \phi_{13,6} = \phi^T_{6,13} = -\ 0.25548 . \tag{82}$$

For the 6th mode, $\bar{k}_6 = 21.5291$ and $\omega_6 = 3.8448$. From (80), we obtain

$$E_6 = 5.862 \times 10^{-4} \quad \text{and} \quad E_6 = 5.894 \times 10^{-4} , \tag{83}$$

which is in qualitative agreement with the direct calculation given in Table 4.

C. Mode Suppression

With the two actuators, a form of mode suppression may be employed. We see from Table 3 that, if we place the actuators at nodes 11 and 15, these actuators will affect only the 1st, 5th, and 6th modes, provided that the actuators produce exactly the same force. (The forces will then cancel on modes 2, 3, 7, 8, 9, and 10, with no effect on the 4th mode.) We will take this then as our objective, that is, to prevent energy going into any of the first 10 modes except for the 1st, 5th, and 6th. The force F_y is still determined from (70) so that we can find u_{11} and u_{15} simply from

$$\begin{aligned} 0.196 &= \bar{u}_{11} + \bar{u}_{15} \\ \bar{u}_{11} &= \bar{u}_{15} . \end{aligned} \tag{84}$$

Since smaller forces are used (the amplitude of each force input is 0.098 N), we expect less energy to go into the two elastic modes. Figures 21a-21e illustrate the FFT spectrum of the displacement obtained at nodes 7, 11, 13, 21,

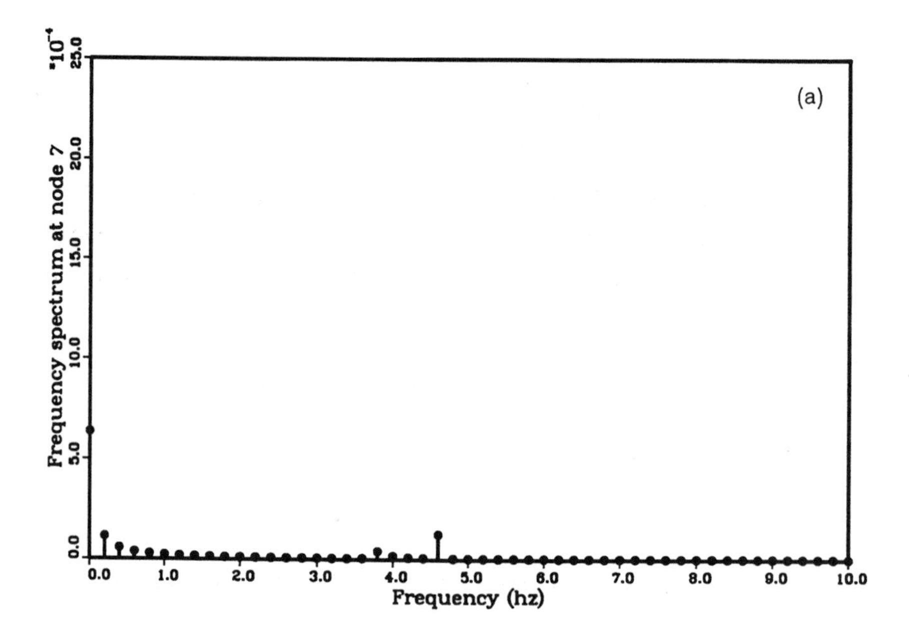

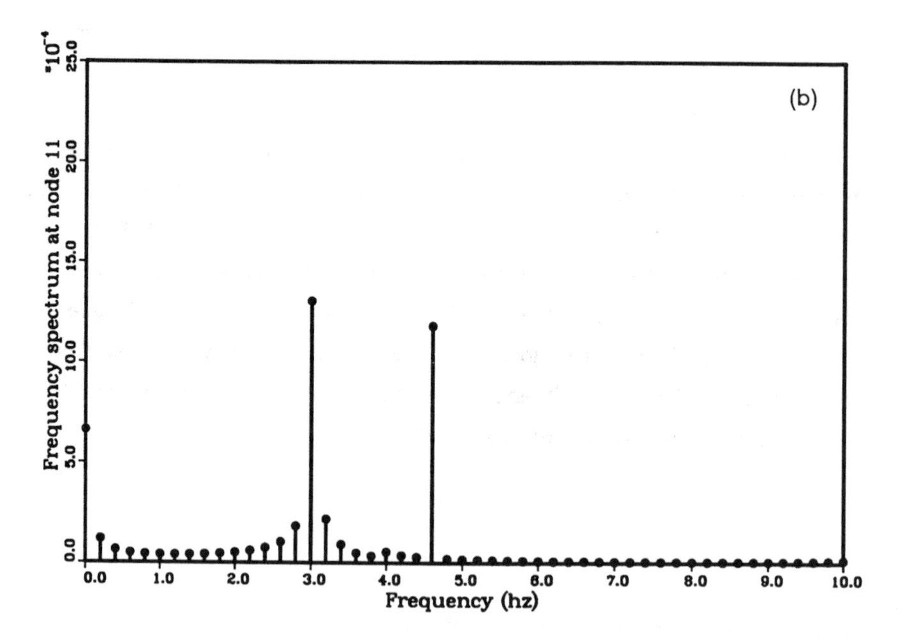

Fig. 21. FFT obtained using two actuators.

Fig. 21.--Continued.

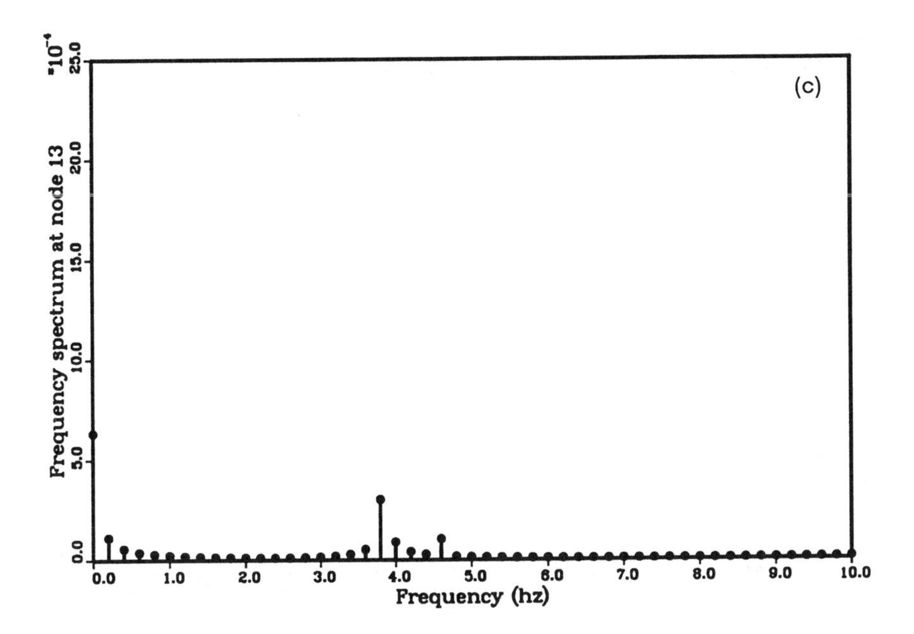

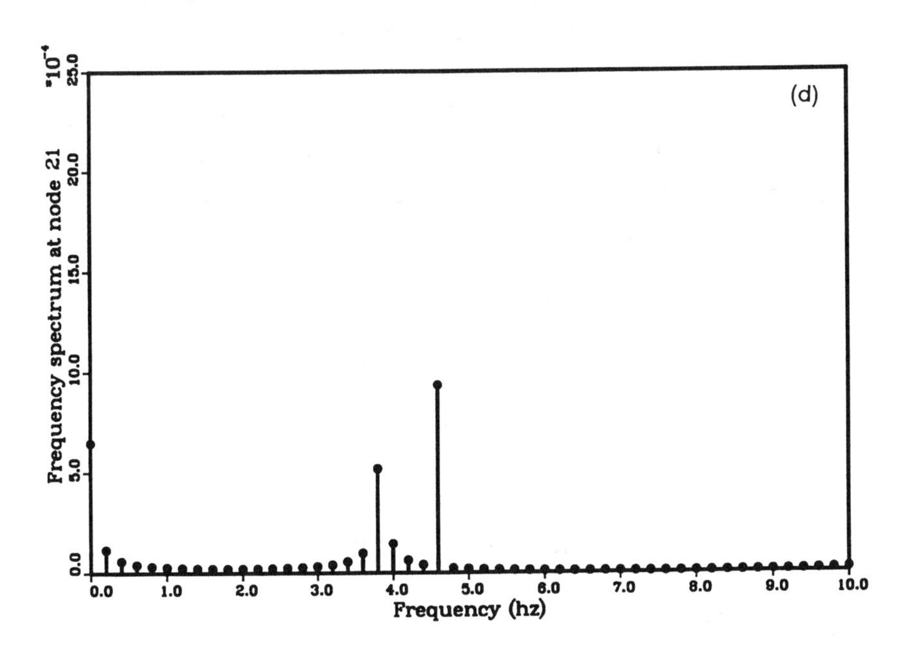

Fig. 21.--Continued.

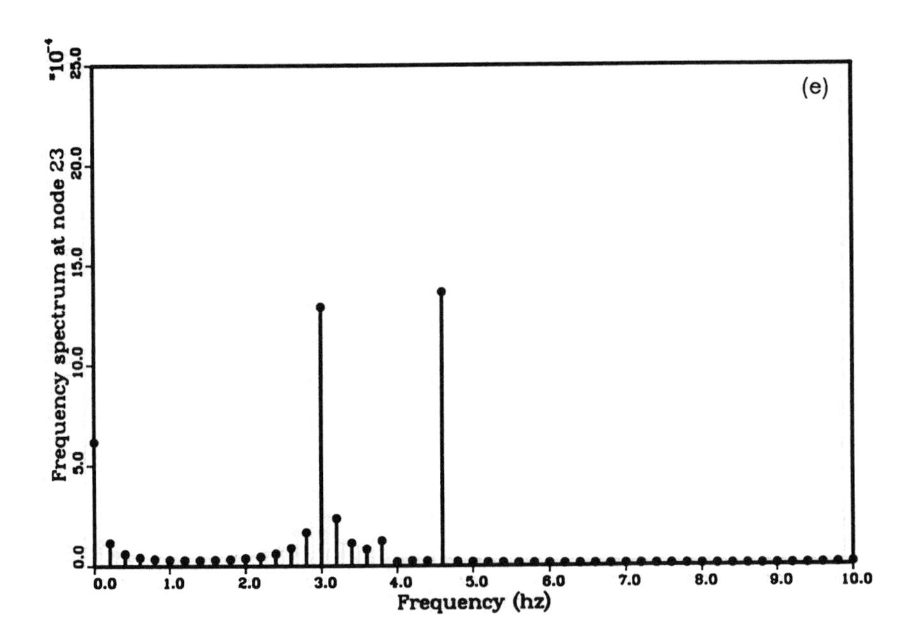

and 23. It is evident from Figures 21a-21e that the expected modes are excited. In every figure, the first mode (0 Hz) and the driving frequency (4.6 Hz) are evident. However, the other two excited modes are not equally represented in every figure. For example, only the 6th mode (3.8 Hz) appears at node 21, whereas the 5th mode (3.0 Hz) is dominate at node 11. This is not unexpected, as the contribution of each mode to the total dipslacement y_j as given by (74) depends on ϕ_{ji}. From Table 3, we have

$$\begin{aligned} \Phi_{21,5} &= \Phi^T_{5,21} = 0 \\ \Phi_{21,6} &= \Phi^T_{6,21} = 0.42944 \ , \end{aligned} \tag{85}$$

which explains the absence of the 5th mode at node 21. Furthermore,

$$\begin{aligned} \Phi_{11,5} &= \Phi^T_{5,11} = 0.30207 \\ \Phi_{11,6} &= \Phi^T_{6,11} = 0.06159 \ , \end{aligned} \tag{86}$$

which explains the results obtained at node 11.

A summary of the average energies for this case are given in Table 4. The majority of elastic energy went into the 5th mode. In this case, 98.2% of the total energy is in the elastic modes, making the two-actuator case a 1.8% mimic. This factor-of-two improvement over the single-actuator case is consistent with the fact that the maximum force produced by each actuator was one-half of that used in the single-actuator case.

Thinking of Figures 20a-20e as experimental data, we can re-derive the energies in the two elastic modes. We obtain $D_{21,6} = 0.5165 \times 10^{-3}$ from Figure 21d and $D_{11,5} = 0.1305 \times 10^{-2}$ from Figure 21b. Corresponding, from Table 3, $\phi_{2,16} = \Phi^T_{6,21} = 0.42944$ and $\phi_{11,5} = \Phi^T_{5,11} = 0.30207$. For the 5th mode, $\bar{k}_5 = 10.2131$ and $\omega_5 = 3.0302$. From (80), we obtain $E_5 = 2.528 \times 10^{-4}$ and (using previously given values for $\bar{k}_6$ and ω_6) $E_6 = 0.3450 \times 10^{-4}$, which is again in qualitative agreement with Table 4.

With three actuators, some additional mode suppression may take place. We see from Table 3 that if we place the actuators at nodes 11, 13, and 15, and if the actuators at 11 and 15 again produce equal forces ($u_{11} = u_{15}$), there will be no input into modes 2, 3, 4, 7, 8, 9, and 10. Because of the requirement that $u_{11} = u_{15}$ and the fact that $\phi^T_{5,13} = 0$, the 5th mode cannot be suppressed (which would require $u_{11} = -u_{15}$). However, the additional actuator may be used to suppress the 6th by choosing the controls to satisfy

$$\begin{aligned} 0.196 &= \bar{u}_{11} + \bar{u}_{13} + \bar{u}_{15} \\ \bar{u}_{11} &= \bar{u}_{15} \\ 0 &= 0.0616\, \bar{u}_{11} - 0.2555\, \bar{u}_{13} + 0.616\, \bar{u}_{15} \,. \end{aligned} \tag{87}$$

These equations yield force amplitudes of 0.079 N, 0.038 N, and 0.079 N for $\bar{u}_{11}$, $\bar{u}_{13}$, and $\bar{u}_{14}$, respectively. Figures 22a-22e illustrate the FFT spectrum of the displacement at nodes 7, 11, 13, 21, and 23. The driving frequency is again at 4.6 Hz. These figures show the complete suppression of the 6th (3.8 Hz) mode.

Note that the impact of the 5th mode varies from node to node. While the 6th mode was suppressed, significant energy still went into the 5th (3 Hz) mode (see Table 3). In this case, 96.9% of the total energy is in the elastic modes, making this three-actuator case a 3.1% mimic.

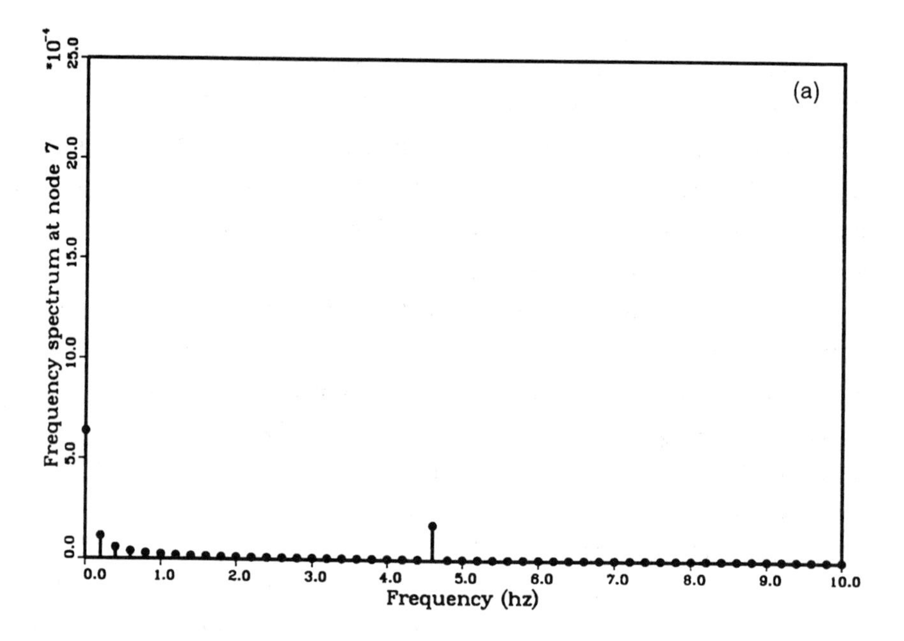

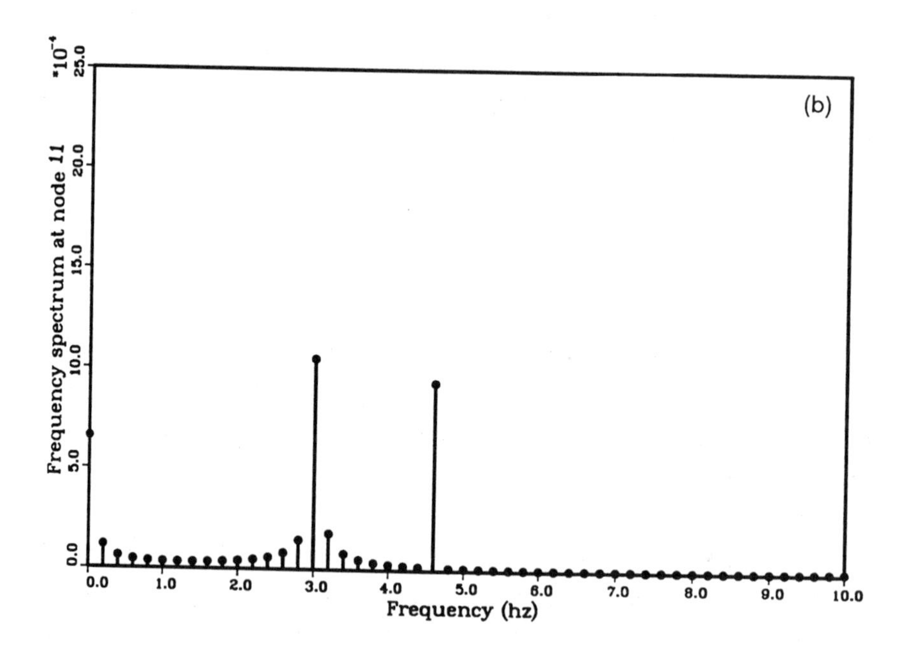

Fig. 22. FFT obtained using three actuators (mode suppression).

Fig. 22.--Continued.

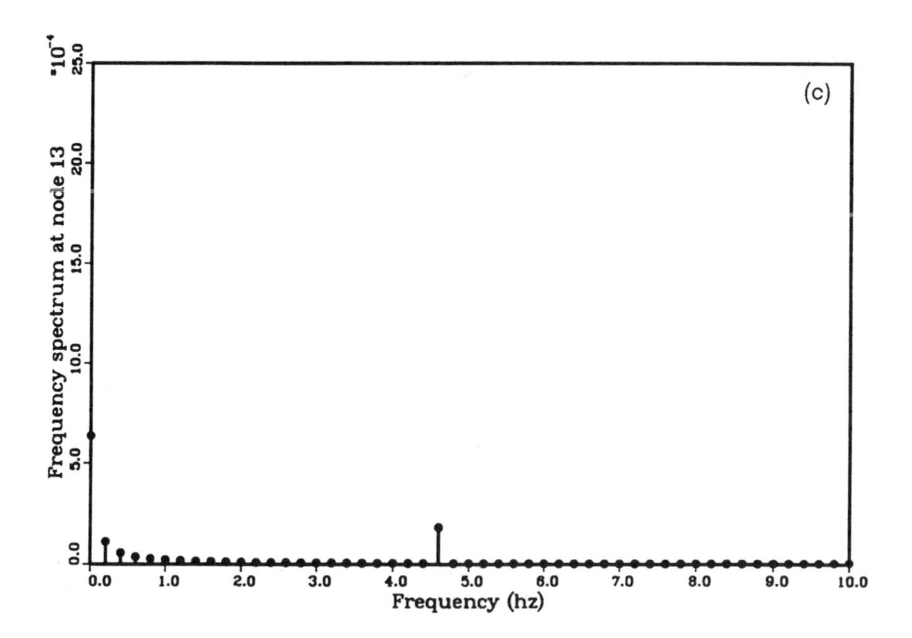

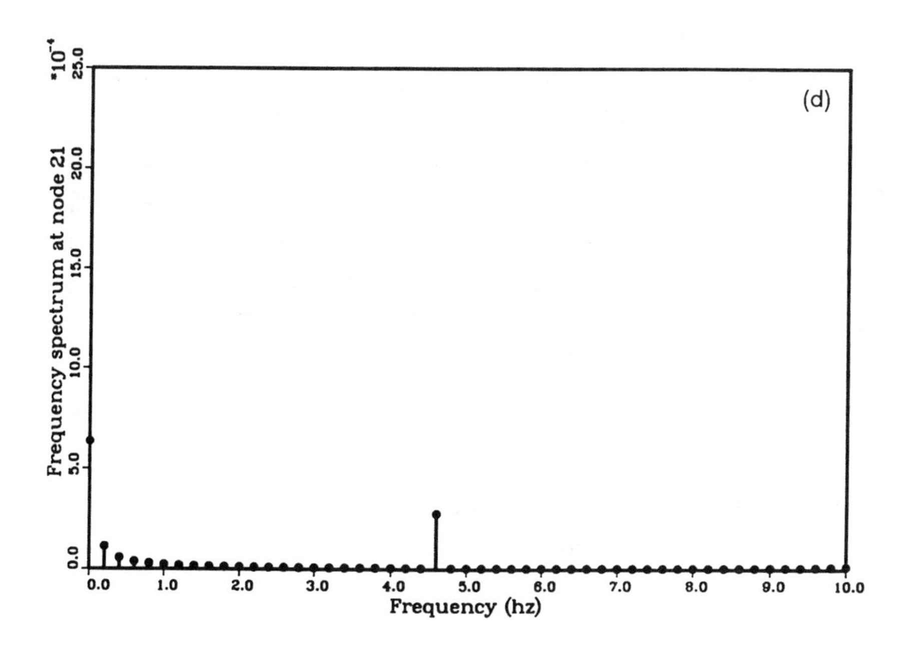

Fig. 22.--Continued.

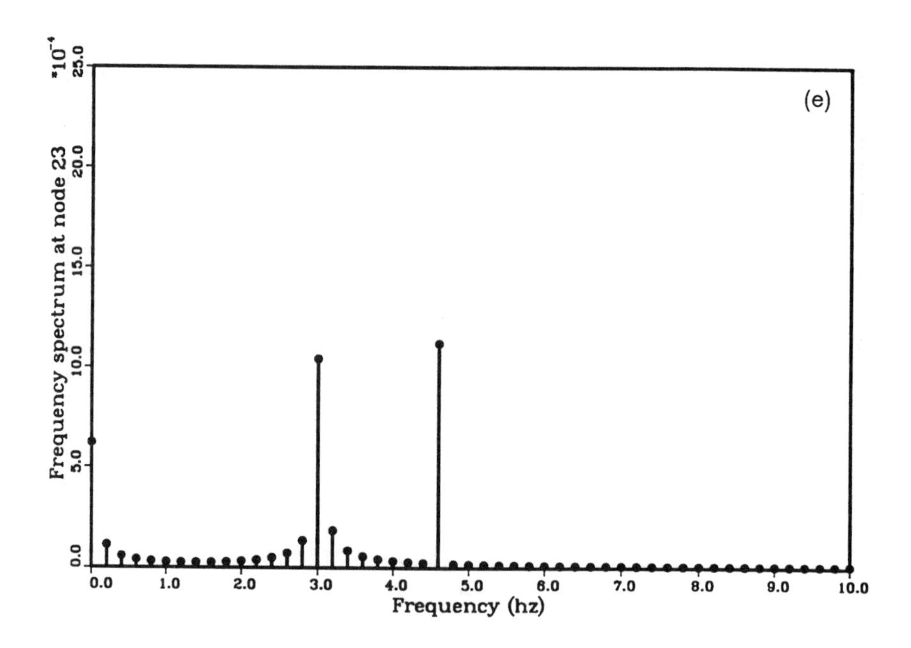

D. Energy Suppression

For those situations where minimizing the total flexible energy is important, the energy suppression technique offers an alternative control strategy. Suppose, again, that we have but three actuators at our disposal. So that we may make a direct comparison with mode suppression, we will again place the actuators at nodes 11, 13, and 15. At these locations, no energy can go into the first flexible (4th) mode (see Table 3). Our objective is to minimize the energy going into the 5th and 6th modes [see (63)],

$$G = \frac{A_5^2}{\tilde{k}_5} + \frac{A_6^2}{\tilde{k}_6} , \tag{88}$$

where, from (62), $\tilde{k}_5 = 13.1044$ and $\tilde{k}_6 = 3.6172$. The function G is to be minimized subject to the constraints [see (64) and (65)],

$$A_5 = -\ 0.3021\ \bar{u}_{11} + 0\ \bar{u}_{13} - 0.3021\ \bar{u}_{15}$$

$$A_6 = 0.0616\ \bar{u}_{11} - 0.2555\ \bar{u}_{13} + 0.0616\ \bar{u}_{15}$$

$$0.196 = \bar{u}_{11} + \bar{u}_{13} + \bar{u}_{15}$$

From this optimization process, we obtain

$$\bar{u}_{11} = 0.0653\ , \qquad \bar{u}_{13} = 0.0654, \qquad \bar{u}_{15} = 0.0653\ .$$

Using this control results in the displacement FFT at nodes 7, 11, 13, 21, and 23, as illustrated in Figures 23a-23e. The 1st (0 Hz), 5th (3 Hz), 6th (3.8 Hz), and driving frequency are represented to varying degrees at each of the nodes. From Table 4, it follows that 96% of the energy goes into the elastic modes in this case. Hence, energy suppression with three actuators at the given locations can provide us with a 4% mimic.

V. DISCUSSION

Sensor and actuator placement has been discussed from a controllability/ observability point of view [9] and from an overall performance point of view [16-19]. The energy suppression idea presented here can also be used to evaluate actuator location. For example, the optimal placement of three actuators (at nodes 7, 13, 19 or 9, 13, 17) will provide a 19.7% mimic for the UA plate under the conditions described above. This was determined by systematically examining all feasible actuator locations and determining the corresponding energy in the first 10 flexible body modes. Different design requirements from those specified above will, in general, result in different optimal locations.

A similar procedure with more actuators can be used to provide additional mode or energy suppression. Introducing a 4th actuator allows for a considerable improvement in performance. We have shown that with four actuators, we can produce a 54.3% mimic using energy suppression.

In positioning a flexible plate by means of actuators, three main problems must be solved: (1) How many actuators should be used, (2) where should they be placed, and (3) what control law should be used. The open-loop approach to positioning a flexible plate discussed here provides an orderly way to investigate these problems. We do this through the use of a force effectiveness matrix. Actuator placement is evaluated in terms of an m% mimic concept for both energy and mode suppression. Both location and number of actuators are shown to have a significant effect on the value of m.

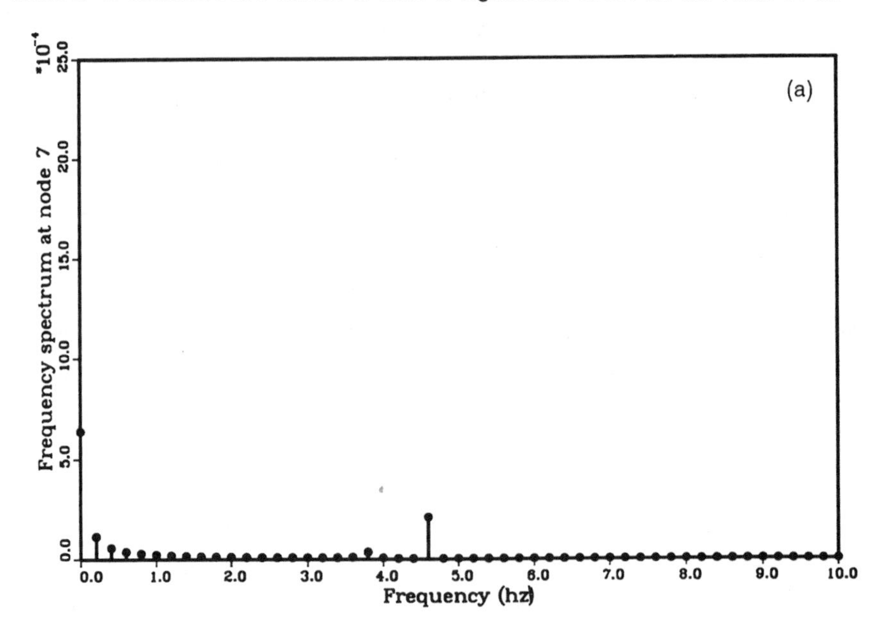

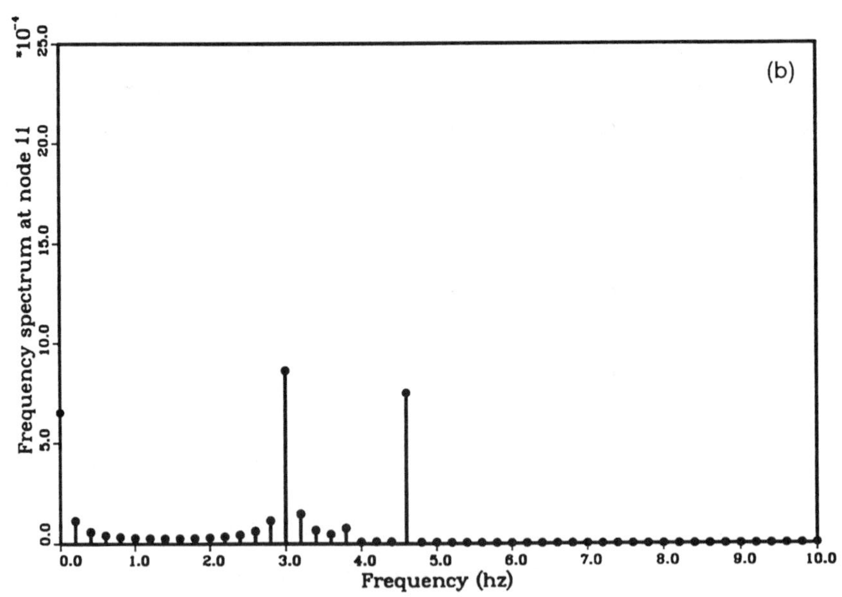

Fig. 23. FFT obtained using three actuators (energy suppression).

Fig. 23.--Continued.

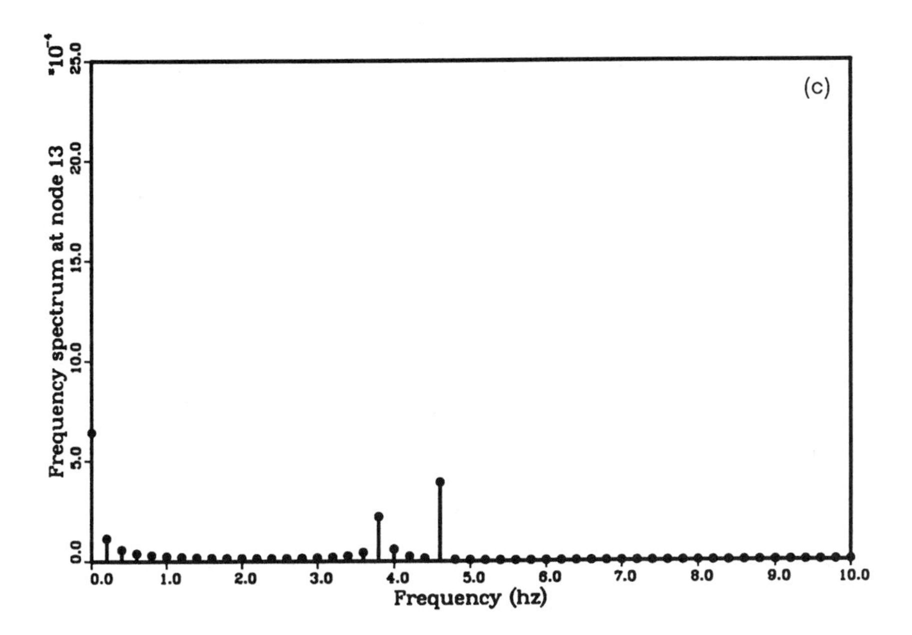

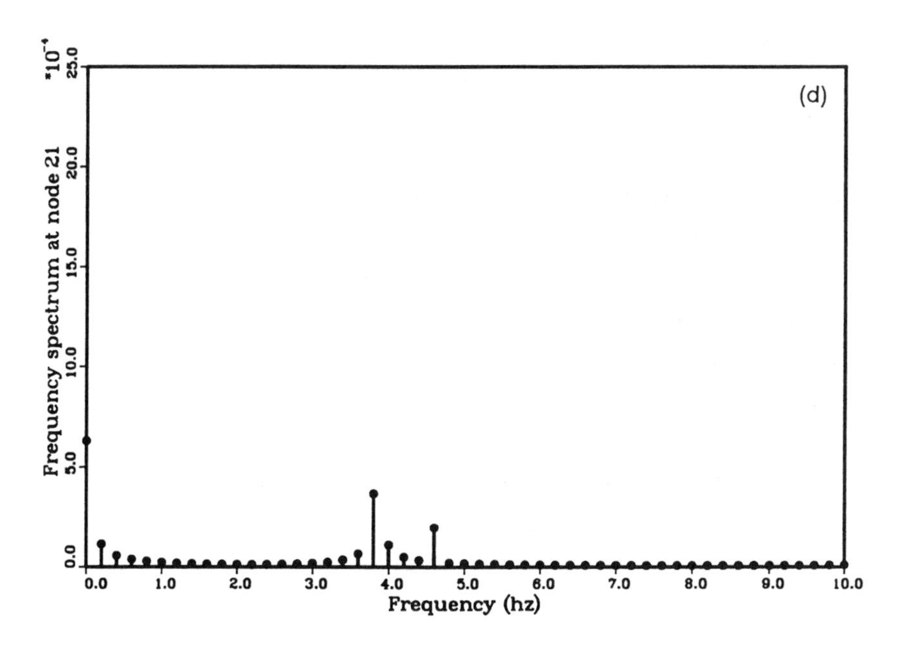

Fig. 23.--Continued.

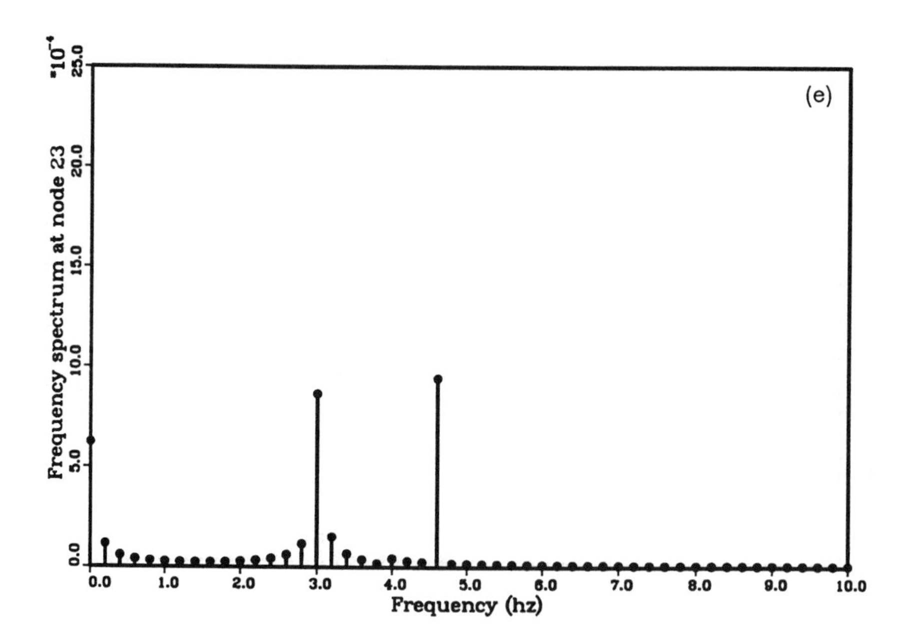

REFERENCES

1. S. P. Joshi, T. L. Vincent, and Y. C. Lin, "Control for Energy Dissipation in Structures," Proceedings AIAA, 1988.

2. T. L. Vincent, S. P. Joshi, and Y. C. Lin, "Positioning and Active Damping of Spring-Mass Systems," *Journal of Dynamical Systems, Measurement, and Control* (to appear).

3. T. L. Vincent, Y. C. Lin, and S. P. Joshi, "Positioning and Active Damping of Flexible Beams," *Journal of Dynamical Systems, Measurement, and Control* (to appear).

4. J. S. Przemmieniecki, "Theory of Matrix Structural Analysis," Dover, New York, 1985.

5. K. J. Bathe, "Finite Element Procedures in Engineering Analysis," Prentice-Hall, Inc., Englewood Cliffs, 1982.

6. D. J. Dawe, "Matrix and Finite Element Displacement Analysis of Structures," Oxford University Press, 1984.

7. L. Meirovitch, "Computational Methods in Structural Dynamics," Sijthoff and Noordhoff, The Netherlands, 1980.
8. L. Meirovitch, and H. Baruh, "The Implementation of Modal Filters for Control of Structures," *Journal of Guidance and Control 8*, 707-716 1985.
9. P. C. Hughes, et. al., "Controllability and Observability for Flexible Spacecraft," *Journal of Guidance and Control 3*, 452-459 1980.
10. S. Timoshenko and S. Woinowksy-Krieger, "Theory of Plates and Shells," McGraw-Hill Book Company, Inc., New York, 1987.
11. D. J. Gorman, "Free Vibration Analysis of Rectangular Plates," Elsevier North Holland, Inc., New York, 1982.
12. T. Y. Yang, "Finite Element Structural Analysis," Prentice-Hall, Inc., Englewood Cliffs, 1986.
13. J. Penzien, "Dynamics of Structures," McGraw-Hill, New York, 1975.
14. R. D. Blevins, "Formulas for Natural Frequency and Mode Shape," Van Nostrand Reinhold, New York, 1979.
15. A. W. Leissa, "The Free Vibration of Rectangular Plates," *Journal of Sound Vibration 31*, 257-293 1973.
16. R. Skelton and D. Chiu, "Optimal Selection of Inputs and Outputs in Linear Stochastic Systems," *J. Astronautical Sciences 31*, 399-414, 1983.
17. R. Skelton and M. L. Delorenzo, "Selection of Noisy Actuators and Sensors in Linear Stochastic Systems," *Large Scale Systems 4*, 109-136, 1983.
18. R. Skelton and M. L. Delorenzo, "Space Structure Control Design by Variance Assignment," *Journal of Guidance and Control 8*, 454-462, 1985.
19. R. Skelton and G. A. Norris, "Selection of Sensors and Actuators in the Presence of Correlated Noise," *Control-Theory and Advanced Technology 4*, 53-71, 1988.

ON THE MODAL STABILITY OF IMPERFECT CYCLIC SYSTEMS

Osita D.I. Nwokah*
Daré Afolabi**
Fayez M. Damra***

*School of Mechanical Engineering
***School of Aeronautics and Astronautics
Purdue University
West Lafayette, IN 47907

**School of Engineering and Technology
Purdue University
Indianapolis, IN 46202

I. Introduction

An important subject in the dynamics and control of structural systems is the behavior of structures under transient or steady state excitations. In this work, we examine the stability of the geometric form of the spatial configuration of structural systems when the structural parameters are subject to small perturbations, and the implications of this instability for frequency response. We show that circularly configured systems which nominally have cyclic symmetry exhibit complicated topological behavior when small perturbations are impressed on them. We further show that the frequency response of a perturbed cyclic system depends considerably on the form of perturbation. On the other hand, a rectilinear configuration of nearly identical subsystems does not ehibit modal instability. Usually, both kinds of systems are implicitly assumed to undergo similar qualitative behavior under a small perturbation whereas, in fact, the cyclic configuration exhibits a very stange behavior, [1].

The distinction between the behavior of cyclic and rectilinear configurations under a perturbation is important because many engineering structures are composed of identical substructures which are replicated either in a *uni-axial* chain, or in a closed *cyclic* formation where modal control is of interest. Examples of the former case of periodicity occur in structures such as space platforms and bridges, which have an obvious periodicity of the uni-axial kind. An example of cyclic periodic systems is a turbine rotor, which consists of a set of nominally identical blades mounted on a central hub, and often referred to as a "bladed disk assembly" [2]. When all the blades are truly identical, then the system is referred to in the literature as a *tuned* bladed disk assembly. Practical realities of manufacturing processes preclude the existence of exact uniformity among all the blades. When residual differences from one blade to another—no matter how small—are accounted for in the theoretical model, the assembly is then termed a *mistuned* bladed disk.

Our primary focus in this investigation is on bladed disk assemblies. However, since we approach the problem from a generalized viewpoint, the conclusions to be drawn will be of relevance to other periodic systems. Therefore, in the sequel, we borrow the 'tuned' and 'mistuned' terminology from the bladed disk literature, and apply it to repetitive systems having cyclic or uniaxial periodicity. Thus, in a tuned periodic system, the nominal periodicity is preserved, whereas it is destroyed in a mistuned system.

If we examine the system matrices of the linear and cyclic chains, we observe a fundamental difference in forms. The dynamical matrix of the linear chain is usually banded. Banded matrices are frequently encountered in structural dynamics. A special form of banded matrices that is of interest here is the tri-diagonal form $a_{ij}=0, |i-j|> 1$. On the other hand, the system matrix of a cyclic chain has a circulant submatrix, or is entirely circulant or block circulant [3]. Circulant matrices usually arise in the study of circular systems. They have interesting properties that set them apart from matrices of other forms [4]. We note that all circulants commute under multiplication, and are diagonalizable by the fourier matrix. One of the most important consequences of the foregoing is that the cyclic chain has a *series* of degenerate eigenvalues, whereas the eigenvalues of the uni-axial chain are all simple.

We know that a tuned circulant matrix, having a multitude of degenerate eigenvalues, lies on a bifurcation set [5]. Thus, the reduction of such matrices to Jordan normal form is an unstable operation [6]. Consequently, if a non-singular deformation due to mistuning is applied to a circulant matrix, then some of the eigenvectors will undergo rapid re-alignment, if the mistuning leads to a crossing of the bifurcation set. If however, no crossing of the bifurcation set takes place, then the tuned system's eigenvectors will be very stable, preserving their alignment under small perturbations. In contrast, the eigenvectors of a tuned banded matrix, being analytically dependent on parameters, are not generally disoriented by mistuning until the eigenvalues are pathologically close [7].

If one examines the literature in structural dynamics, it is observed that some unusual behavior has been reported in the study of perturbed cyclic systems. This has been the case in various studies of rings [8], circular saws [9], and other cyclic structures [10]. But that such anomalous behavior is due to a "geometric instability" inherent in the *cyclicity* of the tuned system has not been previously established in the literature, to our knowledge. Indeed, it is often assumed (see, for instance, [11]) that the linear and cyclic chains would undergo the same qualitative behavior under slight parameter perturbations so that small order perturbations of the system matrix will lead to no more than small order differences in the system response relative to the unperturbed case, if the system has "strong coupling".

In this paper, we show that such an assumption regarding qualitative behavior does not actually hold in the case of cyclic systems; that cyclic systems exhibit a peculiarity of their own under parameter perturbation; that, although a certain amount of mistuning may produce little difference relative to the tuned datum in one case, a considerable change could be induced if a slightly different kind of mistuning is applied to the same cyclic system; that such apparently erratic behavior arises in cyclic system, even when the system has "strong" coupling. In carrying out this work, we borrow from certain developments in differential topology specifically, from Arnold's monumental work in singularity theory [6, 12-16].

II. Topological Dynamics of Quadratic Systems

In mistuned dynamical systems, a major concern is to understand which specific kinds of mistuning parameters, or combinations thereof, lead to unacceptably high amplitude ratios. In this section, we give an indication of the taxonomy of the different consequences of mistuning in the hope of isolating those that lead to high ratios.

Consider a mechanical system under small oscillations with kinetic and potential energies given by:

$$T = \tfrac{1}{2}\dot{x}^* M\dot{x} > 0, \;\; U = \tfrac{1}{2}x^* Kx > 0; \;\; \dot{x}, x \neq 0. \tag{2.1}$$

Under the influence of a forcing function $f(t)$, (2.1) produces the following equations of motion by application of Lagrange's formula:

$$M\ddot{x} + Kx = f; \;\; x, f \in \mathbb{C}^n \tag{2.2}$$

where M and K are symmetric and positive definite. A theorem in linear algebra shows that there exists some non-singular transformation matrix P such that:

$$P^T MP = I, \text{ and } P^T KP = \Lambda \tag{2.3}$$

where Λ is a diagonal matrix of eigenvalues whose elements satisfy the equation:

$$\det(M - \lambda K) = 0 \tag{2.4}$$

Consequently, by putting

$$x = Pq, \tag{2.5}$$

substituting for q in (2.1), and premultiplying every term of the resultant equation by P^T, we obtain a new equation set:

$$\ddot{q} + \Lambda q = f', \tag{2.6}$$

where $f' = P^T f$. Hence:

$$\ddot{q}_i + \lambda_i q_i = f'_i \text{ for } i = 1, 2, \cdots, n. \tag{2.7}$$

Systems which can be reduced to the above form are called quadratic systems. They are called quadratic *cyclic* systems if, in addition, M and K are cyclic or circulant matrices. Our basic aim is to determine the nature of the changes in the dynamical properties of a quadratic system of a given order, under random differential perturbations in M and/or K. Central to this investigation are the topological concepts of structural stability and genericity.

Let N be a set with a topology and an equivalence relation e. An element $x \in N$ is stable (relative to e) if the e-equivalence class of x contains a neighborhood of x.

A property P of elements of N is generic if the set of all $x \in N$ satisfying P contains a subset A which is a countable intersection of open dense sets [17].

Genericity is important in our context because a generic system will in effect display a "typical" behavior. More concretely if a given generic system gives a certain frequency response, all systems produced by differential parameter perturbations about the nominal system will also produce frequency response curves that are not only slight perturbations of the original nominal response but also geometrically (isomorphic) equivalent to it. Such systems are called versal deformations of the nominal system [14]. A versal deformation of a system is a normal form to which it is possible to reduce not only a suitable representation of a nominal system, but also the representation of all nearby systems such that the reduction transformation depends smoothly on parameters. The key to establishing versality, and hence genericity, is the topological concept of transversality.

Let $N \subset M$ be a smooth submanifold of the manifold M. Consider a smooth mapping $f: \Gamma \to M$ of the parameter space Γ into M; and let μ be a point in Γ such that $f(\mu) \in N$.

The mapping f is transversal to N at μ if the tangent space to M at $f(\mu)$ is the sum:

$$TM_{f(\mu)} = f_* \, T\Gamma_\mu + TN_{f(\mu)}$$

Consequently, two manifolds intersect transversally if either they do not intersect at all or intersect properly such that perturbations of the manifolds will neither remove the intersection nor alter the type of intersection.

Lemma 2.1, see ref [14].

A deformation $\mathrm{f}(\mu)$ *is versal if and only if the mapping* $\mathrm{f}: \Gamma \to \mathrm{M}$ *is transversal to the orbit of* f *at* $\mu = 0$.

The above result is crucially important because:

(i) It classifies from the set of all perturbations of a given nominal system, those that do not lead to radically different dynamical properties from the nominal.

(ii) It separates the "good" from the "bad" perturbations and hence enables us to concentrate our study on the bad perturbations. Let Q denote the family of all real quadratic systems in $\mathbb{R}^n$. The set Q has the structure of a vector space of dimension $\frac{1}{2}(n[n+1])$. It can be shown that Q also has the structure of a differentiable manifold [13].

Let Q_v denote the set of quadratic systems having v_2 eigenvalues of multiplicity 2, v_3 eigenvalues of multiplicity 3 etc. Q_v is called the degenerate subfamily of Q.

Theorem 2.1, see ref [13].
The transformation $f{:}\Gamma \to Q$ *is transversal to* Q_v.

Consequently, a generic family of quadratic systems of a given order is given by a transformation, f, of the space of parameters Γ into the space of all quadratic systems Q, such that f is transversal to the space of all degenerate quadratic systems Q_v.

Hence Q_v is the degenerate (bad) set and Q/Q_v is the generic set. Observe that Q/Q_v and Q_v are transversal. Consequently, the fundamental group of the space of generic real quadratic systems is isomorphic to the manifold of systems without degenerate eigenvalues.

The above discussion leads inevitably to the following conclusions:

(i) Radical changes in the dynamical properties of a nominal system occurs under perturbations, when the perturbations take the system across the boundary from Q/Q_v to Q_v and vice-versa.

(ii) Q_v is a smooth semi-algebraic submanifold of Q, and can therefore be stratified into distinct fiber bundles [14]. By a bundle, we mean the set of all systems which differ only by the exact values of their eigenvalues; but for which the number of distinct eigenvalues as well as the respective orders of the degenerate eigenvalues are the same. Within the degenerate set, Q_V, the crossing from one bundle to another can also lead to radical dynamical changes. Each bundle is represented by a specific Jordan block of a certain order. Note that each bundle is also transversal to Q.

Theorem 2.2, ref [14].

Q_V is a finite union of smooth sub-manifolds with codimension satisfying Codim $Q_V \geq 2$.

Theorem 2.2 has the following implications:

(i) Q/Q_V is topologically path connected. This means that by smooth parameter variations, provided that the number of variable parameters is less than the codimension of Q_V, it is possible to smoothly pass from one member of Q/Q_V to another without reaching any singularity; that is, without encountering any member of Q_V. Such parameter variations will typically not lead to radical dynamical changes in Q/Q_V.

(ii) Because codim $Q_V \geq 2$, it follows that a generic one-parameter family of quadratic systems cannot contain any degenerate subfamilies. Therefore under one-parameter deformations of a generic family, some eigenvalue pairs may approach each other but cannot be coincident (i.e. cannot collide). After approaching each other, they must veer away rapidly, giving rise to the so-called eigenvalue loci-veering phenomenon [18], under one-parameter deformations of generic families. This offers a theoretical explanation for the eigenloci veering phenomenon which has been observed in perturbed periodic systems without a corresponding phenomenological base [18, 19]. Furthermore, this phenomenon holds provided the system has a quadratic structure, irrespective of whether the model arose from a continuous or discrete structural system [20].

This rapid eigenloci veering can, under the right conditions, produce the mode localization phenomenon [18]. Since the dynamical properties of any linear constant-coefficient system are totally determined by its eigen-structure (eigenvalues and eigenvectors), and since the eigenvalues are continuous functions of the matrix elements, it follows that radical changes in the dynamical properties of a given system under differential parameter perturbations ensue principally from a large disorientation between the eigenvectors of the tuned (unperturbed) and mistuned (perturbed) systems. We study, in Section IV, the variation of eigenvectors of generic families under differential random parameter perturbations.

III. Bounds on Amplitude Ratios

Consider, again, the equation set for the dynamics of quadratic systems:

$$M\ddot{x}_0 + Kx_0 = f, \tag{3.1}$$

where M and K are positive definite matrices. For tuned cyclic systems, M and K have the additional property of being circulant. Taking the Laplace transform of (3.2) under zero initial conditions, gives:

$$(Ms^2 + K)X_0(s) = F(s), \tag{3.2}$$

or

$$A(s) \cdot X_0(s) = F(s) \tag{3.3}$$

where $A = Ms^2 + K$. Suppressing s in all subsequent calculations leads to:

$$X_0 = A^{-1} \cdot F. \tag{3.4}$$

The positive definite nature of M and K guarantees that A^{-1} exists for all s on the Nyquist contour. Under normal operations of the system, suppose A varies to $A + \Delta A := A_e$. Let X_0 then change to $X_0 + \Delta X := X_e$. Then, for the same excitation force as in the tuned state,

$$X_e = (A + \Delta A)^{-1} \cdot F. \tag{3.5}$$

The physical nature of the system guarantees that $A + \Delta A$ will always remain symmetric but not *necessarily* circulant since a true mistuning destroys cyclicity. Equation (3.5) can be rewritten as:

$$X_e = (A + \Delta A)^{-1} \cdot F = (I + A^{-1}\Delta A)^{-1} \cdot A^{-1}F. \tag{3.6}$$

Substituting (3.4) into (3.6) gives:

$$X_e = (I + A^{-1}\Delta A)^{-1} \cdot X_0. \tag{3.7}$$

Normally ΔA will be a differential perturbation of A, so that:

$$\rho(A^{-1}\Delta A) < 1,$$

where $\rho(\cdot)$ is the spectral radius of $(\cdot)$. Hence

$$(I + A^{-1}\Delta A)^{-1} = \sum_{k=0}^{\infty} (-1)^k (A^{-1}\Delta A)^k. \tag{3.8}$$

Substituting (3.8) into (3.7) gives:

$$X_e = \sum_{k=0}^{\infty} (-1)^k (A^{-1}\Delta A)^k \cdot X_0. \tag{3.9}$$

Taking norms in (3.9) gives:

$$\|X_e\| = \| \sum_{k=0}^{\infty} (-1)^k (A^{-1}\Delta A)^k X_0 \|$$

$$\leq \sum_{k=0}^{\infty} \| A^{-1}\Delta A \|^k \cdot \|X_0\|. \tag{3.10}$$

Let $\|A^{-1}\Delta A\| = r$. Because ΔA is a differential perturbation of A, it follows that $r < 1$. Hence:

$$\| X_e \| \leq \|X_0\| \sum_{k=0}^{\infty} r^k = \|X_0\| \left\{ 1 + r + r^2 + \cdots + r^k \right\}$$

$$\leq \frac{\|X_0\|}{1-r}, \text{ since } r < 1. \tag{3.11}$$

Or:

$$\frac{\|X_e\|}{\|X_0\|} \leq \frac{1}{1-r} = \frac{1}{1 - \|A^{-1}\Delta A\|}. \tag{3.12}$$

Write

$$A = D + C = D(I + D^{-1}C) \tag{3.13}$$

where D is a diagonal matrix of the uncoupled system dynamic matrix and C is the relative coupling dynamic matrix, such that the minimum eigenvalue of $D^{-1}C$ at any frequency gives the coupling index of the system at that frequency [21]. If the norms in (3.12) are H^∞- norms, then, over the frequency interval Ω:

$$\underset{\omega\in\Omega}{\text{ess.sup}}\left\{\frac{\|X_e(\omega)\|_\infty}{\|X_0(\omega)\|_\infty}\right\} \le \underset{\omega\in\Omega}{\text{ess.sup}}\left\{\frac{1}{1-\left|\dfrac{\sigma_{\max}\Delta A(\omega)}{\sigma_{\min}A(\omega)}\right|}\right\} \tag{3.14}$$

where $\sigma_{\max}(\cdot)$ and $\sigma_{\min}(\cdot)$ correspond to maximum and minimum singular values of $(\cdot)$ respectively. Note that all the matrices and vectors considered above are functions of frequency $s=i\omega$.

Because A is symmetric it follows from (3.13) that:

$$\begin{aligned}\sigma_{\min}(A) &= \sigma_{\min}(D[I + D^{-1}C]) \\ &= \lambda_{\min}(D)\cdot\lambda_{\min}(I + D^{-1}C), \\ &= d_{\min}[1 + \lambda_{\min}(D^{-1}C)]\end{aligned} \tag{3.15}$$

by the eigenvalue shift theorem, where $d_{\min}$ is the minimum eigenvalue of D. Let $d_{\min} = a$, and $\lambda_{\min}(D^{-1}C) = \lambda_0$. At any frequency ω, let $\dfrac{\|X_e(\omega)\|_\infty}{\|X_0(\omega)\|_\infty} = \Pi(\omega)$.

Then (3.14) reduces to:

$$\delta_e = \underset{\omega\in\Omega}{\text{ess.sup}}\ \Pi(\omega) \le \frac{1}{1 - \underset{\omega\in\Omega}{\text{ess.sup}}\left\{\left|\dfrac{\sigma_{\max}\Delta A(\omega)}{a(\omega)(1+\lambda_0(\omega))}\right|\right\}} \tag{3.16}$$

where Ω is a frequency interval of interest. In some cases it is possible to define Ω by the semi-open interval $\Omega = [\,0\ \ \infty)$. Here λ_0 is called the coupling index of the system. The system is decoupled when $\lambda_0 = 0$. It is weakly coupled if $\lambda_0 < 1$, and is strongly coupled if $\lambda_0 \geq 1$. In general, $0 \leq \lambda_0 \leq \infty$. Observe that $\lambda_0(k, \omega)$ is a function of both the structural coupling k, and frequency ω. Inequality (3.16) leads to the following conclusions:

(i) The mistuned to tuned amplitude ratio is determined by the maximum peak of the mistuning strength $\sigma_{\max}\Delta A(\omega)$, the minimum strength of the weakest link in the system $a(\omega)$, and the minimum peak of the coupling index (strength) $\lambda_0(\omega)$.

(ii) A variation in rigidity (coupling) affects the ratio of (3.16) monotonically for fixed values of $\sigma_{\max}(\Delta A)$ and a. This is because at any given frequency, λ_0 varies continuously and monotonically as the coupling is varied [13].

(iii) A reduction in a caused by a reduction of mass of the blades, and/or more flexible blades, increases the ratio (3.16) monotonically. More specifically, at any frequency when $\lambda_0 \to 0$, from (3.16):

$$\delta_e \leq \operatorname*{ess.sup}_{\omega \in \Omega} \left\{ \left\| \frac{a(\omega)}{a(\omega) - \sigma_{\max}\Delta A(\omega)} \right\| \right\} > 1, \text{ for } \sigma_{\max}\Delta A(\omega) > 0, \ \forall\ \omega \in \Omega.$$

Hence under weak coupling across the frequency interval, the amplitude ratio depends entirely on the relationship between the frequency response of the mistuning strength and that of the strength of the weakest blade in the assembly. Under these conditions, the maximum amplitude ratio will arise from the blade with the worst mistune [22].

IV. Eigenvector Rotations

In section II, we showed that generic systems Q/Q_v will typically have distinct eigenvalues, while degenerate systems Q_v will typically have repeated eigenvalues. To study eigenvector perturbations for generic systems, regular analytical methods will work, while for eigenvector variations in the system Q_v we require singular perturbations [23]. Let $A \in \mathbb{C}^{n \times n}$ be the dynamic matrix arising from any system $Q_r \in Q/Q_V$. Let Γ represent the parameter space and let $\mu \in \Gamma$ be a p-dimensional parameter vector. If Codim $Q_V \geq r$, then for any $\mu \in \Gamma \in \mathbb{R}^p$, where $p<r$, differential parameter variations in $A(\mu)$ will not lead to eigenvalue degeneracies. Thus, if the eigenvalues of $A(\mu)$, given by $\lambda_1(\mu), \lambda_2(\mu), \cdots, \lambda_n(\mu)$, are distinct when $\mu=0$ they will continue to remain distinct when μ is small, by continuity arguments.

Let

$$A(\delta\mu) = A(0) + \delta A, \tag{4.1}$$

where:

$$\delta A = \delta\mu_k \cdot \frac{\partial}{\partial\mu_k} A(\mu) \Big|_{\mu=0} \tag{4.2}$$

δA can be expanded in Taylor series form as: $\delta A = \Delta A$ + higher terms dependent on μ. To a first order approximation we can write the perturbed matrix as:

$$A = A_0 + \Delta A. \tag{4.3}$$

Write

$$A_0 = U \Lambda U^{-1} \tag{4.4}$$

where U is the modal matrix of A_0, and $V^* = U^{-1}$ where:

$$U = [u_1, u_2, \cdots, u_n]$$

and

$$V^* = [v_1^* \; v_2^* , \cdots , {}_n^*]^*$$

with

$$\Lambda = \text{diag}(\lambda_1, \lambda_2, \cdots, \lambda_n).$$

where $(\cdot)^*$ is the complex conjugate transpose of $(\cdot)$.

Since A_0 is also generic, we can write the perturbed modal expression as:

$$A_0 + \Delta A = [U + \Delta U][\Lambda + \Delta\Lambda][U + \Delta U]^{-1} \tag{4.5}$$

where ΔU is the perturbation in U resulting from ΔA while $\Delta\Lambda$ is the corresponding perturbation in Λ resulting from ΔA. Under eigenvector normalization, $\| u_i \| = 1$, and $\| u_i + \Delta u_i \| = 1$. Equation (4.5) can be solved as:

$$A_0 U + \Delta A U + A_0 \Delta U = U\Lambda + U\Delta\Lambda + \Delta U \Lambda \tag{4.6}$$

where we neglect second order terms like $\Delta U \Delta\Lambda$ and $\Delta A \Delta U$ [24]. As a measure of the eigenvector variations, we begin by writing Δu_i as a linear combination of all the eigenvectors since the eigenvectors span the whole n-dimensional space. Thus:

$$\Delta u_i = \sum_{j=1}^{n} l_{ji}\, u_j \tag{4.7}$$

Or:

$$\Delta U = UL. \tag{4.8}$$

Now solving for $\Delta\Lambda$ in (4.6) gives:

$$\Delta\Lambda = U^{-1} A_0 U + U^{-1}\, \Delta A U + U^{-1} A_0\, \Delta U - \Lambda - U^{-1}\, \Delta U \Lambda. \tag{4.9}$$

Observe that $U^{-1} A_0 U - \Lambda = 0$, so that

$$\Delta\Lambda = U^{-1}\, \Delta A U + \Lambda L - L\Lambda. \tag{4.10}$$

Notice that the diagonal elements of $(\Lambda L - L\Lambda)$ are zero. Hence:

$$\Delta\lambda_i = [\, U^{-1} \Delta A U]_{ii} = v_i^*\, \Delta A u_i.$$

To solve for ΔU, we need L. The off-diagonal elements of L are given by Skelton [24]

$$l_{ji} = (\lambda_j - \lambda_i)^{-1}\, v_i^*\, \Delta A u_j \quad \text{for } i{\neq}j,\ i,j = 1, 2 \cdots n\,,$$

or:

$$l_{ij} = (\lambda_i - \lambda_j)^{-1}\, v_j^*\, \Delta A u_i \quad \text{for } i{\neq}j,\ i,j = 1, 2 \cdots n.$$

To determine l_{ii}, observe that the constraint equation $\| u_i + \Delta u_i \| = 1$ contains l_{ii}. Thus

$$\| u_i + \Delta u_i \| = \{< u_i + \Delta u_i, u_i + \Delta u_i > \}^{1/2} = 1. \tag{4.12}$$

Or:

$$u_i^* u_i + 2u_i^* \Delta u_i + \Delta u_i^* \Delta u_i = 1. \tag{4.13}$$

But $u_i^* u_i = 1$, so that

$$2u_i^* \Delta u_i + \Delta u_i^* \Delta u_i = 0.$$

Therefore:

$$\Delta u_i = \sum_{j=1}^{n} l_{ji} u_j = \sum_{\substack{j=1 \\ j \neq i}}^{n} l_{ji} u_j + l_{ii} u_i \tag{4.14}$$

$$= x_i + l_{ii} u_i$$

where:

$$x_i = \sum_{\substack{j=1 \\ j \neq i}}^{n} l_{ji} u_j. \tag{4.15}$$

Thus:

$$l_{ii}^2 + (2 + 2u_i^* x_i) l_{ii} + (2u_i^* x_i + x_i^* x_i) = 0 \tag{4.16}$$

Letting:

$$z_i = u_i^* x_i$$

and

$$y_i = x_i^* x_i,$$

gives (on accepting the positive solution of the quadratic):

$$l_{ii} = -(1 + z_i) + (1 + z_i^2 - y_i)^{1/2}. \tag{4.17}$$

Since the eigenvectors u_i and $u_i + \Delta u_i$ can be normalized to unity and since each vector is represented by a magnitude m_i and an angle θ_i, the natural measure of modal variations is θ_i since $m_i \equiv 1$ after normalization. Knowing all the elements of L, we can now determine θ_i as:

$$<u_i, u_i + \Delta u_i> = \| u_i \| \; \| u_i + \Delta u_i \| \cos \theta_i . \tag{4.18}$$

But $\| u_i \| = \| u_i + \Delta u_i \| = 1.$

Hence:

$$\cos \theta_i = <u_i, u_i> + <u_i, \Delta u_i>$$

$$= 1 + u_i^* (x_i + l_{ii} \, u_i)$$

$$= 1 + u_i^* x_i + l_{ii}$$

$$= (1 + z_i^2 - y_i)^{1/2} , \quad 0 \le \theta_i \le \pi/2. \tag{4.19}$$

Consequently for the occurrence of no vector rotation under parameter variations, we require:

$$z_i^2 - y_i = 0 \tag{4.20}$$

Or:

$$x_i^* u_i \, u_i^* x_i - x_i^* x_i = x_i^* (u_i u_i^* - I) x_i = 0 \tag{4.21}$$

This implies x_i belongs to the null space of $(u_i u_i^* - I)$, that is:

$$(u_i u_i^* - I) \sum_{\substack{j=1 \\ j \ne i}}^{n} lji \; u_j = 0 \tag{4.22}$$

where:

$$l_{ji} = (\lambda_i - \lambda_j)^{-1} \, v_j^* \, \Delta A \, u_i \qquad i \ne j$$

The nearer the expression (4.20) is to zero, the less the corresponding eigenvector rotation under the given perturbation. Let

$$\alpha_i = \sum_{\substack{j=i \\ j \ne i}}^{n} l_{ji} \, u_j^* \, (u_i u_i^* - I) \sum_{\substack{j=i \\ j \ne i}}^{n} l_{ji} \, u_j , \quad i = 1, 2 \cdots n.$$

Then $\max_i \{\alpha_i\}$ gives the eigenvector with maximum rotation.

The conclusions are the following:

a) If the separation between the eigenvalues is very large, (i.e. $(\lambda_i - \lambda_j)$ is very large for all i,j), then l_{ji} is relatively small and eigenvector rotation will be correspondingly small.

b) If $v_j^* \Delta A u_i \approx 0$, then eigenvector rotation will also be relatively small, provided $l_{ji} \neq \infty$.

For example, if A_0 is Hermitian as is the case in all quadratic systems, and $\Delta A = \alpha I, \alpha \in \mathbb{C}$, then

$$v_j^* \Delta A u_i \equiv 0, \forall\, i,j = 1,2, \cdots n.$$

Thus, identical increases or decreases in the diagonal elements of a quadratic system will not produce unexpected amplitude excursions [25] because it cannot produce eigenvalue splittings in formerly degenerate families. Therefore, such perturbation cannot take a system either across the boundary of the bifurcation set or across different bundles of Q_v. Geometrically, this implies that degenerate eigenvalues in systems belonging to a bundle in Q_v cannot be lifted by perturbations that leave the perturbed system in the same bundle of Q_v. Indeed, define the eigenvector sensitivity matrix of a quadratic system as

$$S = \Delta U U^{-1} = L,$$

from eqn. (4.8). Defining the eigenvector sensitivity metric measure by

$$S_F = \| S \|_2^2 = \sum_{\substack{j=1 \\ i=1}}^{n} l_{ji}^2$$

where S_F is the Frobenius norm of S shows that the maximum eigenvector sensitivity is obtained at the positions of minimum eigenvalue separation, which is not difficult to compute. Alternatively, $(S_F)_{max}$ also corresponds to the position of maximum angular rotation between the tuned and mistuned system eigenvectors. This condition is evidenced by strong eigenloci deformations.

If $A(\omega)$ is a frequency response matrix arising from a generic system, the eigenvalues $\lambda_i(\omega)$ and eigenvectors $u_i(\omega)$ are also continuous functions of frequency. We can therefore plot the frequency response functions $S_F(\omega)$ to determine the frequencies at which maximum deformations take place.

V. Examples

To illustrate the theory so far developed we consider two examples. The first is an interconnected linear chain of oscillators. This has been studied by Arnold [13] and more recently by Pierre [18].

Example 1: *Mode Localization in Generic Periodic Systems.*

Consider a coupled pendulum, as shown in Fig 1, with identical masses but of different lengths l_1 and l_2, where l_2 is a perturbation of l_1, i.e., $l_2 = (l_1 + \Delta l_1)$. If we put $l_1 = l$, then the kinetic energy is given by

$$T = \frac{1}{2} m[l^2 \dot{\theta}_1^2 + (l + \Delta l)^2 \dot{\theta}_2^2] \tag{5.1}$$

while the potential energy is given by

$$U = ml\frac{\theta_1^2}{2} + m(l + \Delta l)\frac{\theta_2^2}{2} + \frac{k}{2}(\theta_1 - \theta_2)^2. \tag{5.2}$$

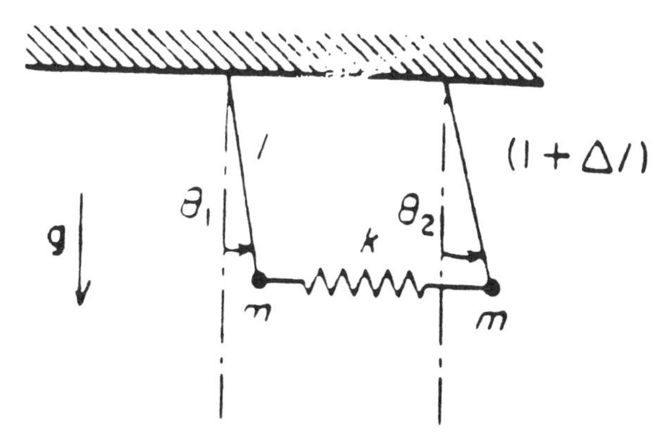

Fig. 1. *Two coupled oscillators.*

Under unit gravitational force, application of Lagrange's equations results in the equation of motion:

$$M \begin{bmatrix} \ddot{\theta}_1 \\ \ddot{\theta}_2 \end{bmatrix} + \begin{bmatrix} K \end{bmatrix} \begin{bmatrix} \theta_1 \\ \theta_2 \end{bmatrix} = 0 \tag{5.3}$$

where

$$M = \begin{bmatrix} ml^2 & 0 \\ 0 & m(l + \Delta l)^2 \end{bmatrix} \tag{5.4}$$

and

$$K = \begin{bmatrix} ml + k & -k \\ -k & m(l + \Delta l)^2 + k \end{bmatrix}. \tag{5.5}$$

The dynamic matrix for the above system is given by

$$A(\omega) = \begin{bmatrix} ml + k - \omega^2 ml^2 & -k \\ -k & m(l + \Delta l) + k - \omega_m^2 (l + \Delta l)^2 \end{bmatrix}. \tag{5.6}$$

Rewrite $A(\omega)$ as:

$$A(\omega) = \begin{bmatrix} a & -k \\ -k & b \end{bmatrix}. \tag{5.7}$$

The characteristic equation of $A(\omega)$ is given by

$$\lambda^2 - (a + b)\lambda + (ab - k^2) = 0$$

Both M and K are symmetric and positive definite. The eigenvalues of $A(\omega)$ are:

$$\lambda_{1,2} = \frac{[(a + b) \pm \sqrt{(a - b)^2 + 4k^2}\,]}{2}. \tag{5.8}$$

Note that $\lambda_{1,2}$ cannot be degenerate. Thus under one-parameter deformations, the eigenvalues can deform but cannot collide.

Indeed,

$$\frac{\partial \lambda_1}{\partial a} = \frac{1}{2}\left[1 + \frac{a - b}{\sqrt{(a - b)^2 + 4k^2}}\right] \tag{5.9}$$

and

$$\frac{\partial\lambda_2}{\partial a} = \frac{1}{2}\left[1 - \frac{a-b}{\sqrt{(a-b)^2 + 4k^2}}\right]. \tag{5.10}$$

Hence:

$$\frac{\partial\lambda_1}{\partial a} + \frac{\partial\lambda_2}{\partial a} = 1 \tag{5.11}$$

$$\left[\frac{\partial\lambda_i}{\partial a}\right] = \frac{1}{2} \text{ when } a=b.$$

The distance between the eigenvalues is given by:

$$d_\lambda = |\lambda_1 - \lambda_2| = \sqrt{(a-b)^2 + 4k^2} \tag{5.12}$$

which assumes its minimum value of $2k$ when $a=b$ or when $\left[\partial\lambda_i/\partial a\right] = ½$ This represents the tuned state of the linear chain. For a fixed mistuning value $(a-b)$, d_λ depends essentially on k. If $(a-b)$ is small, it is clear that $S_F \to \infty$ as $k \to 0$. The modal matrix of the chain is given by

$$U = \begin{bmatrix} -1 & -1 \\ \dfrac{(a-b) - \sqrt{(a-b)^2 + 4k^2}}{2k} & \dfrac{(a-b) + \sqrt{(a-b)^2 + 4k^2}}{2k} \end{bmatrix}. \tag{5.13}$$

Observe that $u_i^* \; u_j \equiv 0, \quad \forall\; k,a,b$. Under tuned conditions, $a=b$, then

$$U_t = \begin{bmatrix} -1 & -1 \\ -1 & 1 \end{bmatrix} \tag{5.14}$$

However, consider the very interesting situation when the mistuning to coupling ratio is rather large. That is to say:

$$\frac{(a-b)}{k} > 1.$$

Then $(a-b)^2 \gg k^2$, and k^2 becomes negligible in the eigenvector expressions. Expanding the term under the radicals and neglecting second and higher order terms gives:

$$[(a-b)^2+k^2]^{1/2}=(a-b)\left\{1+\left[\frac{k}{a-b}\right]^2\right\}^{1/2}$$

$$=(a-b)[1+\frac{1}{2}\left[\frac{k}{a-b}\right]^2+\cdots \tag{5.15}$$

In this case the modal matrix reduces to:

$$U_e=\begin{bmatrix} -1 & -1 \\ -\dfrac{k}{a-b} & \dfrac{a-b}{k} \end{bmatrix}. \tag{5.16}$$

An energy exchange now takes place. The second component of the first mode becomes vanishingly small while the corresponding component of the second mode becomes extremely large. This is an extreme case of classical vibration absorber, and is the *mode localization* phenomenon. We therefore conclude that mode localization (or extreme energy exchange) will occur in a generic system under one-parameter deformations if the following conditions are satisfied:

- at any frequency ω where the system is almost decoupled, i.e., $\lambda(\omega)_0 \to 0$. (Note that $\lambda(\omega)_0 \to 0$ as $k \to 0$).
- when the mistuning to coupling ratio $\dfrac{a-b}{k} \gg 1$.

At the localization stage the eigenvalue and eigenvector sensitivities take on their maximum values, i.e. both $\|\Delta\Lambda\,\Lambda^{-1}\|_F^2$ and $\|\Delta U\,U^{-1}\|_F^2$ have their maximum values. Localized modes always produce:

$$\delta_e=\frac{\|x_e\|_\infty}{\|x_0\|_\infty} \gg 1. \tag{5.17}$$

Example 2: *Cyclic Systems.*

Consider three identical masses, m, arranged in a ring structure and interconnected by identical springs k_c. Assume that all the masses are hinged to the ground by torsional springs of strength k_t, and that the radius of the ring is r; as shown in Fig. 2.

The basic equations of motion of this "ring" is

$$M\ddot{x} + Kx = f \tag{5.18}$$

where

$$M = \begin{bmatrix} m & 0 & 0 \\ 0 & m & 0 \\ 0 & 0 & m \end{bmatrix}, \ K = \begin{bmatrix} 2k_c + \dfrac{k_t}{r} & -k_c & -k_c \\ -k_c & 2k_c + \dfrac{k_t}{r} & -k_c \\ -k_c & -k_c & 2k_c + \dfrac{k_t}{r} \end{bmatrix}. \tag{5.19}$$

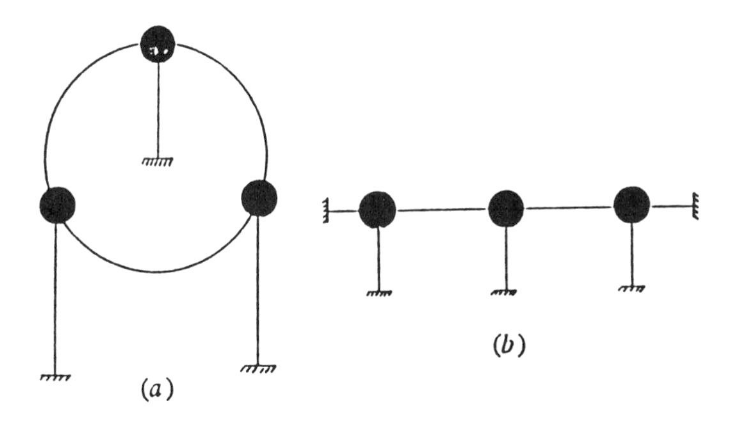

Fig. 2. *Models of (a) the cyclic chain, (b) the linear chain with three degress of freedom.*

Using group theoretic arguments [13], we can easily deduce that the above system has degenerate eigenvalues occurring as doublets, by cyclicity of the corresponding system matrices. Consequently, every quadratic cyclic system $Q_c \subset Q_V$. Furthermore all perturbations of the above system preserving the cyclic structure, leaves the modal geometry invariant [3, 25]. Indeed the eigenvalues of the above system are given as:

$$\lambda_1 = \frac{k_t}{mr}, \quad \lambda_2 = \frac{k_t}{mr} + \frac{3k_c}{m}, \quad \lambda_3 = \frac{k_t}{mr} + \frac{3k_c}{m}. \tag{5.20}$$

Write the dynamic matrix of this system as:

$$A_0(\omega) = \begin{bmatrix} a & -b & -b \\ -b & a & -b \\ -b & -b & a \end{bmatrix} \tag{5.21}$$

where

$$a = 2k_c + k_t - \omega^2 m, \quad b = k_c.$$

We now consider a diagonal perturbation of the form:

$$E = \mathrm{diag}\,(e_1, e_2, e_3). \tag{5.22}$$

Then

$$A_e(\omega) = A_0(\omega) + E. \tag{5.23}$$

This would correspond to the realistic situation where there are slight changes in the values of the ground spring k_t, depending for example on how the blades are coupled to the disk in bladed disk assemblies [22]. The major difference between the behavior of (degenerate) cyclic systems and generic systems are the following:

(i) For generic systems, all the eigenvalues and the distance between adjacent pairs increases as the coupling k_c increases. Consequently the probability of mode localization decreases as k_c and hence $\lambda(\omega)_0$ increases. On the other hand, perturbations which split the degenerate eigenvalue of cyclic systems turn them into generic systems with pathologically close eigenvalues [7]. Hence for previously cyclic systems whose eigenvalues bifurcate under perturbations, S_F is very large. Therefore such systems are susceptible to mode localization, independent of the values of the coupling strength k_c. Recall that

$$S_F = \| L \|_2^2,$$

where

$$l_{ij} = (\lambda_j - \lambda_i)^{-1} v_i^* \, \Delta A u_j, \; i \neq j, \; i,j = 1, 2, \cdots n.$$

(ii) Consequently the only way to avoid large values of S_F in such a situation is if and only if $\| v_i^* \, \Delta A u_j \| \equiv 0$ or in the neighborhood of zero. Perturbations that induce this condition are precisely those that will not induce radical dynamical changes in mistuned cyclic systems. It was already shown that if $\Delta A = \alpha I$, then $\| v_i^* \, \Delta A u_j \| \equiv 0$

(iii) Of the remaining possible perturbations those that have $\| v_i^* \, \Delta A u_j \| = \varepsilon << 1$ will produce minimum dynamical changes. All others for which $\| v_i^* \, \Delta A u_j \|$ is not small will give susceptibility to mode localization, no matter how strong the interblade coupling.

The following numerical example amplifies the above observations. We consider the case of the so-called 'strong coupling', using the following values: k_c=9.5, k_t=1, a = 20, b = 9.5, e_3=0, e_2=–0.1, e_1=0.1. Clearly, the ratio of mistuning to coupling strength is very small. Now, in order to compute the frequency response curves, we need some damping to obtain finite amplitudes at resonance. Assume hysteretic damping of 0.01 for all cases. Without loss of generality, the response to be computed is the direct receptance, i.e. the response of each node to individual excitation. We turn the ring into a linear chain by putting $b = k_{13} = k_{31} = 0$ in equation (5.21). Then A_0 becomes a tridiagonal banded matrix.

The frequency response of the tuned and mistuned systems of the linear chain are shown in Fig 3. The illustration is windowed around one of the resonant frequencies of the coupled system. Notice that, at the tuned state, the amplitudes of nodes 1 and 3 are equal on account of symmetry, while that of node 2 is double that magnitude.

Because the system is now generic, and therefore exhibits modal stability, all nodes have almost the same response patterns and magnitudes as in the tuned system. This is also the case when we change the sign of e_2, from −0.1 to 0.1.

When we repeat exactly the same procedure for the circulant system, a very different picture is obtained. Fig 4 shows the response of individual nodes compared with the tuned case. This case corresponds to a 2-parameter perturbation, with $e_1 = 0.1$, $e_2 = -0.1$, $e_3 = 0$.

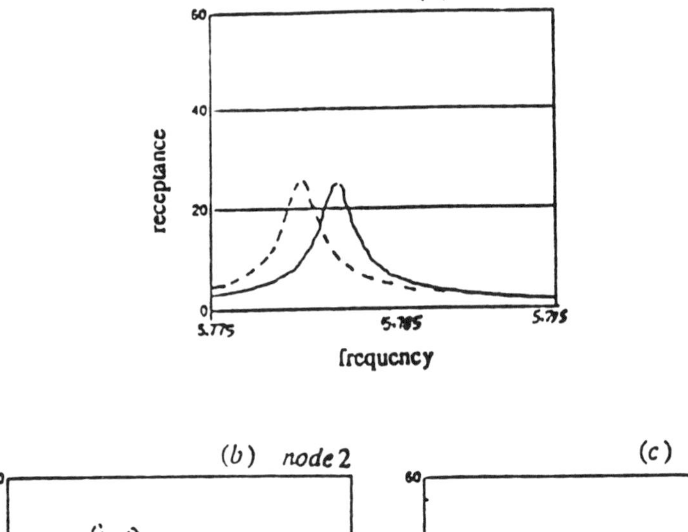

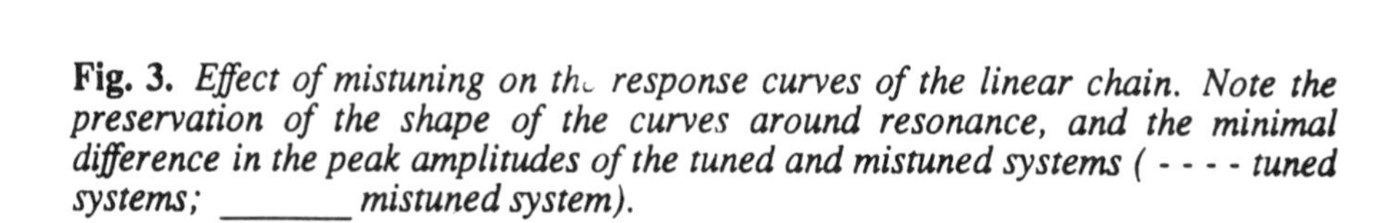

Fig. 3. *Effect of mistuning on the response curves of the linear chain. Note the preservation of the shape of the curves around resonance, and the minimal difference in the peak amplitudes of the tuned and mistuned systems (- - - - tuned systems; _______ mistuned system).*

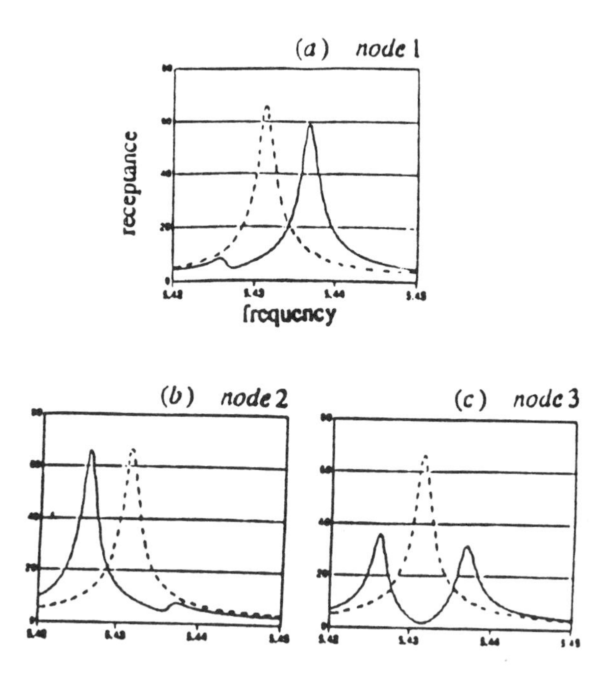

Fig. 4. *Effect of two parameter mistuning on the response curve of the cyclic chain. Note the severe reduction in the amplitude at node 3, which is only 50% of the tuned system (- - - - tuned systems; _______ mistuned system).*

Notice that the node with zero mistuning (mode 3) now has a reduction in amplitude of almost 50%. This extremely unequal amplitude distortion (Fig 4) is the case no matter how small the magnitude of the perturbation is, so long as we keep the *form* of mistuning, and the mistuning does not actually vanish.

If we now change the mistuning matrix in a very small way, by making e_2=0.1, we obtain the response curves in Fig 5. We now notice a substantial difference in the geometry of the curves in Fig 5, compared to those in Fig 4. Thus, a very small change in the perturbation matrix, now results in a considerable difference in the vibration response at the individual nodes. The question of which node will be most responding, or the one having the least amplitude, is now not as easy as one would have expected. In Fig 4, it is node 3, while it is node 2 in Fig 5. In fact, the amplitude of node 3 has been increased by about 100% from Fig 4 to Fig 5, merely by changing only one entry in the system matrix from 19.9

to 20.1, a change of less than 1% !

The foregoing examples, based on a simple 3 degrees of freedom model of a circular ring or disk only, illustrates the instability induced by cyclicity. It is clear that the qualitative conclusions to be drawn from Fig 4 are inconsistent with those from Fig 5, although the difference between the two mistuned matrices is very small indeed. We emphasize that these results, obtained for just a cyclic chain, are not necessarily applicable to bladed disks in all generality, especially those models in which cyclicity is ignored. However, when bladed disk systems are well-modeled to include the effects of blade coupling, blade or disk mistuning and *cyclicity*, similar distortions in the geometry of the frequency response curves can result. The subject is currently under investigation by us.

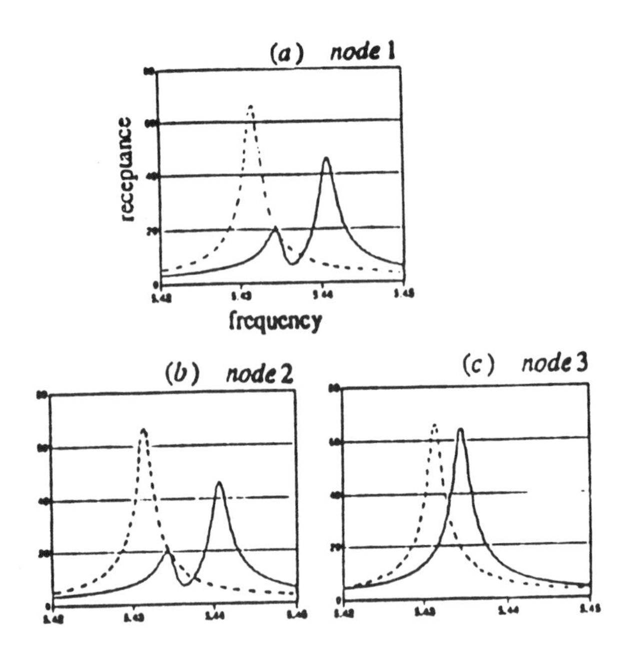

Fig. 5. *Effect of one-parameter mistuning on the response curve of the cyclic chain. Note the symmetrical unfolding of the degenerate singularity (- - - - tuned systems; _______ mistuned system).*

VI. Conclusions

(i) For generic systems, to which linear periodic chains of oscillators belong, differential parameter perturbations are significant for the system dynamics only under weak coupling conditions when the mistuning to coupling ratio exceeds unity (Example 1). Under all other conditions that do not induce eigenvalue degeneracy; small magnitudes of mistuning, or the type of mistuning, is irrelevant to system dynamics.

(ii) For degenerate systems to which a tuned cyclic system with circulant dynamic matrices belongs, it is not just the mistuning to coupling ratio which is significant in the determination of the perturbed system dynamics. The type of mistuning assumes a far greater importance than the mistuning to coupling ratio. All types of mistuning that move the system either across the boundary of the bifurcation set, or from one fiber bundle of the degenerate set to another within Q_v will lead to topological catastrophes [15].

Acknowledgments

This work was supported by the Air Force Office of Scientific Research, Air Force Systems Command, USAF, under Grants AFOSR-89-0002 and AFOSR-89-0014 monitored by Dr. Arje Nachman and Dr. Anthony K. Amos. The US Government is authorized to reproduce and distribute reprints for Governmental purposes notwithstanding any copyright notation thereon.

References

1. Afolabi, D., 1989, *On the Geometric Stability of Certain Modes of Vibration*, NASA Technical Memorandum.
2. Ewins, D. J., 1973, "Vibration Characteristics of Bladed Disc Assemblies", *J. Mechanical Engineering Science* 15, 165-186.
3. Afolabi, D., Nwokah, O.D.I, 1989, 'On the Modal Stability of Mistuned Cyclic Systems,' *Proceeding 2nd University of Southern California Conference on Control Mechanics.*
4. Davies, P. J., 1979, *Circulant Matrices*, John Wiley, New York.
5. Gilmore, R., 1981, *Catastrophe Theory for Scientists and Engineers*, Wiley, New York.
6. Arnold, V. I., 1968, "On Matrices Depending on Parameters", *Russian Mathematical Surveys* vol 26, No 2, pp 29-44.
7. Wilkinson, J. H. 1965, *The Algebraic Eigenvalue Problem*, Clarendon Press, Oxford.
8. Allaei, D., Soedel, W., Yang, T. Y., 1986, "Natural Frequencies and Modes of Rings that Deviate from Perfect Axisymmetry", *Journal of Sound and Vibration*, vol 111, pp 9-27.

9. Mote, C. D. Jr., 1970, "Stability of Circular Plate Subject to Moving Load", *Journal of the Franklin Institute*, vol 290, pp 329-344.

10. Weissenburger, J. T., 1968, "Effect of Local Modifications on the Vibration Characteristics of Linear Systems", *ASME Journal of Applied Mechanics*, vol 35, pp 327-332.

11. Wei, S. T. and Pierre, C., 1988, "Localization Phenomena in Mistuned assemblies with Cyclic Symmetry", *ASME J. Vibration, Acoustics, Stress and Reliability in Design*, vol 110, pp 429-449.

12. Arnold, V. I., 1972, "Lectures on Bifurcation in Versal Families", *Russian Mathematical Surveys*, vol 27, No. 5, pp 54-123.

13. Arnold, V. I., 1978, *Mathematical Methods of Classical Mechanics*, Springer-Verlag, New York.

14. Arnold, V. I., 1981, *Singularity Theory*, Cambridge University Press, Cambridge.

15. Arnold, V. I., 1983, *Catastrophe Theory*, Springer, New York.

16. Arnold, V. I., Gusein-Zade, S. M. and Varchenko, A. N., 1985, *Singularities of Differentiable Maps*, Vol 1, Birkhauser, Boston, MA.

17. Lu, Y.C., 1985, *Singularity Theory and an Introduction to Catastrophe Theory*, Springer-Verlag, New York.

18. Pierre, C., 1988, "Mode Localization and Eigenvalue Loci Veering Phenomena in Disordered Structures", *Journal of Sound and Vibrations*, Vol. 126, pp. 485-502.

19. Cornwell, P.J., Bendiksen, O.O., 1989, "Localization of Vibrations in Large Space Reflectors," *AIAA Journal*, Vol. 27, pp. 219-226.

20. Pierre, C., Cha, P.D., 1989, 'Strong Mode Localization in Nearly Periodic Disordered Structures,' *AAIA Journal*, Vol. 27, pp. 227-241.

21. Nwokah, O. D. I, 1978, "Estimates for the Inverse of a Matrix and Bounds for Eigenvalues", *Linear Algebra and Its Applications* vol 22, pp 283-292.

22. Afolabi, D., 1988, "Vibration Amplitudes of Mistuned Blades", *Journal of Turbomachinery*, vol 110, pp 251-257.

23. Smith, D. R., 1985, *Singular Perturbation Theory*, Cambridge University Press, Cambridge.

24. Skelton, R. E., 1988, *Dynamic Systems Control*, Wiley, New York.

25. Perrin, R., 1971, "Selection Rules for the Splitting of Degenerate Pairs of Natural Frequencies of Thin Circular Rings,"*Acustica*, Vol. 25, pp. 69-72.

CONTROL AND DYNAMIC SYSTEMS, VOL. 35

SIMULTANEOUS STABILIZATION VIA LOW ORDER CONTROLLERS

W.E. SCHMITENDORF

Mechanical Engineering
University of California, Irvine
Irvine, CA 92714

C. WILMERS

Institute for System Dynamics and Control
University of Stuttgart
Stuttgart, West Germany

I. INTRODUCTION

A numerical method for the design of robust low order controllers for single input-single output plants in the frequency domain is developed. To achieve arbitrary pole placement for the closed loop system, the controller must have order $k \geq n-1$, where n is the order of the denominator of the plant. However, it may be possible to achieve stability (but not arbitrary pole assignment) using a controller of order $k < n-1$.

Here we present a numerical technique for determining a low order controller. The algorithm searches for a stabilizing controller whose order is specified by the user. The controller sought is the one that minimizes a function of the distance between the actual poles and some pre-specified set of desired poles. If no stabilizing compensator with the specified order is found, the order can be increased and the numerical procedure repeated.

For the design of a controller which simultaneously stabilizes q plants, the same numerical procedure can be used. The distance between the actual poles and the desired ones is calculated for each plant and the controller is chosen to minimize the maximum distance.

The method is applied to typical examples of robust control problems: a remotely piloted vehicle, and a track guided bus.

II. PRELIMINARIES

Consider the single input-single output system of Figure 1 where P(s) is the plant transfer function and C(s) is the compensator transfer function. Here y, u and w are the scalar output, control and reference input, respectively. The plant is assumed to be rational and strictly proper,

$$P(s) = \frac{N(s)}{D(s)} = \frac{n_0 + n_1 s + \ldots + n_m s^m}{d_0 + d_1 s + \ldots + d_n s^n}, \quad n > m, \tag{1}$$

and the compensator C(s) must be rational and proper with order k,

$$C(s) = \frac{N_c(s)}{D_c(s)} = \frac{\tilde{n}_0 + \tilde{n}_1 s + \ldots + \tilde{n}_k s^k}{\tilde{d}_0 + \tilde{d}_1 s + \ldots + \tilde{d}_k s^k}. \tag{2}$$

The transfer function of the closed loop system is

$$H(s) = \frac{P(s)C(s)}{1 + P(s)C(s)} = \frac{N_c(s)N(s)}{D_c(s)D(s) + N_c(s)N(s)}.$$

The system is asymptotically stable if, and only if, all the poles of H(s) lie in the strict left half of the complex plane, or, equivalently, if and only if all the solutions of

$$\delta(s) \equiv N(s)N_c(s) + D(s)D_c(s) = 0$$

lie in the strict left half of the complex plane. The polynomial $\delta(s)$ has order n+k.

It is shown in [1] that in order to achieve arbitrary pole placement for the closed loop system, the compensator must have order $k \geq n - 1$. However, it may be possible to achieve stability (but not arbitrary pole assignment) using a compensator of order $k < n - 1$. If P(s) is minimum phase, i.e. all solutions of

$N(s) = 0$ lie in the strict left half of the complex plane, then stability can always be achieved with a compensator of order $k = n - m - 1$ [2, 3].

Here we present a numerical technique for determining a low order stabilizing compensator; for a non-minimum phase system we desire $k < n - 1$ and for a minimum phase system $k < n - m - 1$. The algorithm searches for a stabilizing compensator whose order is specified by the user. The compensator sought is the one that minimizes a function of the distance between the actual poles and some pre-specified set of desired poles. This minimization procedure will be made more precise in the next section. If no stabilizing compensator with the specified order is found, the order can be increased and the numerical procedure repeated. No minimum phase assumption is required. Furthermore, we can often find a stabilizing compensator of order less than $n - m - 1$. After presenting our results for a one plant, we extend the results to the problem of finding a single compensator which simultaneously stabilizes several plants.

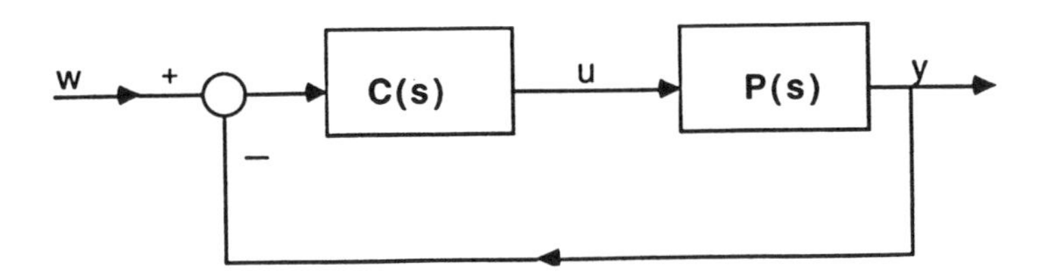

Figure 1: Closed loop control system.

III. DETERMINATION OF THE LOW ORDER COMPENSATOR

For plant (1) with compensator (2), the characteristic polynomial of order $n + k$ for the closed loop system is

$$\delta(s) = N(s)N_c(s) + D(s)D_c(s)$$

$$\equiv \delta_0 + \delta_1 s + \ \ldots \ + \ \delta_{n+k} s^{n+k} \ . \tag{3}$$

Define the closed loop characteristic vector

$$\delta^T \equiv \left[\delta_{n+k} \quad \delta_{n+k-1} \cdots \delta_1 \ \delta_0\right]^T \tag{4}$$

and the (2k + 2) controller parameter vector

$$x^T = \left[\tilde{n}_k \ \tilde{d}_k \cdots \tilde{n}_1 \ \tilde{d}_1 \ \tilde{n}_0 \ \tilde{d}_0\right]^T . \tag{5}$$

For a specific controller parameter vector x, the resulting closed loop characteristic vector δ is obtained from [1, 4]

$$Px = \delta \tag{6a}$$

where the (n + k + 1) x (2k + 2) plant parameter matrix is

$$P = \begin{bmatrix} n_n & d_n & 0 & 0 & . & . & . & 0 & 0 \\ n_{n-1} & d_{n-1} & n_n & d_n & . & . & . & 0 & 0 \\ . & . & . & . & . & . & . & . & . \\ . & . & . & . & . & . & . & . & . \\ . & . & . & . & . & . & . & . & . \\ n_0 & d_0 & n_1 & d_1 & . & . & . & . & . \\ 0 & 0 & n_0 & d_0 & . & . & . & . & . \\ . & . & . & . & . & . & . & . & . \\ . & . & . & . & . & . & . & n_1 & d_1 \\ 0 & 0 & 0 & 0 & . & . & . & n_0 & d_0 \end{bmatrix} \tag{7}$$

where $n_i = 0$, $i > m$ because of the strictly proper assumption on P(s). If we denote the desired characteristic polynomial and desired characteristic vector by

$$\delta^*(s) = \delta^*_0 + \delta^*_1 s + \ \ldots + \delta^*_{n+k} s^{n+k} \tag{8}$$

$$\delta^{*T} = \left[\delta^*_{n+k} \cdots \delta^*_1 \ \delta^*_0\right]^T \tag{9}$$

then we would like to choose x so that $\delta = \delta^*$. If $k = n - 1$, this can be done. However if $k < n - 1$, then x cannot always be chosen such that the two characteristic vectors (4) and (9) match. If exact matching cannot be achieved, then one approach would be to choose x to minimize $\|Px - \delta^*\|$. While the solution to this problem will lead to a characteristic polynomial which is close to the desired one, the resulting poles may not be close to the desired ones. Instead, we try to choose x so that the actual poles are close to the desired poles.

Denote the poles corresponding to a particular choice of controller parameter x by $\{\lambda_i(x)\}$ and the desired ones by $\{\mu_i\}$. In [5], a measure of the separation distance between two sets of poles is given by

$$F(x) = \sum_{i=1}^{n+k} \left|\mu_i - \lambda_i(x)\right| ,$$

$$= \sum_{i=1}^{n+k} \sqrt{\left[\mathrm{Re}(\lambda_i(x) - \mathrm{Re}(\mu_i)\right]^2 + \left[\mathrm{Im}(\lambda_i(x)) - \mathrm{Im}(\mu_i)\right]^2} \tag{10}$$

Prerequisite for this distance evaluation is that both the actual and the desired poles are in an order representing the magnitude of their real parts,

$$\mathrm{Re}(\lambda_1) \le \mathrm{Re}(\lambda_2) \le \ldots \le \mathrm{Re}(\lambda_{n+k}) . \tag{11}$$

Poles with identical real parts are ordered so that the pole with the smaller index corresponds to the pole with the smaller imaginary part.

The summation of the absolute distance has a major drawback: it does not relate this distance to the location of the desired poles. For example, although the absolute distance between $\lambda = -12$ and $\mu = -13$ is the same as between $\lambda = -1.2$ and $\mu = 0.2$, the latter difference causes a much larger change in the dynamics of the system. To avoid this problem the objective function F(x) is computed as the sum of the relative distances

$$F(x) = \sum_{i=1}^{n+k} c_i \frac{\left|\mu_i - \lambda_i(x)\right|}{\left|\mu_i\right|} , \tag{12}$$

We have also introduced weighting constants c_i which are positive penalties used to avoid a controller design which yields an unstable closed loop system, "slow' poles or insufficiently damped poles. The default value is c = 1. The penalty constant for all unstable poles is chosen as c = 1000. Another constant, c = 100, is used for all poles with a real part greater than $-\gamma$. In this way, a required stability margin σ is enforced, see Figure 2. Furthermore the controller design should lead to poles with a sufficiently large damping ratio and the damping ratio should exceed a minimum value.

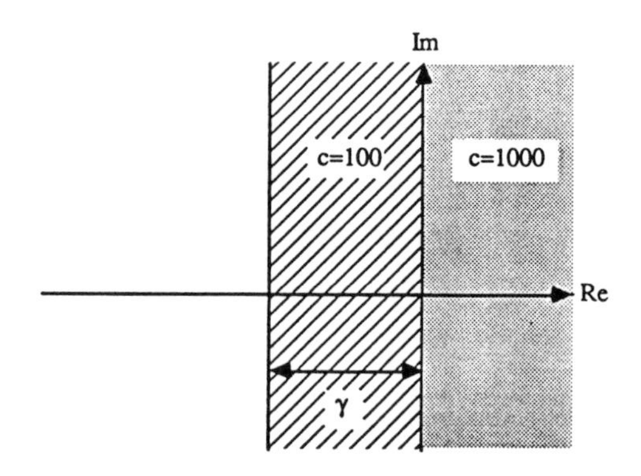

Figure 2: Penalty functions for unstable and slow poles.

$$\zeta_{min} = \sin\alpha \quad . \tag{13}$$

Therefore, for all poles with a damping ratio

$$\zeta_i = \frac{|Re(\lambda_i)|}{\sqrt{Re(\lambda_i)^2 + Im(\lambda_i)^2}} \tag{14}$$

smaller than ζ_{min}, a penalty constant

$$c = \frac{\zeta_{min}}{\zeta} \tag{15}$$

is used. The area for this penalty function is shown in Figure 3.

Combining the three penalty functions in the objective function (12) the controller parameters are computed as solution of

$$\underset{x}{\text{minimize}}\ F(x). \tag{16}$$

A modified simplex optimization method [5] is used for the computational algorithm.

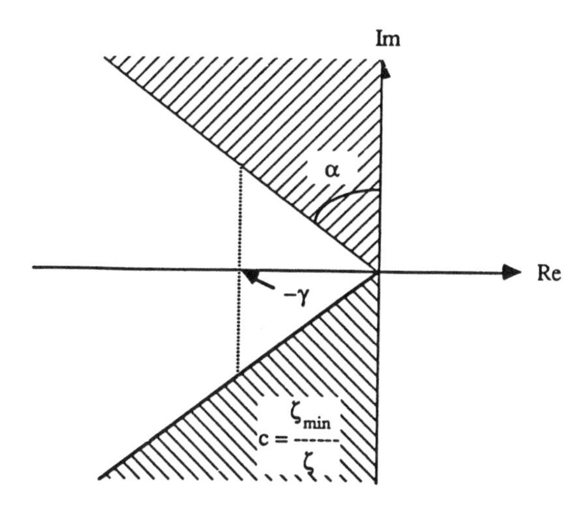

Figure 3: Penalty function for insufficient damped eigenvalues.

To provide initial values for the controller parameters x to start the optimization procedure, either a least squares solution of equation (6) is used or a first estimate for the controller is given by the user. Of course, the choice of the weights c is arbitrary and can be changed. We have found the above choices work quite well. If, for a particular controller order k, no stabilizing controller is found, the order is increased and the algorithm repeated. Thus a pole placement method for the design of a low order controller for one given plant P(s) is realized.

IV. SIMULTANEOUS STABILIZATION

In this section we consider the problem of finding a single compensator which simultaneously stabilizes several plants, $P_i(s)$, $i = 1, 2 \ldots q$. This problem occurs when the true system deviates from the assumed model due to changing parameters or because of inexact modelling. This single controller concept can also be used to design one controller for a system operating at several different conditions rather than having to use several controllers and a gain scheduling procedure. A compensator that stabilizes several different plants will be called a robust controller.

For the design of a robust controller, the technique of the previous section can be used if it is applied to all q plants simultaneously. A controller vector x is specified and the characteristic vector δ_i for each closed loop system is calculated using

$$P_i \; x = \delta_i \; , \; i = 1, \ldots, q \; , \tag{17}$$

where P_i is the coefficient matrix (7) for plant i, and x is the vector for the controller parameters (5). After computing the eigenvalues for all q characteristic polynomials, the relative eigenvalue distances $F_i(x)$, $i = 1, 2, \ldots, q$ are evaluated. The objective for the optimizations procedure is to choose x to minimize the maximum distance for all plants,

$$\underset{x}{\text{minimize}} \; \underset{i}{\text{maximize}} \; F_i(x). \tag{18}$$

An algorithm based on these ideas has been implemented successfully on a personal computer. Several examples of robust controller design using this algorithm are presented in the next section.

V. EXAMPLES

Examples 1. Lateral Autopilot for a Remotely Piloted Vehicle

A fifth order model for the model for the lateral dynamics of a RPV is given [7]. The aileron actuator dynamics are included as a first order lag with a time constant of 0.05. Using the state variables $\mathbf{x} = [\, v, p, r, \phi, \delta_a]^T$, with v as the component of the vehicle velocity parallel to the pitch axis, p as the roll rates, r as the yaw rate, ϕ as the roll angle and δ_a as the aileron deflection together with the demanded aileron deflection as input u, the system equation is,

$$\dot{x} = \begin{bmatrix} -0.85 & 25.47 & -979.5 & 32.14 & 0 \\ -0.339 & -8.789 & 1.765 & 0 & 59.89 + q1.71 \\ 0.021 & -0.547 & -1.407 & 0 & 6.477 + q3.22 \\ 0 & 1 & 0.0256 & 0 & 0 \\ 0 & 0 & 0 & 0 & -20 \end{bmatrix} x + \begin{bmatrix} 0 \\ 0 \\ 0 \\ 0 \\ 20 \end{bmatrix} u, \tag{19}$$

where

$$q = \frac{C_{n\delta a} - 1.99}{1.99} . \tag{20}$$

The unknown aerodynamic coefficient $C_{n\delta a}$ varies in the range

$$-.99 \le C_{n\delta a} \le 2.99, \tag{21}$$

and its nominal value is $C_{n\delta a} = 1.99$, which corresponds to $q = 0$.

With output $y = r$, the transfer functions for three different values of $C_{n\delta a}$ are calculated:

$C_{n\delta a} = 1.99$, $q = 0$:

$$G(s) = \frac{129.54\, s^3 + 604.4\, s^2 + 2167\, s + 2197}{s^5 + 31.046\, s^4 + 272.122\, s^3 + 1419.9\, s^2 + 7930.5\, s + 276.7}, \tag{22}$$

$C_{n\delta a} = -.99$, $q = -1.5$:

$$G(s) = \frac{161.9\, s^3 + 906.9\, s^2 + 2689\, s + 2561}{s^5 + 31.046\, s^4 + 272.122\, s^3 + 1419.9\, s^2 + 7930.5\, s + 276.7}, \tag{23}$$

$C_{n\delta a} = 2.99,\ q = 0.5$:

$$G(s) = \frac{33.1\ s^3 - 297.09\ s^2 + 610.2\ s + 1111.4}{s^5 + 31.046\ s^4 + 272.122\ s^3 + 1419.9\ s^2 + 7930.5\ s + 276.7}, \quad (24)$$

The desired pole locations for the closed loop system are given in [7] as

$$\mu_1 = -40,\ \mu_2 = -10,\ \mu_{3,4} = -2.14 \pm 6.22,\ \mu_5 = -.29\ .$$

In addition, a stability margin $\sigma = .1$ is specified. The design method yields a zeroth order

$$C(s) = 0.48\ , \quad (25)$$

controller and the pole location for the closed loop system using this compensator is shown in the root locus plot in Figure 4. (The pole near -40 is not shown on the plot.) The 0th order controller (25) will stabilize the system for the whole parameter range (21) and can be called robust.

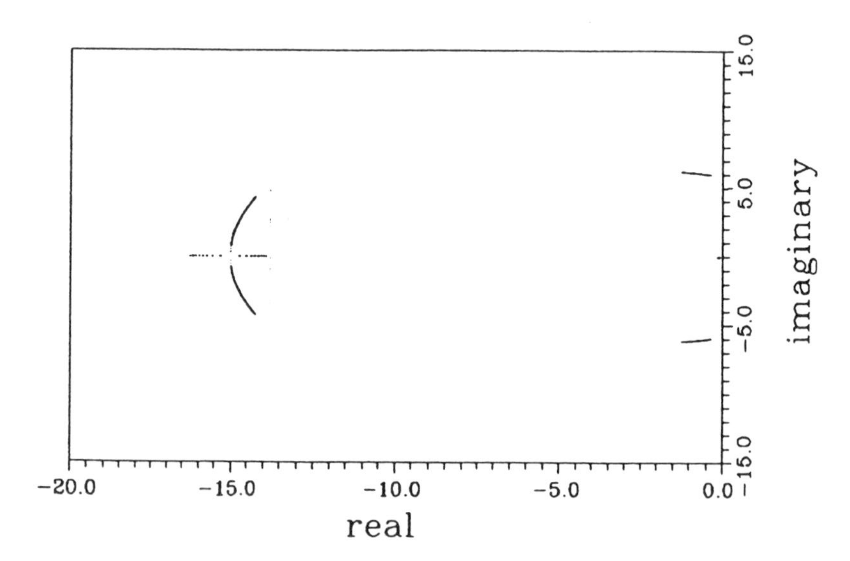

Figure 4: Root locus plot RPV, controller $C(s) = 0.48$, $-0.99 \le C_{n\delta a} \le 2.99$.

In [3], a technique for simultaneous stabilization of single input-single output systems is presented. Their results lead to a controller of order n - m - 1 where n and m are the order of denominator and numerator of the plant, respectively. Furthermore, in [2] it is assumed that all the zeros of the transfer functions lie strictly in the left half of the complex plane (minimum phase assumption). Since the transfer function (24) does not satisfy the minimum phase assumption, the results presented in [2] do not apply. Furthermore, even if they did, a first order controller would result rather than the zeroth order controller obtained here.

To reduce the pole location sensitivity for the closed loop system, we tried a first order controller. The desired eigenvalues were taken as

$$\mu_1 = -40, \mu_2 = -10, \mu_{3,4} = -2.14 \pm 6.22, \mu_{5,6} = -.29 \pm 0.5 .$$

The simultaneous pole placement for all three given transfer functions (22)-(24) yields the controller

$$C(s) = \frac{1.5 + 0.16\,s}{0.3 + s}, \tag{26}$$

which gives the eigenvalue location plotted in Figure 5. (Again, the pole near -40 is not shown.) Comparing the root locus plots in Figures 4 and 5, one can see

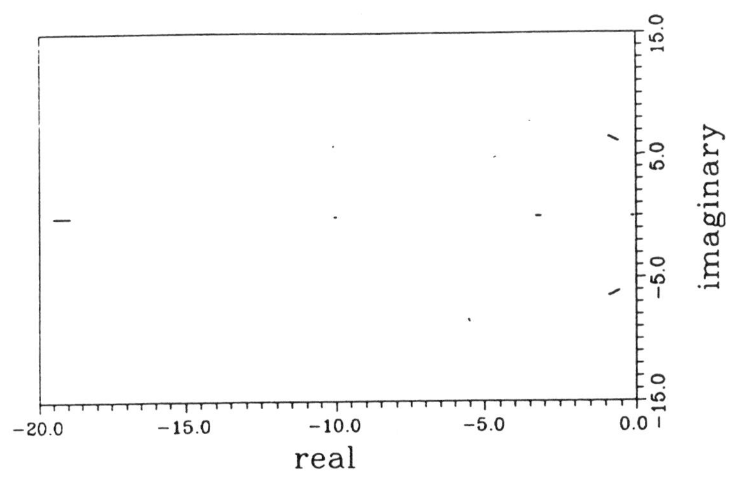

Figure 5: Root locus plot for RPV, 1st order controller, -0.99 ≤ Cnδa ≤ 2.99.

that this controller is superior to the zeroth order controller because the sensitivity of the closed loop pole location to changes in the aerodynamic coefficient $C_{n\delta a}$ is much smaller than that obtained with the zeroth order controller.

Although the actual eigenvalues of the closed loop system do not match the desired ones, both designs, especially the first order one, are good enough to consider for possible application.

Example 2. Track Guided Bus

A novel application of robust control theory is to track guided vehicles whose dynamics vary. Three parameters are mainly responsible for the change of the dynamics during the operation: the velocity v, the mass m, and the friction coefficient μ. The particular model for this example is a Daimler Benz 0305 bus guided by the electric field generated via a wire in the street. The model with its parameters is given in [5]. The five state variables $\mathbf{x}= [\, \alpha, \varepsilon, \varepsilon, y\ \beta\,]^T$ are defined in Figure 6.

The linearized dynamic equations for small deviations of the bus from the guide line are

$$\dot{x}(t) = \begin{bmatrix} -668\ b & -1+181ab^2 & 0 & 0 & 198ab \\ 16.8a & -409ab & 0 & 0 & 67.3a \\ 0 & 1 & 0 & 0 & 0 \\ \frac{1}{b} & 6.12 & \frac{1}{b} & 0 & 0 \\ 0 & 0 & 0 & 0 & 0 \end{bmatrix} x(t) + \begin{bmatrix} 0 \\ 0 \\ 0 \\ 0 \\ 1 \end{bmatrix} u(t), \tag{27}$$

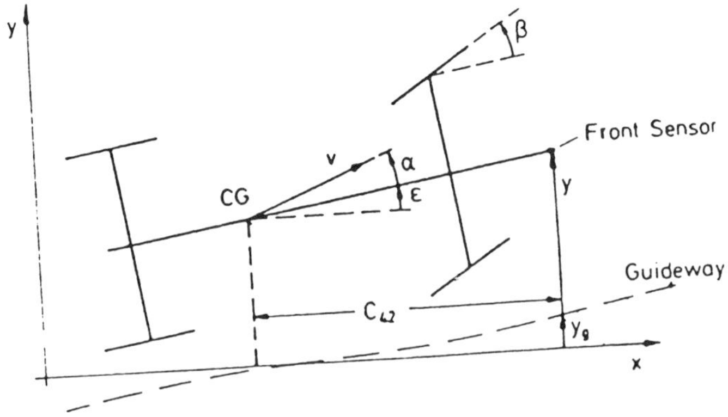

Figure 6: Definition of variables for the track guided bus.

where $a = \frac{\mu}{m}$ and $b = \frac{1}{v}$. The parameters vary in the range

$$0.5 \le \mu \le 1, \quad 9.95 \le m \le 32, \quad 3 \le v \le 20,$$

or

$$\frac{1}{32} \le a \le \frac{1}{9.95}, \; 0.05 \le b \le 0.3333. \tag{28}$$

All possible parameter combinations lie inside a rectangle in the a-b-plane. The four corners of the rectangle are chosen for the controller design, Figure 7.

With the steering angle rate as input u and the displacement y as output, the transfer function is

$$G(s, a, b) = \frac{48282.8a^2 + 388654\, a^2bs + 609\, as^2}{s^3(270316a^2b^2 + 16a + 1077abs + s^2}. \tag{29}$$

The first step for the design of a robust controller is to find a controller for one set of parameters a,b. We start with a pole placement for the heaviest and fastest configuration of the bus, the parameter set 3 ($v = 20$, $m/\mu = 32$). In order

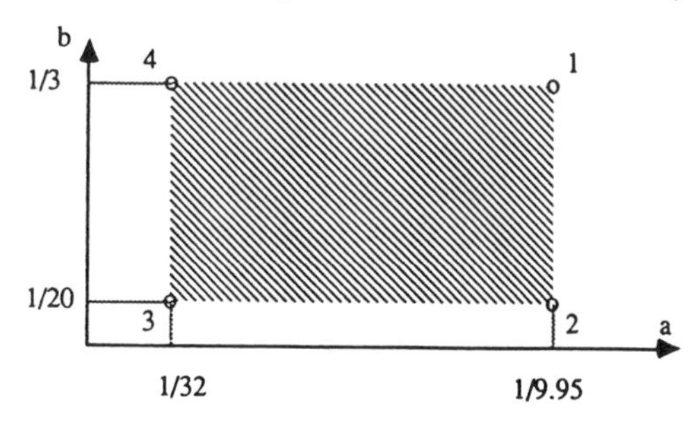

Figure 7: Parameter plane for track guided bus with selected parameter sets 1-4.

to get an idea where to place the poles for the closed loop system, an optimal state feedback controller $u = -k^T x$ is computed with the algebraic Riccati equation,

$$A^T P + P A - P B R^{-1} B^T P + Q = 0 \;, \tag{30}$$

and the gain k is

$$k^T = R^{-1} B^T P \ . \tag{31}$$

For the system (27) with $a = \frac{1}{32}$, $b = \frac{1}{20}$ and the matrices R and Q set equal to the identity matrix, the optimal controller is computed from (30) and (31),

$$k^T = [\,11.57 \quad 5.18 \quad 16.57 \quad 1 \quad 5.47\,] \ . \tag{32}$$

The poles for the corresponding closed loop system are

$$s_1 = -2.97,\ s_{2,3} = -1.63 \pm 2.09j,\ s_{4,5} = -0.49 \pm 1.43j \ . \tag{33}$$

To limit the bandwidth of the system, a pole is placed at s = -20. The desired stability margin is 0.1. Thus, the desired poles are

$$\mu_1 = -20,\ \mu_2 = -3,\ \mu_{3,4} = -1.8 \pm 2j\,,\ \mu_{5,6} = -0.5 \pm j\ , \tag{34}$$

and the computed first order controller is

$$C(s) = \frac{3 + 29.6s}{49.8 + s} \ . \tag{35}$$

The next step is to "robustify" this controller using the simultaneous pole placement method for the four transfer functions with the parameter sets (a, b) specified in Figure 7. Using the controller in (35) as the initial controller for the computer algorithm, we obtain

$$C(s) = \frac{3.33 + 32.8\,s}{47.7 + s} \ . \tag{36}$$

The controller will minimize the distance between the actual pole locations and the desired ones for the corner points of the parameter plane. The sensitivity of the closed loop system with this controller design can be seen in Figure 8. An equidistant 11 by 11 mesh is mapped onto the parameter plane and the poles are

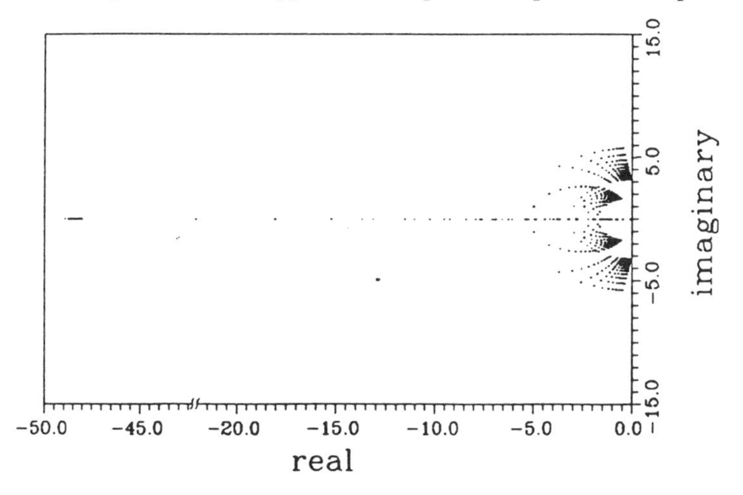

Figure 8: Root locus plot for track guided bus, 1st order Controller.

calculated and plotted for each of the 121 parameter sets (a,b). This graphs shows that the closed loop system remains stable for the whole parameter range (28) and that the maximum real part of an eigenvalue is -0.1, as specified with the stability margin.

The same procedure is performed for a second order controller with the desired eigenvalues

$$\mu_1 = -20, \mu_{2,3} = -3 \pm j, \mu_{4,5} = -1.8 \pm 2j, \mu_{6,7} = -0.5 \pm j. \tag{37}$$

Based only on the plant with parameter set 3, the computed controller is

$$C(s) = \frac{102.3 + 108.4s + 36.1s^2}{178.6 + 24.3\,s + s^2} \quad . \tag{38}$$

This controller defines the starting point for the simultaneous pole placement using the four transfer functions from the parameter sets 1-4. This robustification yields the robust controller

$$C(s) = \frac{92.2 + 127.7s + 13.7s^2}{179.6 + 12.8s + s^2} , \tag{39}$$

with the pole location for the closed loop system plotted in Figure 9 for the entire parameter range. As with the first order controller, the second order controller stabilizes the closed loop system for all parameter values, not just the corner values.

Both low order controllers (36) and (39) are robust. It appears that the second order controller is more sensitive to parameter changes since the closed loop poles vary much more. In [8], a third order controller is presented,

$$C(s) = \frac{9375 + 10937.5s + 2343.75s^2}{15625 + 1250s + 50s^2 + s^3} . \tag{40}$$

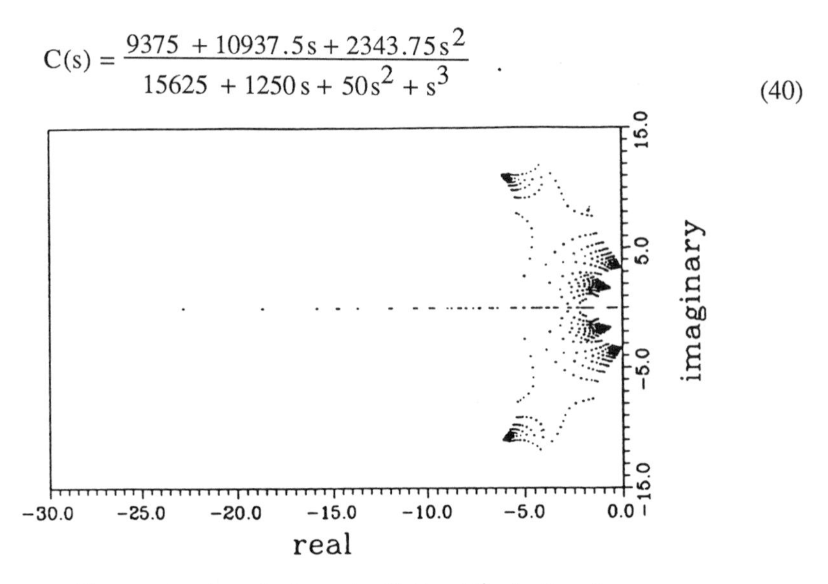

Figure 9: Root locus plot for track guided bus, 2nd order Controller.

The pole locations for all parameters using this controller are plotted in Figure 10. The main difference between this plot and the plots in Figure 8 or 9 is that the pole location with the latter two controllers is concentrated in a smaller area in the complex domain than using the third order controller (40). Whether one of these controllers is superior to the other two designs should be determined with simulations, but it appears that the first order controller is the least sensitive.

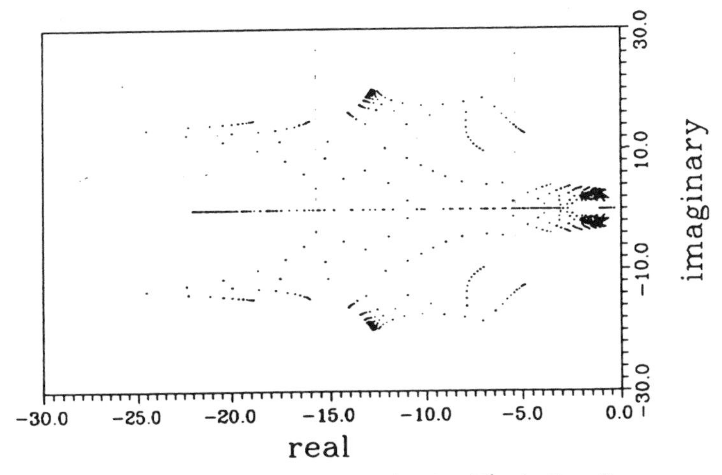

Figure 10: **Root locus plot for track guided bus, 3rd order Controller.**

VI. CONTROLLER DESIGN FOR MISO AND SIMO SYSTEMS

The method developed in the previous sections for single input-single output systems can easily be extended to feedback systems with multiple inputs and a single output (the MISO case). Let the number of inputs be r. Then the plant is described by the 1 x r transfer function row vector

$$P^T(s) = \frac{n^T(s)}{D(s)} \tag{41}$$

where

$$D(s) = d_0 + d_1 s + \ldots + d_n s^n \tag{42}$$

and

$$n^T(s) = n_0^T + n_1^T s + \ldots + n_m^T s^m \tag{43}$$

with n_i, i = 1, 2, ..., m a vector with r components.

The transfer function of the compensator is an rx1 column vector of the form

$$C(s) = \frac{n_c(s)}{D_c(s)} \tag{44}$$

where

$$D_c(s) = \tilde{d}_0 + \tilde{d}_1 s + \ldots + \tilde{d} s^k , \tag{45}$$

$$n_c(s) = \tilde{n}_0 + \tilde{n}_1 s + \ldots + \tilde{n}_1 s^k \quad , \tag{46}$$

and and $\tilde{n},i = 1, 2,.., k$ is a vector with r components.

The closed loop transfer function is

$$H(s) = \frac{n^T(s)\, n_c(s)}{D(s)D_c(s) + n^T(s)n_c(s)} .$$

Similar to the design for SISO systems, the vector for the coefficients of the closed loop characteristic polynomial is denoted by

$$\delta^T = \left[\delta_{n+k} \;\; \delta_{n+k-1} \cdots \delta_0\right]^T \tag{47}$$

and can be calculated from the product

$$\delta = Px . \tag{48}$$

where the (n + k + 1) x [(k +1) (r + 1)] plant parameter matrix P is

$$
P = \begin{bmatrix}
n_n^T & d_n & 0 & 0 & \cdots & 0 & 0 \\
\cdot & \cdot & n_n^T & d_n & \cdots & 0 & 0 \\
\cdot & \cdot & \cdot & \cdot & \cdots & \cdot & \cdot \\
\cdot & \cdot & \cdot & \cdot & \cdots & \cdot & \cdot \\
\cdot & \cdot & \cdot & \cdot & \cdots & n_n^T & d_n \\
\cdot & \cdot & \cdot & \cdot & \cdots & \cdot & \cdot \\
n_0^T & d_0 & \cdot & \cdot & \cdots & \cdot & \cdot \\
\cdot & \cdot & n_0^T & d_0 & \cdots & \cdot & \cdot \\
\cdot & \cdot & \cdot & \cdot & \cdots & \cdot & \cdot \\
\cdot & \cdot & \cdot & \cdot & \cdots & \cdot & \cdot \\
0 & 0 & 0 & 0 & \cdots & n_0^T & d_0
\end{bmatrix} \tag{49}
$$

If $k \geq \frac{n}{r} - 1$ and P has full rank, there are more unknowns than equations and exact pole placement can be achieved. For $k < \frac{n}{r} - 1$, an optimization routine can be used to minimize the distance between the desired poles and the actual ones. A controller for the MISO simultaneous stabilization problem can be obtained in a similiar fashion to the SISO case in Section IV.

Finally, for the single input-multiple output problem (SIMO), the closed loop transfer function is an r x 1 matrix (r denotes the number of outputs)

$$
H(s) = \frac{n(s)n_c^T(s)}{D(s)D_c(s) + n_c^T(s)n(s)} ,
$$

whose denominator is the same as the denominator for the MISO problem. Thus the simultaneous stabilization problem for the SIMO problem can also be handled by the technique in Sec. IV. The problem of simultaneous stabilization for multiple input-multiple output problem remains an open problem and further research is needed.

One other open problem is that of determining low order controllers for plants where the order of the plant can change. The theory in [2] applies to such problems, but the results presented here have not yet been generalized to such problems.

REFERENCES

1. C.T Chen, Linear System Theory and Design, Holt, Rinehart and Winston, New York, 1984.
2. B.R. Barmish, and K.H Wei, "Simultaneous Stabilizability of Single Input-Single Output Systems", Proceedings 7th International Symposium on Mathematical Theory of Networks and Systems, Stockholm, 1985.
3. M.C. Smith and K.P. Sondergeld, "On the Order of Stable Compensators", Automatica, Vol. 22, No. 1, 1986.
4. R.M. Biernacki, H. Hwang, and S.P. Bhattacharya, "Robust Stability with Structured Real Parameters Peturbations", IEEE Transactions on Automatic Control, Vol. AC-32, No. 6, 1987.
5. E. Shapiro, D.A. Fredricks, R.H. Rooney, and B.R. Barmish, "Pole Placement with Output Feedback", J. Guidance, Control and Dynamics, Vol. 4, No. 4, 1981.
6. J.A. Nelder and R. Mead, "A Simplex Method for Function Minimization", The Computer Journal, Vol. 7, No. 4, 1964.
7. A. Vinkler and L.J. Wood, "Multistep Guaranteed Cost Control of Linear Systems with Uncertain Parameters", J. Guidance, Control and Dynamics, Vol. 2, No. 5, 1979.
8. J. Ackermann and R. Muench, "Robustness Analysis in a Plant Parameter Plane", Proceedings of the 1987 IFAC Congress, Munich, 1987.

CONTROL AND DYNAMIC SYSTEMS, VOL. 35

USE OF LIAPUNOV TECHNIQUES FOR COLLISION-AVOIDANCE OF ROBOT ARMS

R.J. STONIER

Department of Mathematics and Computing
Capricornia Institute of Advanced Education
Rockhampton, Australia 4702

I. INTRODUCTION

The problem of real time pathfinding in multi-robot systems (robot navigation) is one of current interest. Freund and Hoyer [1] examine this problem using a systematic design method with a hierarchical controller utilising the full dynamics of all robots involved. The hierarchical controller was designed for real time collision avoidance, where the collision avoidance strategy is based upon an analytically described avoidance trajectory that served for collision detection as well as avoidance. The design of this avoidance trajectory was described in the case of two PR robot arms with a common workspace in the plane.

In this paper we propose an alternative approach for the path-finding problem of two PR manipulators in the horizontal plane based upon established Liapunov theory in the control and differential game literature for capture within a target, and for avoidance of antitargets, cf. [2]-[13].

The theory we present does not rely upon the use of a fictitious robot arm [1] for the establishment of an avoidance trajectory. As illustration, consider Figure 1.

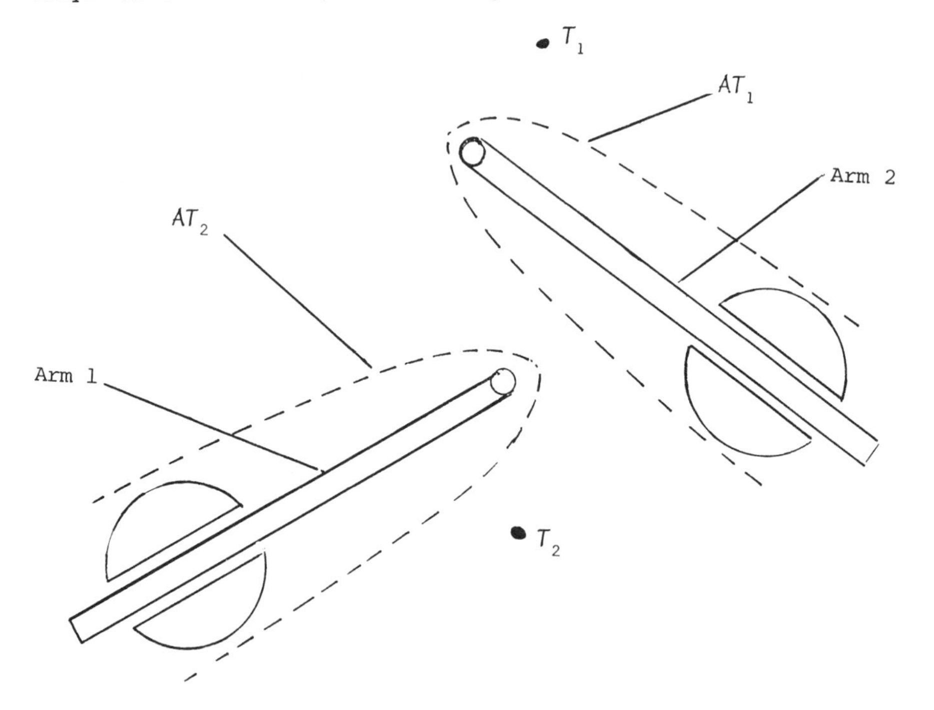

FIGURE 1: Two robot arm configuration with targets and antitargets

The control objective may be described as follows:

(i) to control the movement of arm 1 to position the gripper at target T_1 whilst ensuring the entire arm 1 avoids target T_2 of the arm 2 and an established secure avoidance region AT_1 about arm 2;

(ii) to control the movement of arm 2 to position the gripper at target T_2 whilst ensuring the entire arm 2 avoids target T_1 and an avoidance region AT_2 about arm 1.

Our aim is to develop a method which yields analytically defined nonlinear controllers for each arm using the knowledge of the current state of each arm, target/antitarget configurations and constraints without the imposition of any optimisation calculations.

The trajectories of the pathfinding problem will consequently be nonunique but may be later re-examined with the imposition of a desired performance index. Essentially this is the position of qualitative control, cf. [7].

In order to illustrate the method, we shall firstly look at the movement of two point masses in the plane; then an analysis of a single RP manipulator moving to a target in the presence of state constraints. The method applied to two robot arms in a common planar workspace will appear in a future publication.

II. CONTROL OF TWO POINT MASSES MOVING ON A PLANE

Let the dynamics of two point objects (1) and (2) be given by the state equations:

$$\begin{aligned} \dot{x}_1 &= x_2 & \dot{y}_1 &= y_2 \\ \dot{x}_2 &= u_1 & \dot{y}_2 &= v_1 \\ \dot{x}_3 &= x_4 & \dot{y}_3 &= y_4 \\ \dot{x}_4 &= u_2 & \dot{y}_4 &= v_2 \end{aligned}$$

or in vector form

$$\dot{\underset{\sim}{x}} = \underset{\sim}{f}_1(\underset{\sim}{x},\underset{\sim}{u}) \qquad (1)$$

and

$$\dot{\underset{\sim}{y}} = \underset{\sim}{f}_2(\underset{\sim}{y},\underset{\sim}{v}) \qquad (2)$$

where the vectors $\underset{\sim}{x}$, $\underset{\sim}{y}$, $\underset{\sim}{f}_1$ and $\underset{\sim}{f}_2$ are defined in the appropriate way. Here $\underset{\sim}{u} = [u_1 \ u_2]^T$ and $\underset{\sim}{v} = [v_1 \ v_2]^T$. With no restrictions upon the control vectors $\underset{\sim}{u}$ and $\underset{\sim}{v}$ each system is completely controllable and the position on the plane for objects (1) and (2) are $[x_1\ x_3]^T$ and $[y_1\ y_3]^T$ respectively.

Define target sets and antitarget sets in the plane as follows:

$$T_1 = \{\underset{\sim}{z} : (z_1 - p_1c_1)^2 + (z_3 - p_1c_2)^2 \le rp_1^2\} ,$$

$$T_2 = \{\underset{\sim}{z} : (z_1 - p_2c_1)^2 + (z_3 - p_2c_2)^2 \le rp_2^2\} ,$$

$$AT_1^1 = T_2 ,$$

$$AT_2^1(t) = \{\underset{\sim}{z} : (z_1 - y_1(t))^2 + (z_3 - y_3(t))^2 \le rap_1^2\} ,$$

$$AT_1^2 = T_1 ,$$

$$AT_2^2(t) = \{\underset{\sim}{z} : (z_1 - x_1(t))^2 + (z_3 - x_3(t))^2 \le rap_2^2\} .$$

Fixed circular targets T_1 and T_2 are defined for each object, with centres $[p_1c_1\ p_1c_2]^T$ and $[p_2c_1\ p_2c_2]^T$, and radii rp_1 and rp_2. The two antitargets for object (1) are AT_1^1 (object (2)'s target set) and a moving target $AT_2^1(t)$, which is a circular disc of radius rap_1 about the current time position of object (2).

The two antitargets for object (2) are $AT_1^2 = T_1$ and $AT_2^2(t)$ which is a circular disc of radius rap_2 about the current time position of object (1).

To begin, suppose our objective is that described in the introduction, ignoring the physical structure of the arm, the robot arm (or robots) being idealised as just two point objects in the plane.

Furthermore, we impose the requirement that the speed of each object be made 'sufficiently small' as it approaches the designated target.

Consider now object (1). We form the following four functions defined as follows:

(i) $V_0(\underset{\sim}{x}) = ((x_1 - p_1c_1)^2 + (x_3 - p_1c_2)^2 - rp_1^2)/2 + (x_2^2 + x_4^2)/2$.

V_0 is the sum of two functions, one measuring the distance squared from the target and the other measuring the object's speed squared. It is positive outside T_1 .

(ii) $V_1(\underset{\sim}{x}) = ((x_1 - y_1(t))^2 + (x_3 - y_3(t))^2 - rap_2^2)/2$.

V_1 measures the distance squared of $\underset{\sim}{x}$ from the anti-target of radius rap_2 about the position $[y_1(t)\ y_3(t)]^T$ of object (2). Here $\underset{\sim}{y}(t)$ is recognised as the state at time t , being the output of Eqs. (2).

(iii) $V_2(\underset{\sim}{x}) = ((x_1 - p_2c_1)^2 + (x_3 - p_2c_2)^2 - rp_2^2)/2$.

V_2 measures the distance squared of $\underset{\sim}{x}$ from the fixed antitarget AT_2^1 .

(iv) Suppressing arguments, let

$$V = V_0 + \beta_{11}/V_1 + \beta_{12}/V_2$$

where β_{11} and β_{12} are chosen positive constants.

With this definition V is well defined on R^4 and

$V(\underset{\sim}{x}) > 0$ for $\underset{\sim}{x} \in R^4 \backslash (T_1 \cup AT_1^1 \cup AT_2^1)$.

To establish a coordination control algorithm, we intuitively reason as follows:

We seek a strategy $\underset{\sim}{u}^*$ for which $\dot{V} \le 0$ (or < 0) , and expect that the state would under such strategy through Eq. (1) 'move down' through decreasing levels of V and, if β_{ij} is sufficiently small, 'move down' through decreasing levels of V_0 to pass into the target T_1 whilst avoiding AT_1^1 and AT_2^1 and reduce the objects speed upon approach to T_1 . V would have to increase if the state moved towards either of the two antitargets.

This construction in forming V is an extension of that used for defining the Liapunov function in Example 4.1 of [8]. It also follows the use of penalty functions in the Fiacco and McCormick minimisation algorithm in nonlinear optimisation theory, cf. [13].

In the numerical simulation performed and illustrated in the following diagrams, a fourth order Runge-Kutta was used to integrate Eqs. (1) and (2).

For convenience, let us assume the state equation (1) is integrated first over the interval $[t, t+\delta t]$. It is assumed that information on states $\underset{\sim}{x}$ and $\underset{\sim}{y}$ are known at time t. In the calculation of the control strategy $\underset{\sim}{u}$ from $\dot{V}$, the value of $\underset{\sim}{y}(t)$, or rather the 'state position' components $y_1(t)$ and $y_3(t)$, are held constant over $[t, t+\delta t]$. Provided δt is sufficiently small this last assumption is not unrealistic if the systems are relatively slow moving.

Now with this assumption, we find

$$\dot{V} = [(x_1 - p_1c_1) - \beta_{11}(x_1 - y_1)/V_1^2 - \beta_{12}(x_1 - p_2c_1)/V_2^2]x_2 + x_2u_1$$
$$+ [(x_3 - p_1c_2) - \beta_{11}(x_3 - y_3)/V_1^2 - \beta_{12}(x_3 - p_2c_2)/V_2^2]x_4 + x_4u_2 .$$

Then

$$\dot{V} = -p_1\gamma_2x_2^2 - p_1\gamma_4x_4^2 \leq 0 \quad \text{on} \quad R^4\backslash(T_1 \cup AT_1^1 \cup AT_2^1) ,$$

if we make the selection

$$u_1 = -(x_1 - p_1c_1) + \beta_{11}(x_1 - y_1)/V_1^2 + \beta_{12}(x_1 - p_2c_1)/V_2^2 - p_1\gamma_2x_2 , \tag{3}$$

$$u_2 = -(x_3 - p_1c_2) + \beta_{11}(x_3 - y_3)/V_1^2 + \beta_{12}(x_3 - p_2c_2)/V_2^2 - p_1\gamma_4x_4 , \tag{4}$$

where $p_1\gamma_2$ and $p_1\gamma_4$ are selected positive constants.

Given β_{11} and β_{12} sufficiently small, we observe that when these control strategies are implemented in the state equation, an equilibrium state of Eq. (1) is 'close to' $[p_1c_1\ 0\ p_1c_2\ 0]^T$. The maximal set E for which $\dot{V}(t) = 0$ includes this state, so we would hope that the above conditions upon V and $\dot{V}$ ensure an 'asymptotic convergence' to within the target T_1. With the adequate selection of parameters in the following examples, this is seen to be true, particularly in the movement of the single robot arm in the presence of constraints.

A symmetric discussion can now be made for object (2). Take

$$W_0(\underset{\sim}{y}) = ((y_1 - p_2c_1)^2 + (y_3 - p_2c_2)^2 - rp_2^2)/2 + y_2^2/2 + y_4^2/2$$

$$W_1(\underset{\sim}{y}) = ((x_1(t) - y_1)^2 + (x_3(t) - y_3)^2 - rap_1^2)/2$$

$$W_2(\underset{\sim}{y}) = ((y_1 - p_1c_1)^2 + (y_3 - p_1c_2)^2 - rp_1^2)/2 ,$$

and form

$$W = W_0 + \beta_{21}/W_1 + \beta_{22}/W_2 .$$

As discussed previously, on $[t, t+\delta t]$, it is assumed object (1) has 'right of way' in the execution of the numerical simulation. So with the knowledge of $\underset{\sim}{x}(t+\delta t)$, the development of the control algorithm is based upon $x_1(t+\delta t)$ and $x_3(t+\delta t)$ being held at constant value in W_1 while $\underset{\sim}{y}(t+\delta t)$ is calculated from $\underset{\sim}{y}(t)$.

With this assumption we find

$$\begin{aligned}\dot{W} &= [(y_1 - p_2c_1) + \beta_{21}(x_1 - y_1)/W_1^2 - \beta_{22}(y_1 - p_1c_1)/W_2^2]y_2 + y_2v_1 \\ &\quad + [(y_3 - p_2c_2) + \beta_{21}(x_3 - y_3)/W_1^2 - \beta_{22}(y_3 - p_1c_2)/W_2^2]y_4 + y_4v_2 \\ &= -p_2\gamma_2y_2^2 - p_2\gamma_4y_4^2 \le 0 \quad \text{on} \quad R^4\backslash(T_2 \cup AT_1^2 \cup AT_2^2) ,\end{aligned}$$

if we make the strategy selection,

$$v_1 = -(y_1 - p_2c_1) - \beta_{21}(x_1 - y_1)/W_1^2 + \beta_{22}(y_1 - p_1c_1)/W_2^2 - p_2\gamma_2y_2 , \tag{5}$$

$$v_2 = -(y_3 - p_2c_3) - \beta_{21}(x_3 - y_3)/W_1^2 + \beta_{22}(y_3 - p_1c_2)/W_2^2 - p_2\gamma_4y_4 , \tag{6}$$

where $p_2\gamma_2$ and $p_2\gamma_4$ are selected positive constants.

The same comments regarding convergence again apply here with respect to the movement of the state to target T_2.

Based upon the algorithm described above, with the defined control laws, a number of numerical simulations were performed for various selections of control and convergence parameters, initial states, and target/antitarget constants. The computer program was run over a prescribed time interval. If one object entered its target, the program was allowed to run using the same control laws until the second object entered its target or the final time was reached.

EXAMPLE 1

In the first simulation, the following initialisation was chosen:

Time interval: [0,20] .

RK4 step size: 0.01 .

Target centres: $p_1c_1 = 12.0$, $p_1c_2 = 0.0$,

$p_2c_1 = -12.0$, $p_2c_2 = 0.0$.

Target and antitarget radii: $rp_1 = rp_2 = rap_1 = rap_2 = 6.0$.

Control and convergence parameters:

$$\beta_{11} = \beta_{12} = 5.0 , \quad \beta_{21} = \beta_{22} = 5.0 ,$$

$$p_1\gamma_2 = p_1\gamma_4 = p_2\gamma_2 = p_2\gamma_4 = 5.0 .$$

Initial states:

Object (1) $[-20.0 \quad 1.0 \quad 5.0 \quad 5.0]^T$,

Object (2) $[\ 20.0 \quad -1.0 \quad 2.0 \quad 2.0]^T$.

The plane trajectories of each object are shown in Figure 2. Object (1) entered target T_1 at time 11.1, whilst object (2) entered target T_2 at time 12.2. Figure 3 shows the control graphs. The minimum and maximum of each control component was found to be:

	Minimum	*Maximum*
u_1	-143.68	26.94
u_2	- 29.96	81.19
v_1	- 29.31	339.20
v_2	- 12.01	135.62

□

EXAMPLE 2

Consider again Example 1 with initial states

Object (1) $[-20.0 \quad 1.0 \quad -2.0 \quad 2.0]^T$

Object (2) $[\ 20.0 \quad -1.0 \quad 2.0 \quad 2.0]^T$

and step size reduced to 0.005.

It was found that object (1) entered target T_1 at time 9.7, whilst object (2) entered T_2 at time 10.27. The plane trajectories of each object are shown in Figure 4. The minimum and maximum of each control component in this case was found to be

	Minimum	*Maximum*
u_1	-370.99	32.66
u_2	- 99.01	7.90
v_1	- 29.45	314.40
v_2	- 12.06	125.98

Figure 5 shows the control graphs. With the smaller step size, both objects are seen to avoid collision with the opposing object's target, but control components become quite large as the anti-target is approached and avoided. □

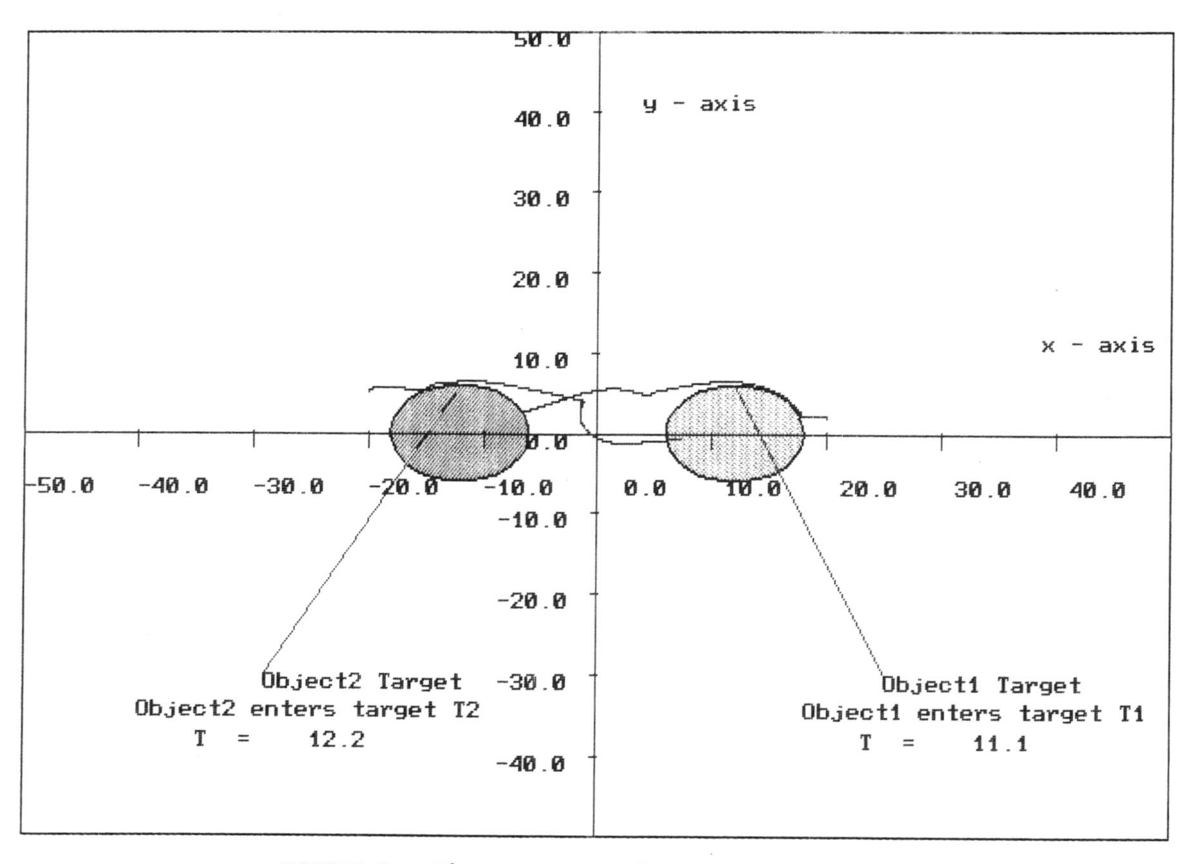

FIGURE 2: Plane trajectories, Example 1

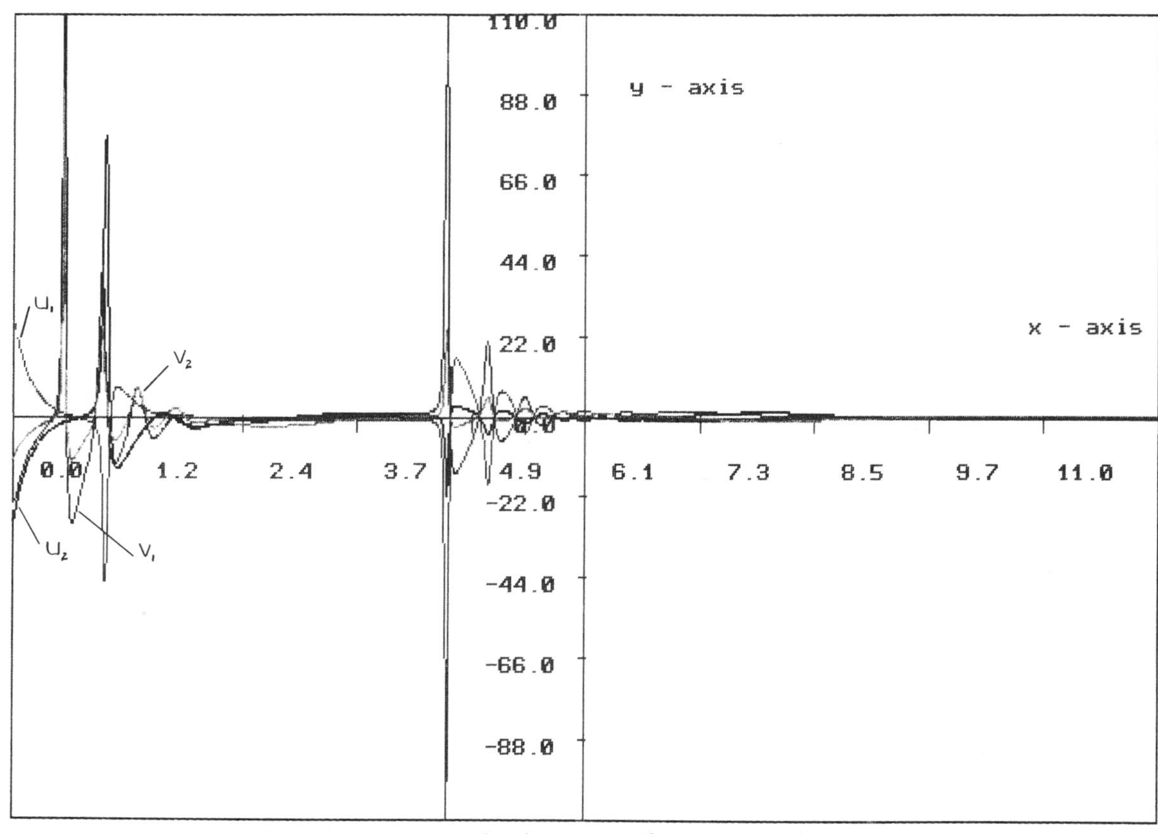

FIGURE 3: Control plots $\underset{\sim}{u}$ *and* $\underset{\sim}{v}$, *Example 1*

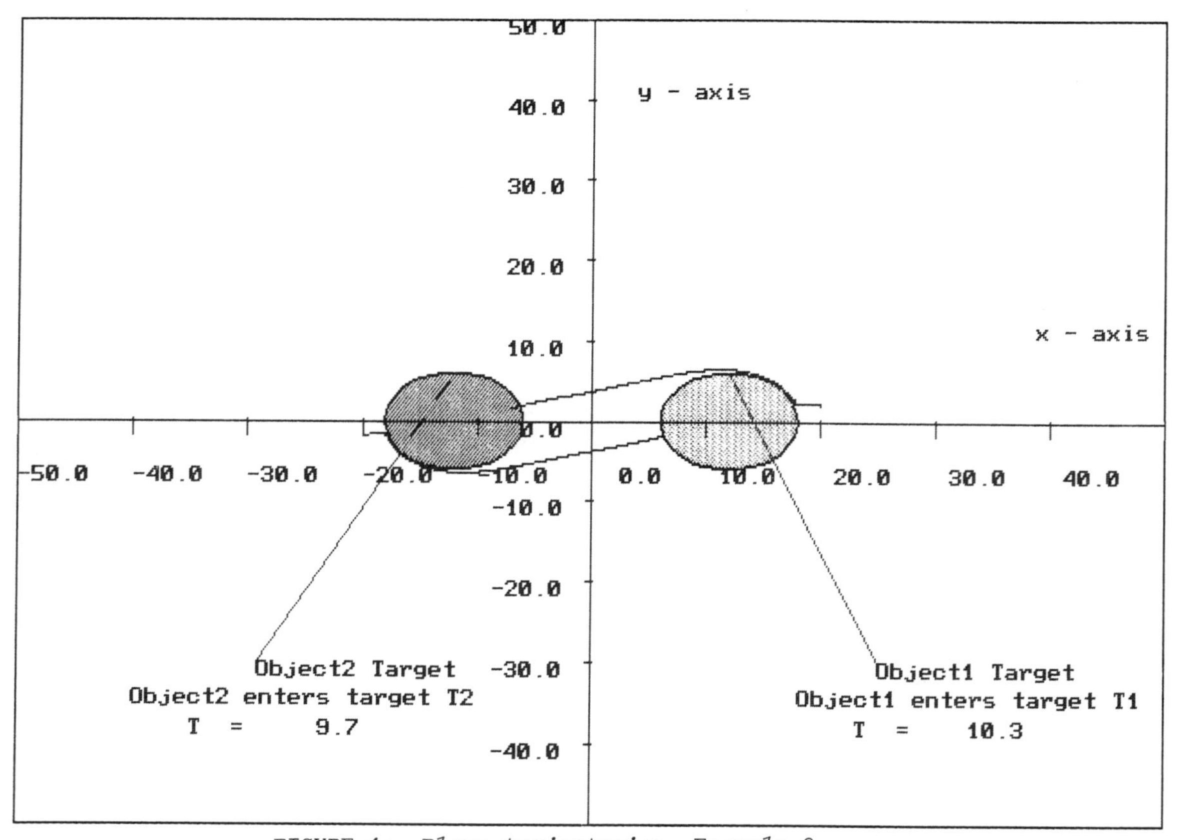

FIGURE 4: Plane trajectories, Example 2

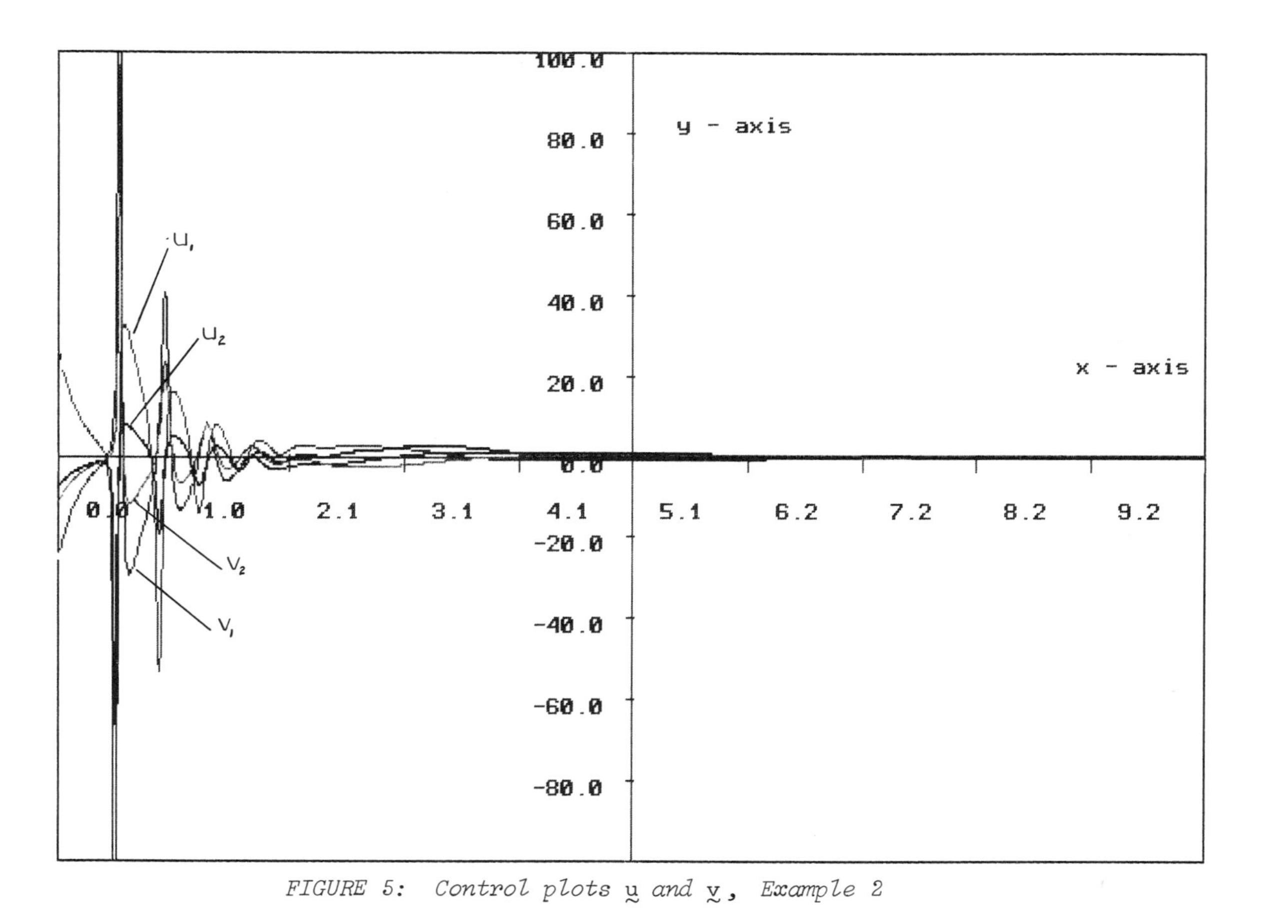

FIGURE 5: *Control plots* $\underset{\sim}{u}$ *and* $\underset{\sim}{v}$, *Example 2*

EXAMPLE 3

A head-on situation is depicted in Figure 6. The program was run over the interval [0,30] with the following initialisation of parameters and states

RK4 step size: 0.0015

Target centres: $p_1c_1 = 20.0$, $p_1c_2 = 0.0$

$p_2c_1 = -20.0$, $p_2c_2 = 0.0$

Target and antitarget radii: $rp_1 = rp_2 = 5.0$

$rap_1 = rap_2 = 6.0$

Control and convergence parameters:

$$\beta_{11} = \beta_{12} = \beta_{21} = \beta_{22} = 5.0$$

$$p_1\gamma_2 = p_1\gamma_4 = p_2\gamma_2 = p_2\gamma_4 = 10.0$$

Initial states:

Object (1) $[-10.0 \quad 2.0 \quad 0.0 \quad 4.0]^T$

Object (2) $[10.0 \quad -2.0 \quad 0.0 \quad -1.0]^T$

Each object entered its target at around $t = 22.0$. The minimum and maximum of each control component was found to be:

	Minimum	*Maximum*
u_1	-59.48	14.37
u_2	-40.00	3.92
v_1	-14.37	59.52
v_2	- 3.92	10.00

The features of the control graph are essentially the same as the other two cases, but the largest magnitude of the control components is considerably less as both objects manoeuver around each other initially. □

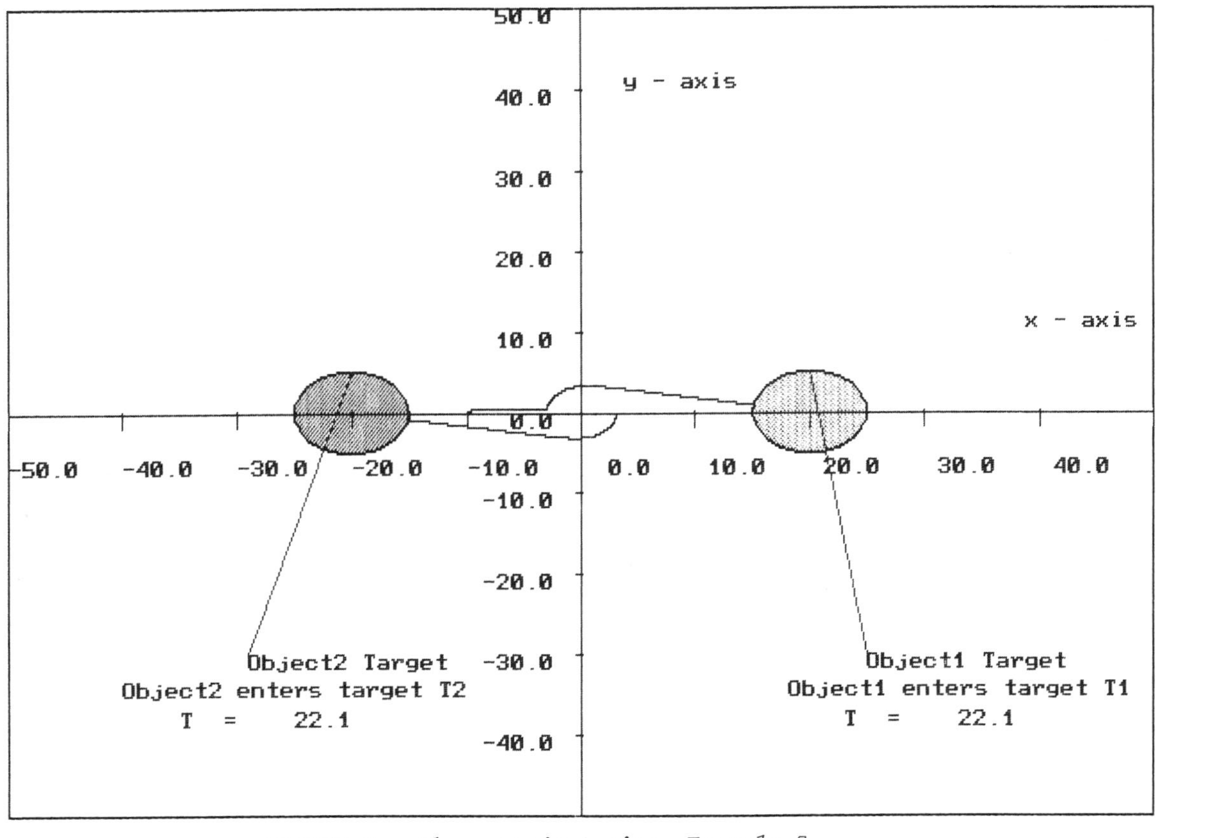

FIGURE 6: Plane trajectories, Example 3

EXAMPLE 4

This last illustration highlights the step selection necessary to obtain a 'good' result. The following initialisation of parameters and states gave the plane trajectory graphs of Figure 7.

Time interval: $[0,15]$

RK4 step size: 0.015

Target centres: $p_1c_1 = 5.0$, $p_1c_2 = 0.0$
$p_2c_1 = -8.0$, $p_2c_2 = 0.0$

Target and antitarget radii: $rp_1 = 1.0$, $rp_2 = 2.0$
$rap_1 = 6.0$, $rap_2 = 6.0$

Control and convergence parameters:

$$\beta_{11} = \beta_{12} = \beta_{21} = \beta_{22} = 1.0$$
$$p_1\gamma_2 = 3.0\ , \quad p_1\gamma_4 = 4.0$$
$$p_2\gamma_2 = 2.0\ , \quad p_2\gamma_4 = 3.0$$

Initial states:

Object (1) $[-10.0 \quad 2.0 \quad 20.0 \quad -2.0]^T$

Object (2) $[\ 10.0 \quad 2.0 \quad 20.0 \quad 3.0\ ]^T$

As seen in Figure 7, object (2) is 'repelled' from object (1) quite violently as they approach each other. If the step size is reduced by a factor of 10 to 0.0015, the detection of the opposing antitarget is more efficient and we obtain the graph given in Figure 8. It is observed in this case that object (2) has effectively reached the centre of T_2 before object (1) has reached its designated target. □

Only a few of many simulations have been presented here. It was found that to achieve the desired objective for each point object so that the antitargets were not entered required careful selection of the step size and the constant control and convergence parameters. If the convergence parameters $p_i\gamma_j$ were too high,

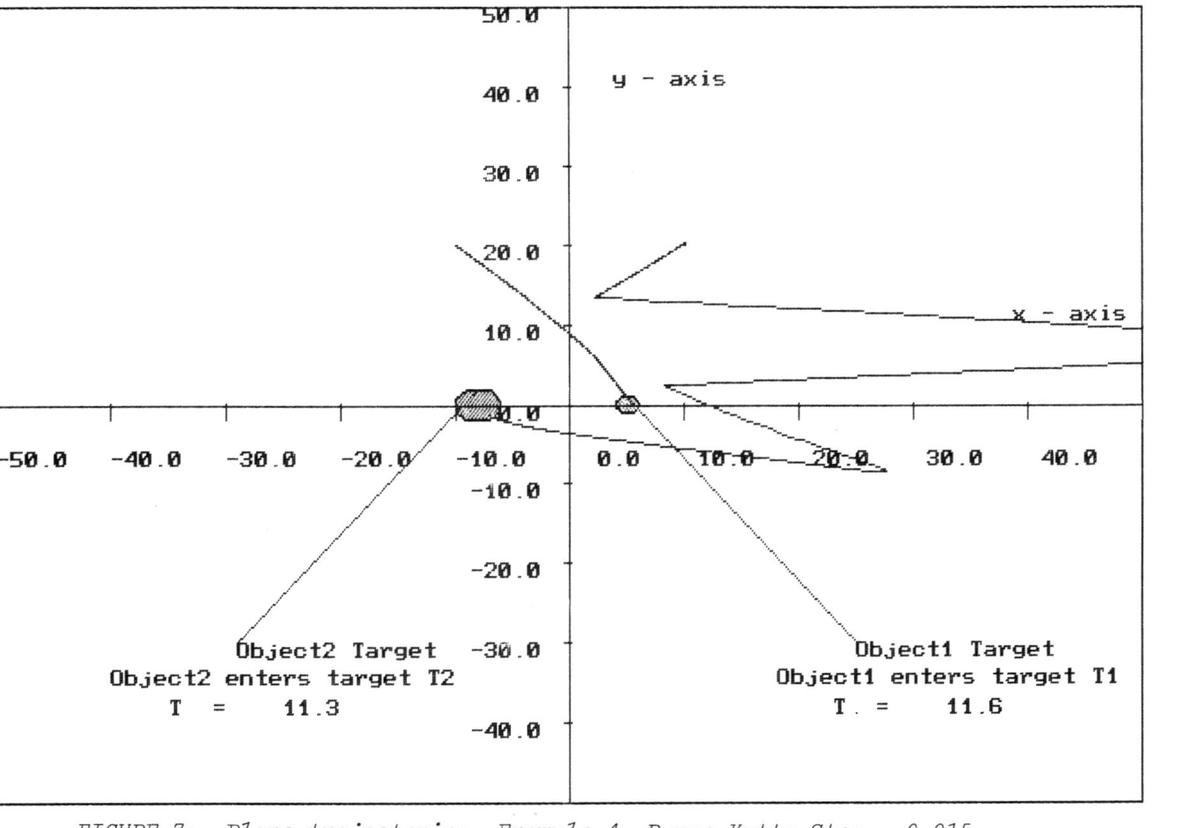

FIGURE 7: Plane trajectories, Example 4, Runge-Kutta Step = 0.015

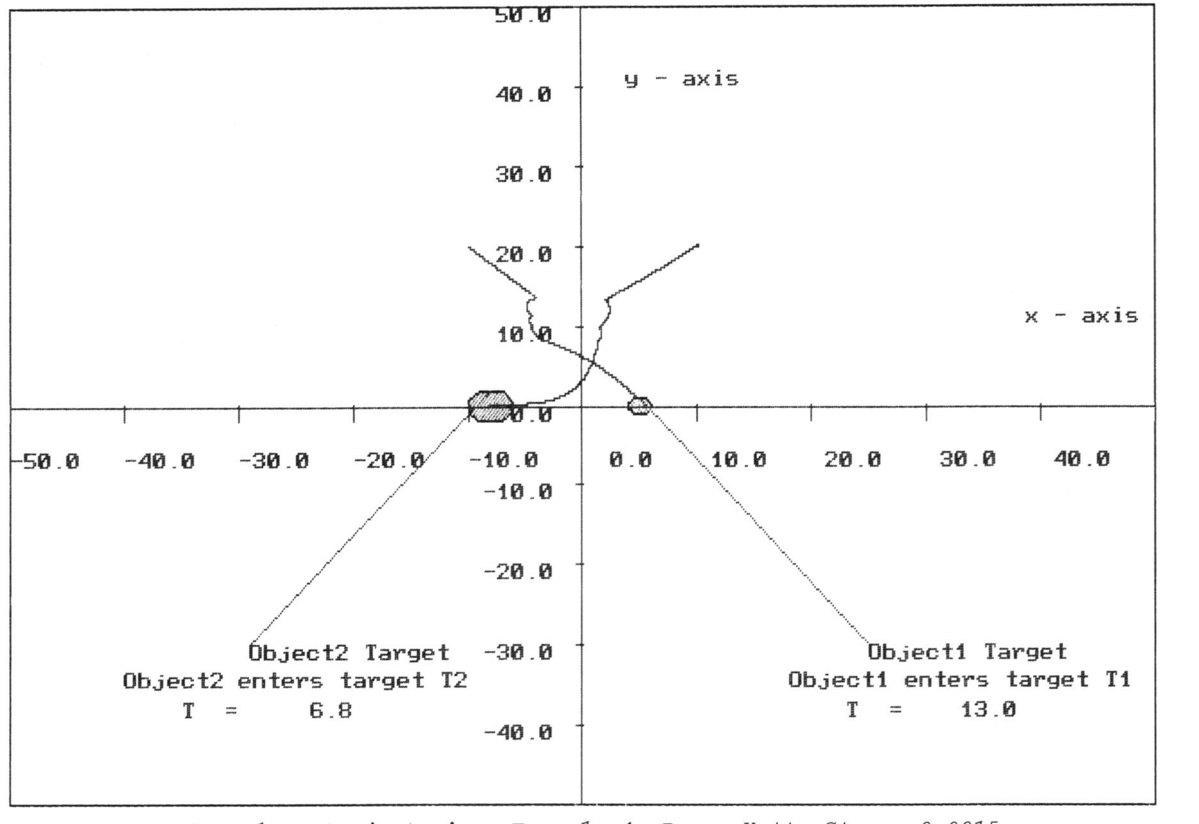

FIGURE 8: Plane trajectories, Example 4, Runge-Kutta Step = 0.0015

then a trajectory for object (1) starting 'behind' object (2)'s target would collide with the target unless the step size was reduced. If the β_{ij} were large, and the targets T_1 and T_2 small, the convergence parameters had to be increased to obtain the attraction required to the targets.

It is clear that an improvement in performance could be achieved by making the β_{ij} and $p_i\gamma_j$ dependent upon the state of each object. For example, the $p_i\gamma_j$ may be reduced by formula if an object is closing in on an antitarget.

III. CONTROL OF A SINGLE ROBOT ARM IN THE PRESENCE OF CONSTRAINTS

Let us now consider the PR manipulator with its schematic representation using equivalent masses shown in Figure 9.

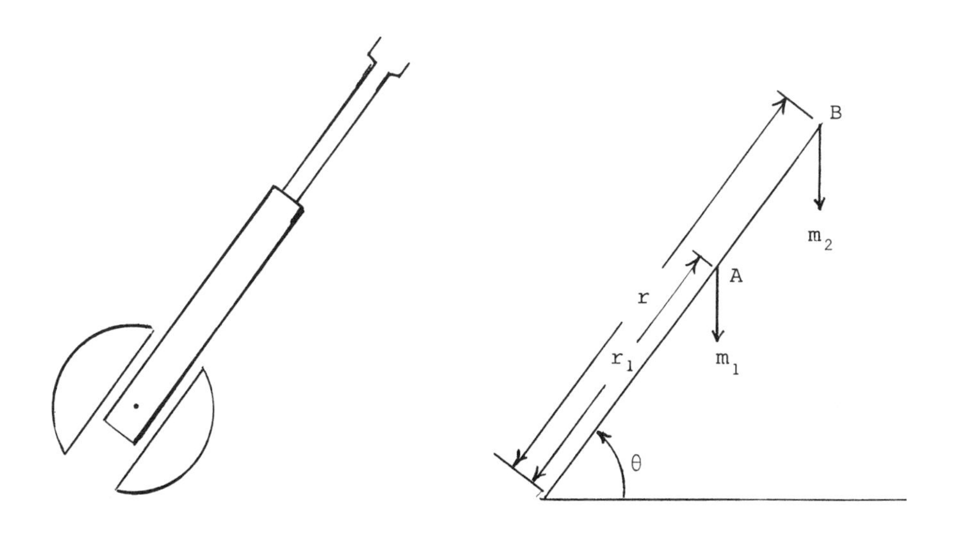

FIGURE 9: Robot arm and schematic representation

Using Lagrangian analysis we find

$$m_1 r_1^2 \ddot{\theta} + m_2 r^2 \ddot{\theta} + 2m_2 r \dot{r} \dot{\theta} + g \cos\theta (m_1 r_1 + m_2 r) = T_\theta$$

and

$$m_2 \ddot{r} - m_2 r \dot{\theta}^2 + m_2 g \sin\theta = F_r$$

where T_θ is the torque about the rotational actuator, F_r is the force applied by the linear actuator, mass of the manipulator m_1 is all centered at A and the payload has mass m_2 at B.

Setting $x_1 = \theta$, $x_2 = \dot{\theta}$, $x_3 = r$ and $x_4 = \dot{r}$ and letting $u_1 = T_\theta$ and $u_2 = F_r$, the state equations of the system are

$$\dot{x}_1 = x_2$$

$$\dot{x}_2 = (u_1 - 2m_2 x_3 x_4 x_2) / (m_1 r_1^2 + m_2 x_3^2)$$

$$\dot{x}_3 = x_4$$

$$\dot{x}_4 = (u_2 + m_2 x_3 x_2^2) / m_2 .$$

We shall consider the movement of the arm in the plane given constants

$$m_1 = 7 \text{ kg}, \quad m_2 = 1 \text{ kg}, \quad r_1 = 1 \text{ metre},$$

and constraints

$$-1 < \dot{\theta} = x_2 < 1, \quad \text{(rad/sec)}$$

$$1 < r = x_3 < 2, \quad \text{(m)}$$

$$-1 < \dot{r} = x_4 < 1, \quad \text{(m/sec)}$$

$$\max |\dot{x}_2| = \max |\ddot{\theta}| = 1,$$

$$\max |\ddot{x}_4| = \max |\ddot{r}| = 1 .$$

Our objective is to determine analytical control laws to move the arm to a desired target in the workspace in the presence of the above defined constraints.

Let us suppose we wish to position the gripper in a circle,

centre $[p_1c_1 \ p_1c_2]^T$ with radius rp_1, in the $x_1 - x_3$ plane and that we require the speeds x_2 and x_4 tend towards zero as the physical target is reached. The physical target set is thus

$$T_1 = \{\underset{\sim}{x} \, ; \, (x_3 \cos x_1 - p_1 c_1^2) + (x_3 \sin x_1 - p_1 c_2)^2 \le rp_1^2\}$$

$$= \{\underset{\sim}{x} \, ; \, x_3^2 - 2x_3(p_1 c_1 \cos x_1 + p_1 c_2 \sin x_1) + p_1 c_1^2 + p_1 c_2^2 \le rp_1^2\} \, .$$

The antitargets are defined from the constraint inequalities:

$$AT_1 = \{\underset{\sim}{x} \, ; \, x_2 + 1 \le 0\} \, ,$$

$$AT_2 = \{\underset{\sim}{x} \, ; \, x_2 - 1 \ge 0\} \, ,$$

$$AT_3 = \{\underset{\sim}{x} \, ; \, x_3 - 1 \le 0\} \, ,$$

$$AT_4 = \{\underset{\sim}{x} \, ; \, x_3 - 2 \ge 0\} \, ,$$

$$AT_5 = \{\underset{\sim}{x} \, ; \, x_4 + 1 \le 0\} \, ,$$

$$AT_6 = \{\underset{\sim}{x} \, ; \, x_4 - 1 \ge 0\} \, .$$

With similar reasoning to that given in Section II, we form the Liapunov functions

$$V_0 = x_3^2/2 - x_3(p_1 c_1 \cos x_1 + p_2 c_2 \sin x_1) + (x_2^2 + x_4^2)/2$$
(Positive on $C(T_1)$),

$V_1 = x_2 + 1$ (Positive on $C(AT_1)$),

$V_2 = 1 - x_2$ (Positive on $C(AT_2)$),

$V_3 = x_3 - 1$ (Positive on $C(AT_3)$),

$V_4 = 2 - x_3$ (Positive on $C(AT_4)$),

$V_5 = x_4 + 1$ (Positive on $C(AT_5)$),

$V_6 = 1 - x_4$ (Positive on $C(AT_6)$),

and

$$V = V_0 + \sum_{i=1}^{6} \beta_1(i)/V_i$$

where β_1 is an array of selected positive constants.

Selection of the following control laws,

$$u_1^* = (m_1r_1^2 + m_2x_3^2)[-p_1\gamma_2x_2^2 - x_2x_3(p_1c_1 \sin x_1 - p_1c_2 \cos x_1)]V_1^2V_2^2$$
$$/[x_2V_1^2V_2^2 - \beta_1(1)V_2^2 + \beta_1(2)V_1^2] + 2m_2x_3x_4x_2$$

$$u_2^* = m_2V_5^2V_0^2[-p_1\gamma_4x_4^2 - x_4(x_3 - p_1c_1 \cos x_1 + p_1c_2 \sin x_1) - \beta_1(3)V_3^{-2}$$
$$+ \beta_1(4)V_4^{-2}]/(x_4V_5^2V_6^2 - \beta_1(5)V_0^2 + \beta_1(6)V_5^2) - m_2x_3x_2^2$$

where $p_1\gamma_2$ and $p_1\gamma_4$ are positive constants, yields the Liapunov derivative

$$\dot{V} = -p_1\gamma_2x_2^2 - p_2\gamma_4x_4^2 .$$

The Liapunov derivative is negative semi-definite in the workspace so the system is Lagrange stable. We shall see that the control laws u_1^*, u_2^* are indeed sufficient for asymptotic attraction to the target in the following examples.

To incorporate bounding constraints on the accelerations which may be broken at a given instant, we define the control laws u_1 and u_2 by:

If $\dot{x}_2 > 1.0$ then $u_1 = (m_1r_1^2 + m_2x_3^2) + 2m_2x_3x_4x_2$;

If $\dot{x}_2 < -1.0$ then $u_1 = -(m_1r_1^2 + m_2x_3^2) + 2m_2x_3x_4x_2$;

Else $u_1 = u_1^*$;

If $\dot{x}_4 > 1.0$ then $u_2 = m_2 - m_2x_3x_2^2$;

If $\dot{x}_4 < -1.0$ then $u_2 = -m_2 - m_2x_3x_2^2$;

Else $u_2 = u_2^*$;

to be calculated and used in the next integration interval.

The following examples illustrate the numerical computation for different selections of control and convergence parameters $\beta_1(i)$ $(i = 1,\ldots,6)$, $p_1\gamma_2$ and $p_1\gamma_4$ and the step size in the Runge-Kutta algorithm.

EXAMPLE 5

This example illustrates a 'good' selection of parameters to effect a transition of the gripper from an initial state to the target.

The characteristics of the system were:

Time interval: [0,20]

RK4 step size: 0.01

Target centre: $p_1c_1 = 1.5$, $p_1c_2 = 0.0$

Target radius: $rp_1 = 0.1$

Control and convergence parameters:

$$\beta_1(i) = 1.0\ , \quad i = 1,\ldots,6$$

$$p_1\gamma_2 = p_1\gamma_4 = 5.0$$

Masses: $m_1 = 7$ kg , $m_2 = 1$ kg

Radius: $r_1 = 1$ m

Initial state: $[3.14 \quad 0.1 \quad 1.5 \quad -0.1]^T$

The gripper reaches the target at $t = 15.52$ and the path from initial state to target is shown in Figure 10. The minimum and maximum of the control components were found to be:

	Minimum	*Maximum*
u_1	-1.13	0.61
u_2	-0.75	0.54

Plots of the control components with respect to time over the interval [0,15.5] are shown in Figure 11. □

EXAMPLE 6

Consider the system with the following characteristics:

Time interval: [0,20]

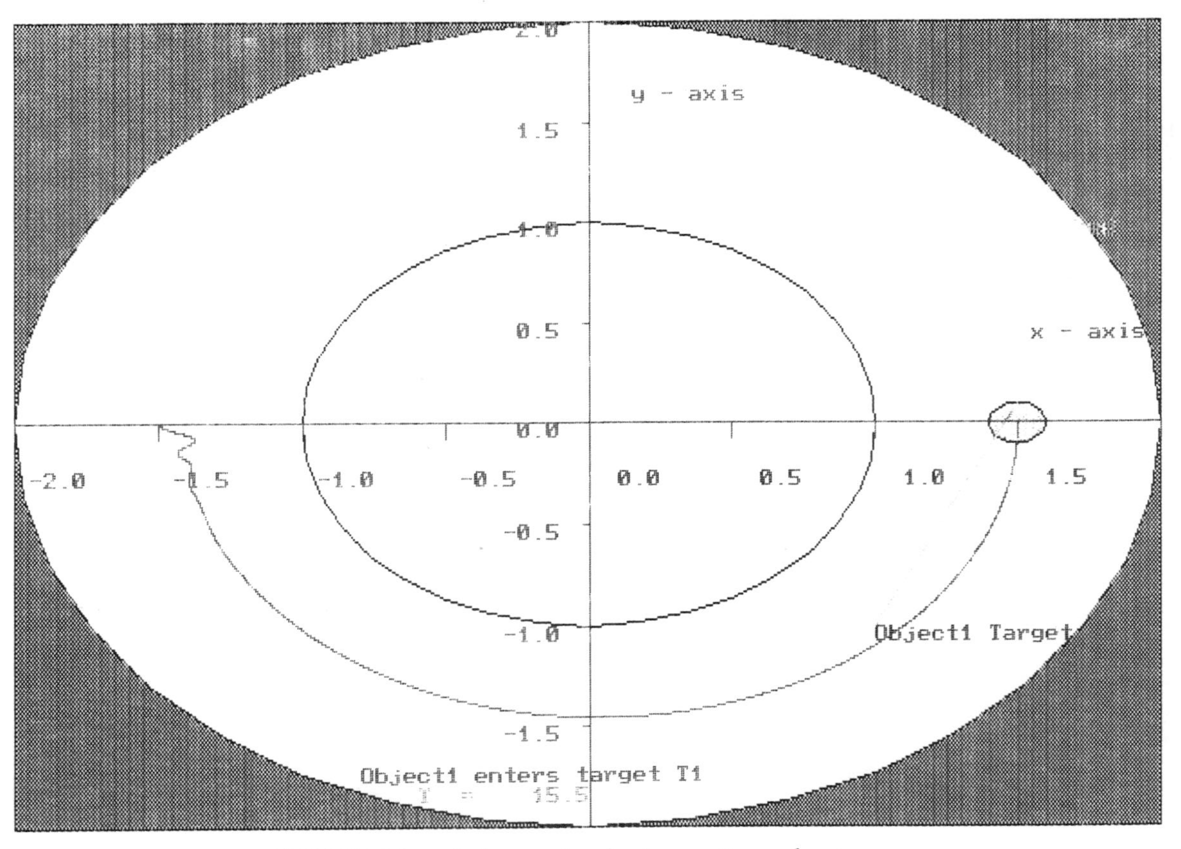

EXAMPLE 10: Gripper trajectory, Example 5

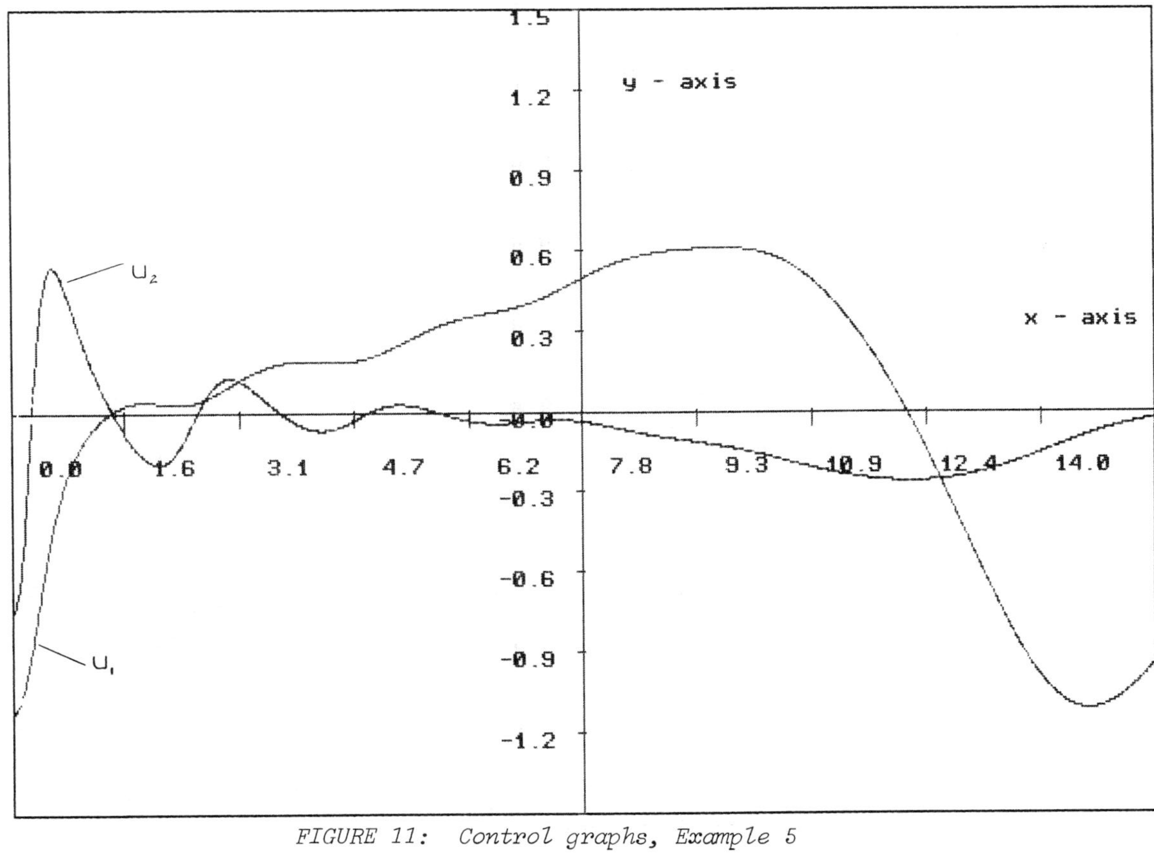

FIGURE 11: Control graphs, Example 5

RK4 step size: 0.01

Target centre: $p_1c_1 = 1.5$, $p_1c_2 = 0.0$

Target radius: $rp_1 = 0.1$

Control and convergence parameters:

$$\beta_1(i) = 5.0 , \quad i = 1,\ldots,6$$

$$p_1\gamma_2 = p_1\gamma_4 = 10.0$$

Masses: $m_1 = 7$ kg , $m_2 = 1$ kg

Radius: $r_1 = 1$ m

Initial state: $[3.14 \quad -0.1 \quad 1.1 \quad 0.1]^T$

We see in Figure 12 that the changes in control and convergence parameters, and increase in step size, yield unacceptable transfer of the gripper from initial state to the target. The gripper enters the target at $t = 15.52$. The minimum and maximum control components u_1 and u_2 graphed in Figure 13 were found to be:

	Minimum	*Maximum*
u_1	-1.43	1.97
u_2	-1.08	1.00

□

IV. CONCLUSION

In this paper we have presented a technique using Liapunov theory for capture and avoidance in control theory, to determine analytic forms of control laws for the planar movement of two point objects in a plane and movement of a single robot arm in a plane.

The method is parameter dependent and requires careful selection of control and convergence parameters to ensure an acceptable movement of the object to the target whilst avoiding antitargets in the form of constraints and obstacles, fixed or moving. The illustration in Section III shows how state constraints may be

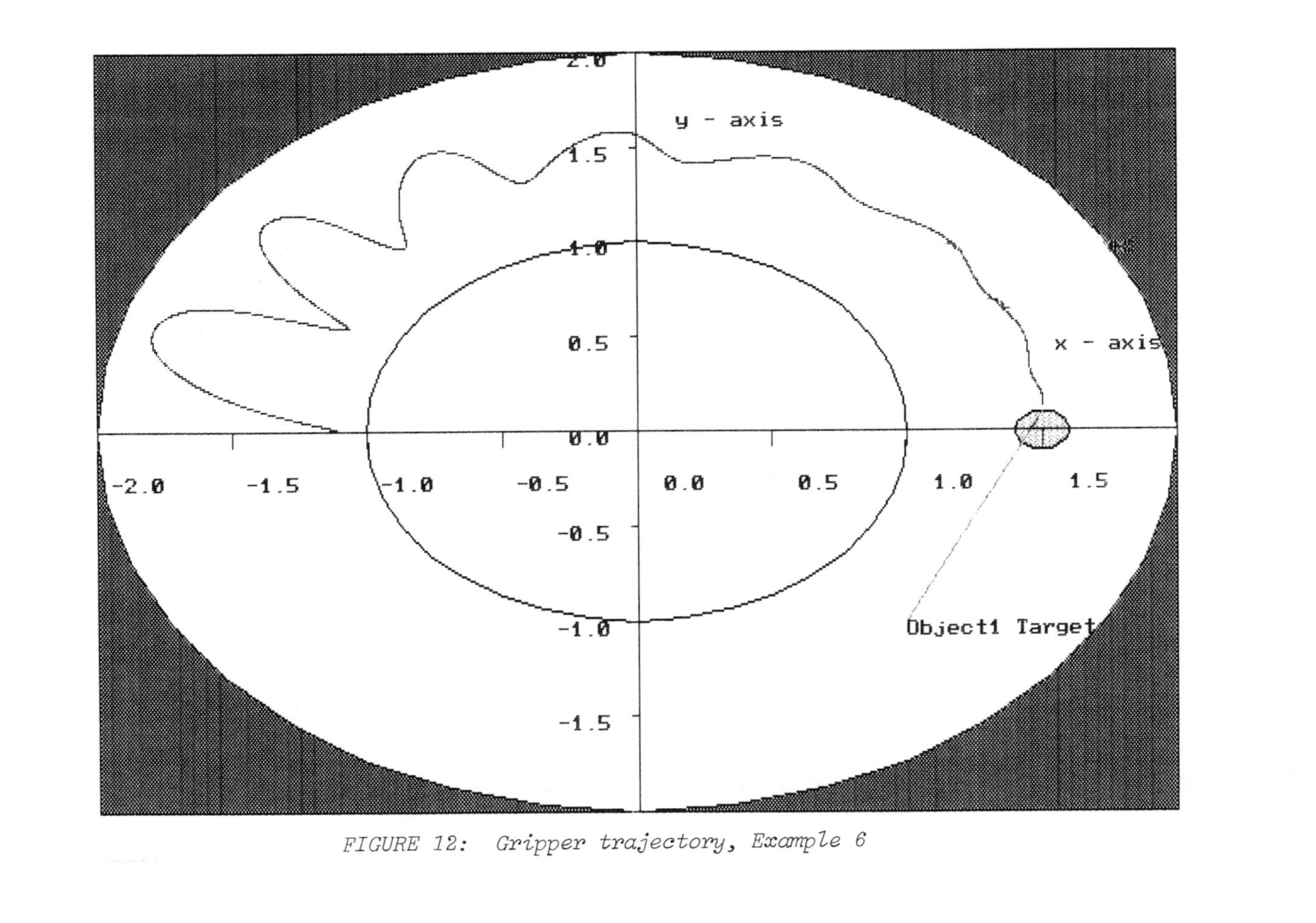

FIGURE 12: Gripper trajectory, Example 6

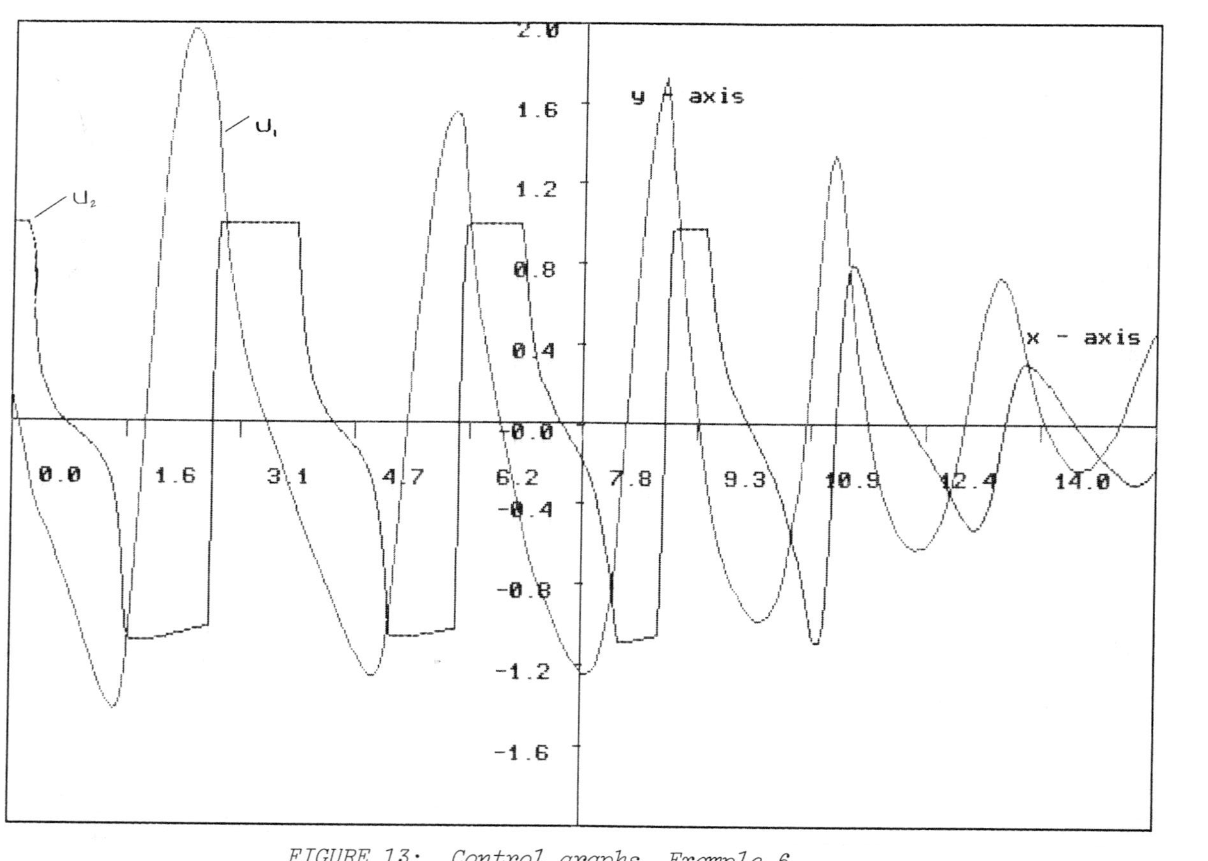

FIGURE 13: Control graphs, Example 6

handled provided an analytical form for the boundary of avoidance region may be defined. The illustration in Section II shows how avoidance of moving objects may be handled.

Improvement in the ability to avoid antitargets and to converge to the prescribed target may be obtained by selecting functional forms for the control and convergence parameters. For example, β_i may be increased to an appropriate value if the 'state' is close to an antitarget, and reduced to a small value when away from the antitargets and close to the target. Switching to a repulsive force such as an inverse square force, when close to a circular avoidance region, can also be considered and implemented easily.

With respect to the application of the method to the movement of two robot arms in a common workspace in the plane, consideration must be given to the construction of appropriate analytic forms of the boundaries of AT_1 and AT_2 shown in Figure 1. This will be the subject of a future paper.

ACKNOWLEDGEMENT

The author wishes to thank Professor Jan Skowronski (Mechanical Engineering, University of Southern California) for suggesting this topic of research and for his helpful comments.

REFERENCES

1. E. Freund and H. Hoyer, "Real-time Pathfinding in Multirobot Systems including Obstacle Avoidance," *International Journal of Robotics Research*, Vol. 7, 1, 42-70, 1988.

2. D.J. Sticht, T.L. Vincent and D.G. Schultz, "Sufficiency Theorems for Target Capture," *Journal of Optimisation Theory and Applications*, Vol. 1, 5/6, 523-542, 1975.

3. W.M. Getz and G. Leitmann, "Qualitative Differential Games with Two Targets," *Journal of Mathematical Analysis and Applications*, Vol. 68, 421-430, 1979.

4. T.L. Vincent and J.M. Skowronski, "Controllability with Capture," *Journal of Optimisation Theory and Applications*, Vol. 29, 1, 1979.

5. J.M. Skowronski, "Collision with Capture and Escape," *Israel Journal of Technology*, Vol. 18, 70-75, 1981.

6. J.M. Skowronski and T.L. Vincent, "Playability with and without Capture," *Journal of Optimisation Theory and Applications*, Vol. 36, 111-128, 1982.

7. R.J. Stonier, "Liapunov Reachability and Optimisation in Control," *Journal of Optimisation Theory and Applications*, Vol. 39, 3, 403-416, 1983.

8. R.J. Stonier, "On Qualitative Differential Games with Two Targets," *Journal of Optimisation Theory and Applications*, Vol. 41, 4, 587-598, 1983.

9. J.M. Skowronski and R.J. Stonier, "Barrier in a Pursuit Evasion Game with Two Targets," *in* International Journal of Computers and Mathematics in Applications, Special Issue on Pursuit-Evasion Games, Pergamon Press, Vol. 13, 1/3, 37-45, 1987.

10. G. Leitmann and J.M. Skowronski, "Avoidance Control," *Journal of Optimisation and Applications*, Vol. 23, 4, 581-591, 1977.

11. G. Leitmann, "Guaranteed Avoidance Strategies," *Journal of Optimisation Theory and Applications*, Vol. 32, 4, 569-576, 1980.

12. M. Corless, G. Leitmann and J.M. Skowronski, "Adaptive Control for Avoidance or Evasion in Uncertain Environments," *in* International Journal of Computers and Mathematics with Application, Special Issue on Pursuit-Evasion Games, Pergamon Press, Vol. 13, 1/3, 1-11, 1987.

13. A.V. Fiacco and G.P. McCormick, "Computational Algorithm for the Sequential Unconstrained Minimisation Technique for Nonlinear Programming," *Management Science*, 10, 601-617, 1964.

AVOIDANCE CONTROL OF A TWO-POINT MECHANICAL SYSTEM

GEORGE BOJADZIEV

Mathematics and Statistics Department
Simon Fraser University
Burnaby, B.C., Canada V5A 1S6

I. INTRODUCTION

We consider a controlled mechanical system with two degrees of freedom consisting of two interacting material points (bodies) moving along a close curve, for instance a circle, located in a plane.

The system without control modelled by the differential equations

$$\begin{aligned} x_1'' + R_1(x_1,x_1') + f_1(x_1) &= \gamma_1 F(x_2'-x_1') + k_1\,\psi(x_2-x_1)\;, \\ x_2'' + R_2(x_2,x_2') + f_2(x_2) &= -\,\gamma_2 F(x_2'-x_1') - k_2\psi(x_2-x_1)\;, \end{aligned} \tag{1}$$

has been investigated qualitatively by Serebrijakova and Barbashin [1], and Barbashin and Tabueva [2].

The two material points P_1 and P_2 with angular coordinates x_1 and x_2 correspondingly, are assumed to move along a circle c_α in a vertical plane α with coordinate frame $0\xi\eta$ (see Fig. 1); $x_s = \angle(0\eta, 0P_s)$, s=1,2, are positive in the counterclockwise direction. The functions $f_s(x_s)$, s=1,2, are the restoring forces applied at the points, $R_s(x_s,x_s')$ are damping forces due to the resistance of the surrounding media, and the functions $F(x_2'-x_1')$ and $\psi(x_2-x_1)$ reflect the existence of connections between the two points (bodies). The coefficients γ_s and k_s, $s=1,2$, are positive constants.

In this paper we extend the Eqs. (1) by introducing control forces acting on the material points. The study of the controlled system is based on a Liapunov stability methodology developed by Leitmann and Skowronski [3]. We find conditions under which the control results in avoidance. We refer also to the monograph book [4] by Skowronski on control of robot manipulators and the paper [5] by the author on controlled chains of pendulums. To illustrate the results numerical simulation is applied to a particular two-point system.

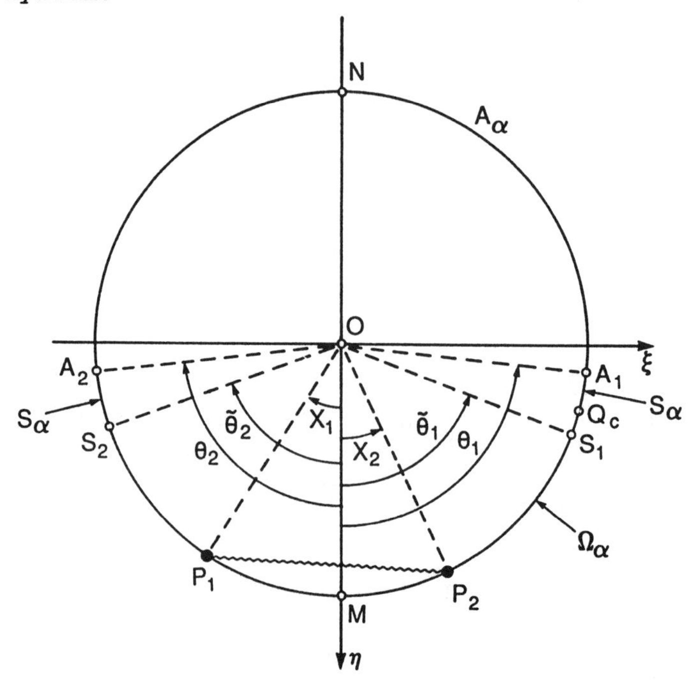

Fig. 1. The two-point system.

II. THE CONTROLLED SYSTEM

In the model (1) we introduce control forces u_1 and u_2 acting on the material points P_1 and P_2 correspondingly, and obtain

$$\begin{aligned} x_1'' + R_1(x_1,x_1') + f_1(x_1) &= \gamma_1 F(x_2'-x_1') + k_1\psi(x_2-x_1) + u_1, \\ x_2'' + R_2(x_2,x_2') + f_2(x_2) &= -\gamma_2 F(x_2'-x_1') - k_2\psi(x_2-x_1)+u_2, \end{aligned} \tag{2}$$

The control $u(t) = (u_1(t), u_2(t))^T \in U \subset R^2$ and the control objectives which will result in avoidance will be specified later.

To simplify the study we assume here that

$$\gamma_1 = \gamma_2 = \gamma, \qquad k_1 = k_2 = k, \tag{3}$$

$$f_s(x_s) = \beta_s f(x_s), \qquad s=1,2, \quad \beta_s = \text{const} > 0. \tag{4}$$

In [1] it is assumed (3) and (4) with $\beta_s = 1$.

With (3) and (4), and setting $x_s' = y_s$, $s=1,2$, the model (2) becomes in the phase space R^4

$$\begin{aligned} x_1' &= y_1, \\ x_2' &= y_2, \\ y_1' &= -R_1(x_1,y_1) - \beta_1 f(x_1) + \gamma F(y_2-y_1) + k\psi(x_2-x_1) + u_1, \\ y_2' &= -R_2(x_2,y_2) - \beta_2 f(x_2) - \gamma F(y_2-y_1) - k\psi(x_2-x_1) + u_2. \end{aligned} \tag{5}$$

For $u_1 = u_2 = 0$ in (5) we obtain the uncontrolled model (UCM).

Further we adopt some of the assumptions made in [1] and [2].

(a) The functions $R_s(x_s,y_s)$, $f(x_s)$, $s=1,2$, $\psi(z)$, and $F(\omega)$, $z = x_2 - x_1$, $\omega = y_2 - y_1$, are C^1 functions in the neighbourhood of the equilibrium points of the UCM (5).

(b) The functions $R_s(x_s,y_s)$, $f(x_s)$, and $\psi(z)$ are 2π periodic in the arguments x_s and z.

(c) The functions $\psi(z)$, $F(\omega)$, and $f(x)$ are odd functions, i.e.

$$\psi(z) = -\psi(-z) \Rightarrow \psi(0) = 0 , \tag{6}$$

$$F(\omega) = -F(-\omega) \Rightarrow F(0) = 0 , \tag{7}$$

$$f(x) = -f(-x) \Rightarrow f(0) = 0 . \tag{8}$$

(d) $$\psi(\pi) = \psi(-\pi) = \psi(0) = 0 , \tag{9}$$

$$f(\pi) = f(-\pi) = f(0) = 0 . \tag{10}$$

(e) $$y_sR_s(x_s,y_s) > 0 , \qquad R_s(x_s,0) = 0 , \tag{11}$$

$$\omega F(\omega) > 0 \ \forall \ |\omega| > 0 . \tag{12}$$

From (b) and (c) it follows that

$$z\psi(z) > 0 \ \forall \ |z| \in (0,\pi) , \tag{13}$$

$$xf(x) > 0 \ \forall \ |x| \in (0,\pi) . \tag{14}$$

III. CONTROL OBJECTIVES

Assume that there is no danger of collision between the moving material points P_1 and P_2 on the circle c_α when the angular coordinates are equal ($x_1 = x_2$), but there is limitation on the motion so that P_1 and P_2 cannot enter a certain arc on c_α.

To formulate the control objectives more precisely we introduce the following definitions.

Definition 1. Admissible arc Ω_α,

$$\Omega_\alpha \triangleq \text{arc}(A_2MA_1) \in c_\alpha ; \qquad A_1, A_2 \notin \Omega_\alpha; \tag{15}$$

Definition 2. Avoidance arc A_α ,

$$A_\alpha \triangleq \mathrm{arc}(A_1 N A_2) \in c_\alpha , \quad A_1, A_2 \in A_\alpha ; \tag{16}$$

Definition 3. Safety arc S_α ,

$$S_\alpha \triangleq \mathrm{arc}(S_1 A_1) \cup \mathrm{arc}(A_2 S_2) \in c_\alpha, \quad A_1, A_2 \notin S_\alpha , \tag{17}$$

$$M \notin \mathrm{arc}(S_1 A_1), \mathrm{arc}(A_2 S_2), \quad S_1, S_2 \in S_\alpha .$$

Definition 4. Desirable arc $\tilde{\Omega}_\alpha$,

$$\tilde{\Omega}_\alpha \triangleq \mathrm{arc}(S_2 M S_1) = \Omega_\alpha \backslash S_\alpha . \tag{18}$$

The points A_s and S_s, s=1,2, fixed on c_α, are determined by the angles $\theta_s = \angle MOA_s$, $\tilde{\theta}_s = \angle MOS_s$; the difference $\theta_s - \tilde{\theta}_s$ is as small as desired for a particular case study. For simplicity we assume that $|\theta_1| = \theta_2 = \theta > 0$, $|\tilde{\theta}_1| = \tilde{\theta}_2 = \tilde{\theta} > 0$, i.e. the arc $(S_2 A_2)$ is symmetrically located to the arc $(S_1 A_1)$ with regard to $O\eta$.

We say that the two-point mechanical system enters into the arc A_α if at least one of the two connected points P_1 and P_2 enters into A_α.

Control objectives. We wish to design a strategy according to which the control force $u(t)$ secures that the two-point (P_1 and P_2) mechanical system subjected to the forces f_s, R_s, F, and ψ, s=1,2, does not enter into the avoidance arc A_α but preferably moves along the desirable arc $\tilde{\Omega}_\alpha$ (Fig. 1).

In a real-world process there are always small disturbances which makes meaningful the introduction of the safety arc S_α whose role is to prevent the points P_1 and P_2 of getting dangerously close to the boundary points A_1 and A_2 of the avoidance arc A_α.

The motion of the controlled two-point mechanical

system along $c_\alpha \subset R^2$ is modelled by the dynamical system (5) in R^4 whose solution for a given $\mathbf{u}$ is called response to $\mathbf{u}$; we denote it by $\mathbf{z}[t] = \mathbf{z}(t, \mathbf{x}(t_0), \mathbf{y}(t_0), \mathbf{u})$, $\mathbf{z} \in R^4$, $\mathbf{x} = (x_1, x_2)^T$, $\mathbf{y} = (y_1, y_2)^T$, $z_s = x_s$, $z_{2+s} = y_s$, s=1,2, $(\mathbf{x}(t_0), \mathbf{y}(t_0)) \in R^4$ is the initial state, $t_0 \geq 0$.

In order to achieve the control objectives we find conditions for $\mathbf{u}$ under which the dynamical system (5) is controllable for avoidance of a certain region $A \in R^4$, i.e. the response $\mathbf{z}[t]$ of (5) does not enter A. This facilitates the design of a strategy under which the two-point mechanical system modelled by (2) does not enter arc A_α.

IV. EQUILIBRIA OF THE UNCONTROLLED MODEL

The equilibria of the UCM (5) play an important role in this study. They are located in the hyperplane $(x_1, x_2) \subset R^4$ and are solutions of the system

$$\begin{aligned} \beta_1 f(x_1) &= k\psi(x_2 - x_1) , \\ \beta_2 f(x_2) &= - k\psi(x_2 - x_1) , \end{aligned} \tag{19}$$

which is obtained from the UCM (5) by setting $x' = 0$, $y'_s = 0$, s=1,2, and taking into account (8) and (11).

Due to (9) and (10), the system (19) has solutions of the type $(p\pi, p\pi, 0, 0) \in R^4$, where p is integer. Here we only deal with the following zeros of (19):

$$\begin{aligned} &E_0(0,0,0,0), \quad E_1(\pi,0,0,0), \quad E_2(-\pi,0,0,0), \\ &\qquad E_3(0,\pi,0,0), \quad E_4(0,-\pi,0,0), \end{aligned} \tag{20}$$

which are located in the region

$$\{(x_1,x_2,y_1,y_2):\ x_1^2 + x_2^2 \le \pi^2 , \qquad y_1 = y_2 = 0\} . \qquad (21)$$

Note that the set (20) does not represent all solutions of (19). Following [2] we show the existence of solutions of the type $(x_1^s,x_2^s,0,0)$ in the region (21) in the case of small k, or more precisely

$$k \ll \max|f(x_s)| . \qquad (22)$$

From (19) with (22) it follows that x_1^s and x_2^s should be close to the zeros of the function $f(x)$, i.e. close to $-\pi$, 0, and π.

The addition of equations (19) gives

$$\beta_1 f(x_1^s) + \beta_2 f(x_2^s) = 0 \Rightarrow f(x_1^s)\ f(x_2^s) < 0 .$$

Hence x_1^s and x_2^s belong to intervals in which the sign of $f(x)$ is different. Two cases are possible.

(i) $-\pi < x_2^s < 0 , \quad 0 < x_1^s < \pi \Rightarrow x_2^s - x_1^s < 0 .$

From (19) we also obtain

$$f(x_1^s) > 0 \Rightarrow \psi(x_2^s - x_1^s) > 0 .$$

Since $\psi(z)$ is 2π periodic, we conclude that

$$-2\pi < x_2^s - x_1^s < -\pi .$$

Hence x_1^s and x_2^s cannot both be close to 0.

(ii) $-\pi < x_1^s < 0, \quad 0 < x_2^s < \pi \Rightarrow x_2^s - x_1^s > 0 ,$

then from (19) we get

$$f(x_2^s) > 0 \Rightarrow \psi(x_2^s-x_1^s) < 0 ,$$

which leads to

$$\pi < x_2^s - x_1^s < 2\pi ,$$

i.e. x_1^s and x_2^s cannot both be close to 0.

Hence the UCM (5) has also in the region (21), with the assumption (22), the following equilibria

$$E_5(\pi-\delta_1^1, -\delta_2^1, 0, 0), \quad E_6(-\pi+\delta_1^2, 0, 0),$$
$$E_7(\delta_1^3, -\pi+\delta_2^3, 0, 0), \quad E_8(-\delta_1^4, \pi-\delta_2^4, 0, 0), \tag{23}$$

where δ_s^j, s=1,2, j=1,...,4, are small positive constants.

From (20) and (23) we conclude that the equilibria E_s, s=1,...,8, of the UCM (5) are not close to E_0, hence there is a neighbourhood $\mathcal{B}$ of E_0 which does not contain inside any other equilibrium of the UCM besides E_0.

V. THE LIAPUNOV FUNCTION

Consider the function

$$V(\mathbf{x},\mathbf{y}) = \sum_{s=1}^{2} y_s^2 + 2\sum_{s=1}^{2}\int_0^{x_s} \beta_s f(\tau)d\tau + 2k\int_0^{x_2-x_1} \psi(\tau)d\tau, \tag{24}$$

In $\mathcal{B}$ the function V has the following properties:

(i) $V(\mathbf{0},\mathbf{0}) = 0, \quad \mathbf{0} = (0,0)^T$;

(ii) $V(x,y)$ is positive definite in $\mathcal{B}\backslash E_0$;

(iii) $\dfrac{dV}{dt} = \sum_{s=1}^{2}\left(\dfrac{\partial V}{\partial x_s}x_s' + \dfrac{\partial V}{\partial y_s}y_s'\right) < 0,$

x_s', y_s' are the right hand sides of the UCM (5). Property (i) is obvious; property (ii) follows from (13) and (14); the inequality in property (iii) can be represented in the form

$$- 2 \sum_{s=1}^{2} y_s R_s(x_s, y_s) - 2\gamma(y_2 - y_1) F(y_2 - y_1) < 0 ; \tag{25}$$

it follows from assumptions (11) and (12).

Hence the function (24) is a Liapunov function for the UCM (5) in $\mathcal{B}$.

The function (24) describes in $V \times \mathcal{B} \subset R^5$ a "cup-shaped" hypersurface with minimum at E_0.

The equation

$$V(x,y) = h , \qquad h = \text{const} > 0 , \tag{26}$$

represents a family of level surfaces in $V \times \mathcal{B}$ whose orthogonal projection onto $\mathcal{B} \subset R^4$ generates hypersurfaces $\mathcal{H}_h$ in R^4 with the following properties (see [2] and [4]): (a) They are closed; (b) Do not intersect; (c) Contain inside only the equilibrium E_0; (d) If $h_1 < h_2$, the hypersurface $\mathcal{H}_{h_1}$ is inside of $\mathcal{H}_{h_2}$. The largest hypersurface $\mathcal{H}_{h^*}$ from the set $\mathcal{H}_h$ with the above properties (a)-(d) passes through one or more equilibria of the UCM, (20) and (23) (E_0 exempted), whose coordinates minimize h. We define

$$h^* = \min h_s = \min V(E_s), \; s=1,\ldots,8, \tag{27}$$

$$\mathcal{H}_{h^*} = \{(\mathbf{x},\mathbf{y}) \in R^4 : V(\mathbf{x},\mathbf{y}) = h^*\} . \tag{28}$$

The function $V(\mathbf{x},\mathbf{y})$ possesses the properties (i)-(iii) in larger regions than $\mathcal{B}$. The largest region $\mathcal{D} \subset R^4$ with properties (i)-(iii) is

$$\mathcal{D} \triangleq \{(\mathbf{x},\mathbf{y}) \in R^4 : V(\mathbf{x},\mathbf{y}) \le h^*\} \tag{29}$$

with boundary $\partial D = \mathcal{H}_{h^*}$.

VI. AVOIDANCE CONTROL

Making use of the properties of the Liapunov function (24) in the region $\mathcal{D}$ given by (29) and the level surfaces (26) we define sets Ω, A, S, and $\tilde{\Omega}$ in R^4 which correspond to certain extent to the arcs Ω_α, A_α, S_α, and $\tilde{\Omega}_\alpha$ introduced in Section III.

First using (27) we determine the values h_A and h_S,

$$h_A = h^* - \varepsilon, \quad h_S = h^* - \delta, \quad h_A > h_S , \tag{30}$$

where $0 \le \varepsilon < \delta$; ε and δ are as small as desired for a particular study.

Then we define the following regions in R^4.

Definition 5. Admissible set Ω,

$$\Omega \triangleq \{(\mathbf{x},\mathbf{y}) \in R^4 \colon V(\mathbf{x},\mathbf{y}) < h_A\} , \quad h_A \le h^* , \tag{31}$$

Definition 6. Avoidance set A,

$$A \triangleq \mathcal{D}\backslash\Omega = \{(\mathbf{x},\mathbf{y}) \in R^4 \colon h_A \le V(\mathbf{x},\mathbf{y}) \le h^*\}, \partial A = \mathcal{H}_{h_A} , \tag{32}$$

Definition 7. Safety zone S,

$$S \triangleq \{(\mathbf{x},\mathbf{y}) \in R^4 \colon h_S \le V(\mathbf{x},\mathbf{y}) < h_A\} , \quad \partial S = \mathcal{H}_{h_S} \tag{33}$$

Definition 8. Desirable region $\tilde{\Omega}$,

$$\tilde{\Omega} \triangleq \Omega\backslash S = \{(\mathbf{x},y) \in R^4 \colon V(\mathbf{x},\mathbf{y}) < h_S\} . \tag{34}$$

In order to design a strategy that guarantees avoidance we introduce the following definition.

Definition 9. The region A defined by (32) is avoidable if there is a control $\mathbf{u} \in U$ such that for all states $(\mathbf{x}(t_1), \mathbf{y}(t_1)) \in S$, $t_1 \ge t_0$, S is the safety zone (33), the response $\mathbf{z}[t] = \mathbf{z}(t, \mathbf{x}(t_1), \mathbf{y}(t_1), \mathbf{u})$ of the controlled model (5) cannot enter A, i.e.

$$\mathbf{z}(t, \mathbf{x}(t_1), \mathbf{y}(t_1), \mathbf{u}) \cap A = \phi \ \forall \ t . \tag{35}$$

Sufficient conditions for the avoidance of A are given in the following theorem (see [5]).

Theorem. The response $\mathbf{z}[t]$ of the dynamical system (5) is controllable in $\Omega \subset \mathcal{D}$ for avoidance of the region A if there is a control $\mathbf{u} \in U$ and a Liapunov function $V(\mathbf{x},\mathbf{y})$ given by (24) such that

$$\frac{dV}{dt} = \sum_{s=1}^{2} \left(\frac{\partial V}{\partial x_s} x_s + \frac{\partial V}{\partial y_s} y_s \right) \le 0 , \tag{36}$$

where x'_s and y'_s are determined by (5); Ω and A are defined correspondingly by (31) and (32).

Proof. Following [3] and [5]) let us assume that the region A is not avoidable, i.e. (35) does not hold. Then for some state $(\mathbf{x}(t_1), \mathbf{y}(t_1)) \in S$ (safety zone) and time $t > t_1$, the response $\mathbf{z}[t, \mathbf{x}(t_1), \mathbf{y}(t_1), \mathbf{u}(t_1)]$ enters A. This implies that there is a $t_2 > t_1$ such that $\mathbf{z}(t_2, \mathbf{x}(t_1), \mathbf{y}(t_1), \mathbf{u}(t_1)) \in \partial A$, the boundary of A. The point $(\mathbf{x}(t_1), \mathbf{y}(t_1))$ lies on a hypersurface which is enclosed between ∂S and ∂A. Hence $V(\mathbf{x}(t_1), \mathbf{y}(t_1)) < V(\mathbf{x}(t_2), \mathbf{y}(t_2))$ from which follows that the Liapunov function $V(\mathbf{x},\mathbf{y})$ in increasing, thus contradicting (36) according to which $V(\mathbf{x},\mathbf{y})$ is non-increasing along every response of (5).

The sufficient condition (36) for avoidance can be expressed in explicit form. Using (5) and (24), for (36) we obtain

$$\sum_{s=1}^{2} y_s u_s - \sum_{s=1}^{2} y_s R_s(x_s, y_s) - \gamma(y_2 - y_1) F(y_2 - y_1) \le 0. \tag{37}$$

The inequality (37) establishes a relationship between the control $\mathbf{u}$, the damping forces $R_s(x_s, y_s)$, the force $F(y_2-y_1)$ connecting the points P_1 and P_2, and the velocities $y_s = x_s'$ of the angular coordinates x_s, s=1,2. If (37) is compared to (25), we see that the role of the control force is to balance the effect of the velocities. The control $\mathbf{u}$ must satisfy (37) which does not impose a strict restriction on U. We can assume that the control components u_s can be chosen from R,

$$u(t) \in U = \{u_s: u_s \in [-r,r] \subset R,\ r = \text{const}, s=1,2\} \tag{38}$$

as long as (37) is satisfied.

VII. CONTROL STRATEGY

Now we design an avoidance control strategy for the two-point mechanical system modelled by (2) based on the avoidance control results derived in Section VI concerning the response of the dynamical system (5) in $\mathcal{D} \subset R^4$.

Let us assume that the points $P_1, P_2 \in \text{arc } \tilde{\Omega}_\alpha$, the desirable arc (18), have initial angular coordinates $x_1(t_0)$, $x_2(t_0)$, initial velocities $y_1(t_0)$, $y_2(t_0)$, and are subjected to fixed control $\mathbf{u}(t_0)$, $t_0 \le 0$. As the points move along arc $\tilde{\Omega}_\alpha \in c_\alpha \subset R^2$, the corresponding response $\mathbf{z}[t]$ of (5) moves in R^4. In order to guarantee that P_1 and P_2 will avoid entering into arc A_α, we monitor simultaneously: (i) The value $V(\mathbf{z}[t])$ which we wish to be smaller than h_S (see (30)), i.e. we wish $\mathbf{z}[t] \in \tilde{\Omega}$; (ii) The angular coordinates $x_1(t)$ and $x_2(t)$ which should be such that $P_1, P_2 \in \text{arc } \tilde{\Omega}_\alpha$. The following two situations may occur.

A. P_1, $P_2 \in$ arc $\tilde{\Omega}_\alpha$ at time $t_1 > t_0$ while $h_S < V(\mathbf{z}[t_1]) < h_A$, i.e. the response $\mathbf{z}[t] = \mathbf{z}(t, \mathbf{x}(t_0), \mathbf{y}(t_0), \mathbf{u}(t_0))$ crossing the boundary ∂S enters into the safety zone S, hence is dangerously close to the avoidance region A with boundary ∂A; at $t = t_1$ the response is at the state $Q(\mathbf{x}(t_0), \mathbf{y}(t_0)) \in S$. Our strategy at time t_1 is to change the control $\mathbf{u}(t_0)$ to a new control $\mathbf{u}(t_1) \subset U$ which satisfy the control law (37). The response $\mathbf{z}[t]$ subjected to this strategy would "move down" through decreasing levels of $V(\mathbf{x}(t), \mathbf{y}(t)) < V(\mathbf{z}[t_1])$.

B. Assume that one of the points, say P_2, enters the safety arc S_α, say arc(S_1A_1), and at time $t_1 > t_0$ coincides with the point $Q_c \in$ arc(S_1A_1) (see Fig. 1), while the response $\mathbf{z}[t]$ at the same time reaches the state $\mathbf{z}[t_1] \equiv Q(\mathbf{x}(t_1), \mathbf{y}(t_1)) \in \tilde{\Omega}$ (desirable region), i.e. $V(\mathbf{z}[t_1]) < h_S$. To prevent P_2 of entering into the avoidance arc A_α, a new control vector $\mathbf{u}(t_1) \in U$ which obeys (37) has to be selected at the point Q (switching point). The response $\mathbf{z}(t, \mathbf{x}(t_1), \mathbf{y}(t_1), \mathbf{u}(t_1))$ to the control $\mathbf{u}(t_1)$ will be steered "down" through decreasing levels of $V(\mathbf{x}(t), \mathbf{y}(t)) = h < h_S$. The point P_2 will leave the position $Q_c \in$ arc(S_1A_1) and either immediately or very shortly after will move back into the desirable arc $\tilde{\Omega}_\alpha$.

Note 1. If at time $t = t_1$ one of the points, or both, enter arc S_α, and also the response $\mathbf{z}[t]$ enters S, similarly to A and B, we choose a new control $\mathbf{u}(t_1) \in U$ which satisfies (37).

Note 2. As time progresses, say at $t = t_2 > t_1$, it is possible under the control $\mathbf{u}(t_1)$, one of the points to enter again arc S_α, or the response to enter S. In such a case we apply the strategy described in A or B, selecting a new control $\mathbf{u}(t_2) \in U$ according to (37).

VIII. CASE STUDY

Consider the following particular selection of constants and functions in the model (5):

$$\beta_1 = \beta_2 = \gamma = 1, \quad \psi(x_2-x_1) = 0, \quad f(x_s) = \sin x_s,$$
$$F(y_2-y_1) = (y_2-y_1)^3, \quad R_s(x_s,y_s) = y_s, \quad s=1,2,$$

which then reduces to

$$\begin{aligned} x_1' &= y_1 \\ x_2' &= y_2 , \\ y_1' &= - y_1 - \sin x_1 + (y_2-y_1)^3 + u_1 , \\ y_2' &= - y_2 - \sin x_2 - (y_2-y_1)^3 + u_2 . \end{aligned} \tag{39}$$

The meaning of $\psi=0$ is that there is no direct connection between the two points P_1 and P_2. The system (19) with $\psi=0$ and $f(x_s) = \sin x_s$ has only the zeros (equilibria) given by (20) in the region (21). The Liapunov function (24) now is

$$V(\mathbf{x},\mathbf{y}) = y_1^2 + y_2^2 + 2[(1-\cos x_1) + (1-\cos x_2)] \tag{40}$$

and the sufficient condition for avoidance control (37) takes the form

$$\sum_{s=1}^{2} y_s u_s - \sum_{s=1}^{2} y_s^2 - (y_2-y_1)^4 \le 0 . \tag{41}$$

From (27) with (20) we find $h^* = \min V(E_s) = 4$.

Further we select $\theta = 1.571$ (90°) and $\tilde{\theta} = 1.309$ (75°), $\varepsilon = .1$, $\delta = .5$. Then from (30) we get $h_A = 3.9$, $h_S = 3.5$.

Consider Eqs. (39) at time $t_0 = 0$ with initial state $\mathbf{z}(0) = (-1.04720, .52360, 1, -1)$ and control

$\mathbf{u}(0) = (1,1)^T$. A fourth order Runge-Kutta was used on time interval [0,20] with step .01 to integrate (39), i.e. to find the response $\mathbf{z}(t)$.

Following the control strategy established in Section VII, at each time step we calculate the components x_1, x_2, y_1, y_2 of the response $\mathbf{z}$ and the value $V(\mathbf{x},\mathbf{y})$. As long as $x_s < \tilde{\theta} = 1.309$, s=1,2, and $V(\mathbf{x},\mathbf{y}) < h_S = 3.5$, we continue the calculations. However at time $t = 3.31$, $x_1 = 1.30558 < \theta$, $x_2 = 1.30993 > \tilde{\theta}$, $y_1 = .27898$, $y_2 = .16667$, and $V(\mathbf{x},\mathbf{y}) = 3.0653 < h_S$. Since $x_2 > \tilde{\theta}$, the point P_2 entered arc S_α and according to case (b) we have to change the control at the switching point

$$Q_1(1.30558,\ 1.30993,\ .27898,\ .16667)\ . \tag{42}$$

The new control can be selected from (38) in many ways as long as inequality (41) is satisfied (we are dealing with qualitative control). Two control selections are illustrated below.

A. RANDOM CONTROL

We select from U at the switching point Q_1 the new control by random, $\mathbf{u} = (-1.45, -1.22)^T$. At time $t=3.50$, $x_1 = 1.31290 > \tilde{\theta}$, $x_2 = 1.30164 < \tilde{\theta}$, hence the point P_2 is returned back from arc S_α to arc $\tilde{\Omega}_\alpha$ but the point P_1 is now at arc S_α. However, at time $t = 3.79$ we have $P_1, P_2 \in \Omega_\alpha$, but $V(\mathbf{x}(3.79), \mathbf{y}(3.79)) = 3.50306 > h_S$, i.e. the response $\mathbf{z}[t]$ enters the safety zone S. According to case (a) we have to change the control again at the new switching point $Q_2(1.17384, 1.15767, -.74299, -.72609)$. We select by random $\mathbf{u} = (.72, 1.11)^T$ which "moves down" the V-level to 2.63120 at $t=4.00$.

The oscillations of the angular coordinates $x_1(t)$ and $x_2(t)$ versus time are shown in Fig. 2 and the projection of the response $z[t]$ in the phase plane $(x_1,x_2) \subset R^4$ in Fig. 3. On the interval $t \in [0, 20]$ there are five switching points Q_s, $s=1,\ldots,5$. The point Q_2 is associated with case (a) which requires lowering the level of V. The other four switching points illustrate situation (b)

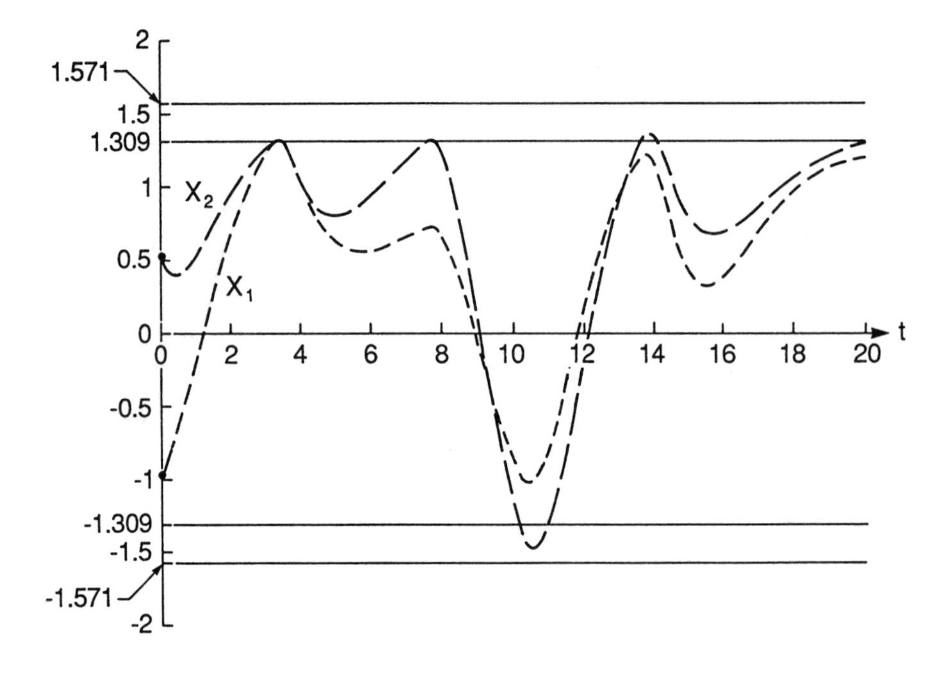

Fig. 2. Case A. Oscillations of x_1 and x_2.

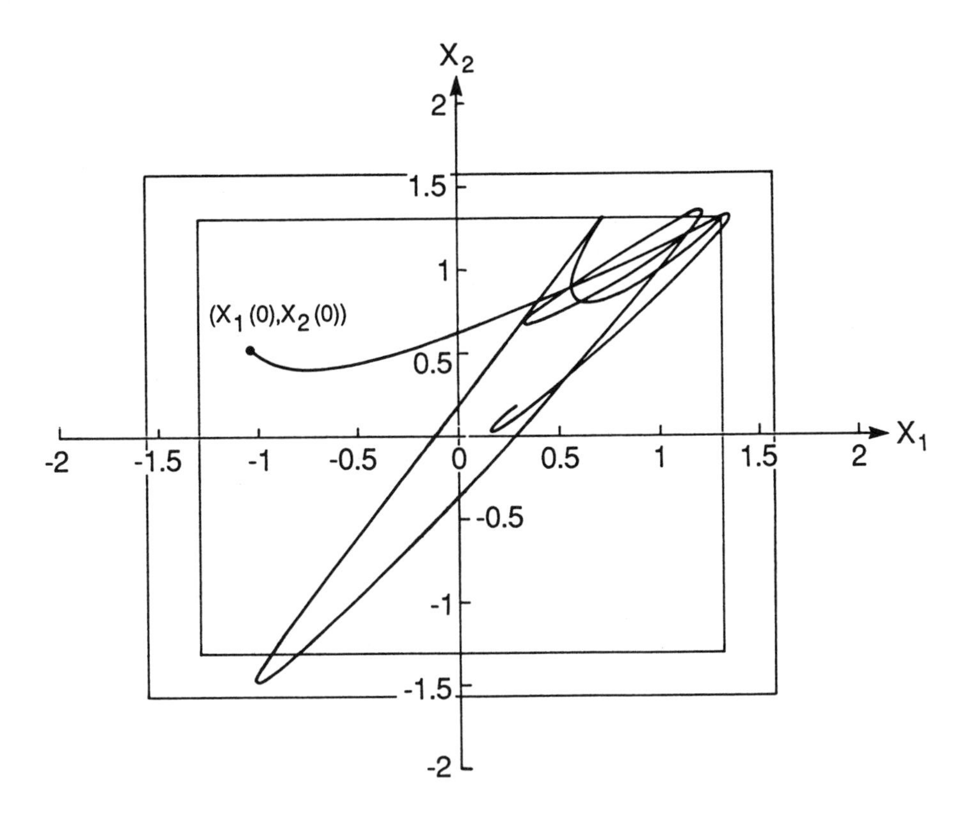

Fig. 3. Case A. Projection of **z** on the plane (x_1,x_2).

B. CONTROL $\mathbf{u} = (y_1,y_2)^T$

Now we make the selection

$$\mathbf{u} = (u_1,u_2)^T = (y_1,y_2)^T \; . \tag{43}$$

Then the sufficient condition for avoidance (41) reduces to $-(y_2-y_1)^4 \leq 0$ which is always satisfied.

At the switching point Q_1 given by (42) we select the new control according to the control law (43), $\mathbf{u} = (.27898,\ .16667)^T$. The point P_2 is not pushed back immediately since at time $t = 3.50$, $x_1 = 1.34213 > \tilde{\theta}$, $x_2 = 1.32520 > \tilde{\theta}$. Now both,

P_1, $P_2 \in$ arc S_α. We can change the control using (43) at that time. However, in order to test the strength of the control we do not do that. At time $t = 4.00$ we get $x_1 = 1.31148 > \tilde{\theta}$, $x_2 = 1.23993 < \tilde{\theta}$ which shows that the point P_2 is in arc $\tilde{\Omega}_\alpha$, out of arc S_α, while the point P_1 is still in arc S_α but moving down. At time $t = 4.50$ both P_1 and P_2 are in arc $\tilde{\Omega}$. There are not other switching points. The oscillations of $x_1(t)$ and $x_2(t)$ versus time presented on Fig. 4 have damped character. The projection of the response z[t] on the (x_1, x_2) plane is shown on Fig. 5.

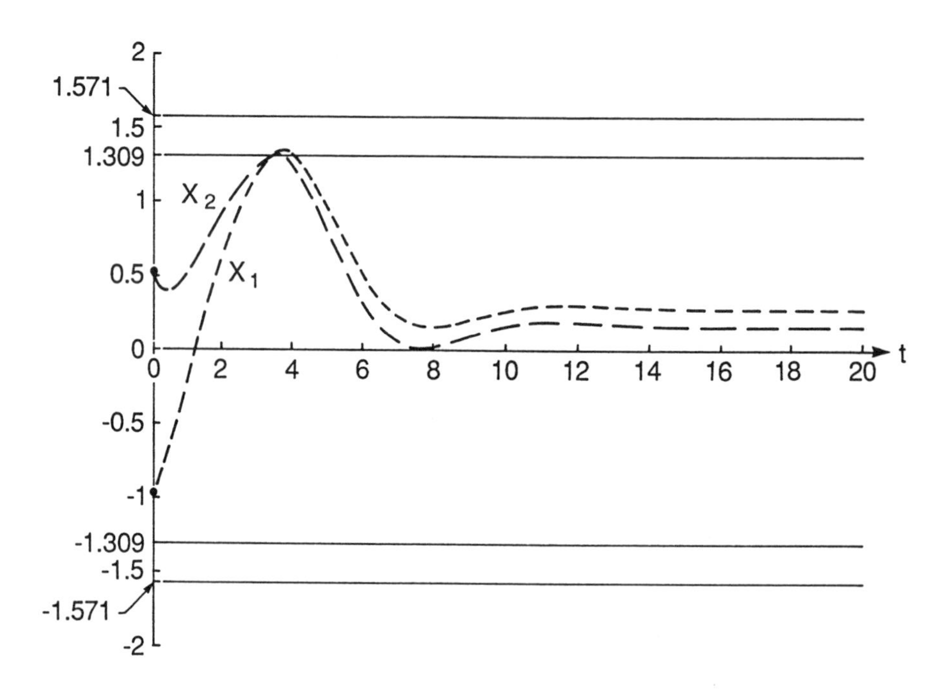

Fig. 4. Case B. Oscillations of x_1 and x_2.

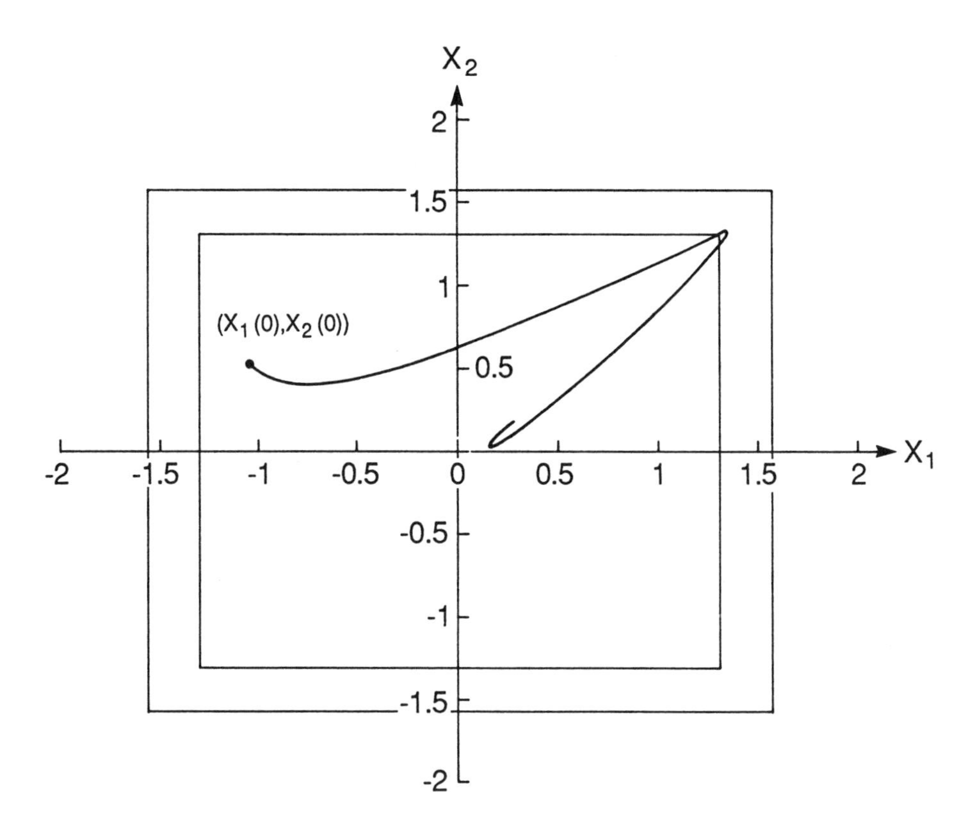

Fig. 5. Case B. Projection of **z** on the plane (x_1, x_2).

REFERENCES

1. B. Serebrijakova and E. Barbashin, "Qualitative study of equations describing the motion of interacting points on a circle", (in Russian). Izv. VUZ, Mathematics, 2 (1961).
2. E. Barbashin, and B. Tabueva, "Dynamical systems with cylindrical phase space" (in Russian), Izd. Nauka, Moskow, 1969.
3. G. Leitmann, and J. Skowronski, "Avoidance control", J. Optim. Theory and Appl.", 23, 581-591 (1977).
4. J. Skowronski, "Control dynamics of robotic manipulators", Academic Press, New York, 1986.
5. G. Bojadziev, "Controlled chain of pendulums", Am. Society Mech. Eng., J. Dyn. Syst. Meas. Control (to appear) (1989).

COORDINATION CONTROLLERS FOR MULTI -ARM MANIPULATORS – A CASE STUDY

M. D. Ardema

Santa Clara University
Santa Clara, CA.

J. M. Skowronski

University of Southern California
Los Angeles, CA.

Abstract

For a pick and place robot with two entirely independent arms, we obtain controllers that quarantee colllision-free regions in the same workspace, by applying a game-theoretic method. A barrier between dextrous regions for each of the arms is determined. Although the scenario is highly simplified, the method may be applied to a large class of multi-arm robots.

1 Introduction

The manufacturing industry assumes an organized distribution of manipulative activities along the production line. These activities are carried out by a distributed set of robotic arms in a proper arrangement. In this sense a single arm robotic system can be considered a handicapped operator. Apart from controlling each arm as a highly complex, high precision and fast working machine, which itself is difficult, the designer of present manufacturing systems must look at the problem of *coordinating control* for the work of a sequence of such machines. The latter includes all the requirements for controlling individual arms, but much more besides, including guaranteeing the satisfactory coexistance of the machines.

There are a number of studies on master-slave or follow-the-leader (hierarchical) techniques, in which the manipulator arms have a specified relationship to each other, see [1,2]. Other studies have used closed chain propositions, see [1,3-8] and symmetry assumptions, see [1,2]; for a review see [9]. In all of these approaches, the second arm motion is specified by ›nstraints. It seems, however, that when both arms are controlled indepen-lently to operate on the same work piece that a differential game formulation ; more appropriate for developing control algorithms. Two formulations are of interest. First, in normal operation it is desired that the manipulator arms operate cooperatively to optimize the manufacturing operation – a cooperative differential game. Second, it is desired that collision of the arms be avoided in any event, and this leads to a formulation of a noncooperative differential game. It is this latter case we consider here.

For example, in an untended manufacturing space it would be desirable to design a controller for one manipulator arm such that it would continue to function safely, at least at some reduced level of effectiveness, even in the event of a catastrophic control failure of a nearby arm. This situation can be formulated as the problem of reaching one target set in the combined manipulator state space while avoiding another set (anti target). This describes what has been called a game of combat [10–12].

Such a scenerio describes what we usually call a qualitative semi-game for each arm and the interface of the semi-games for the two arms concerned gives the designer of the coordinating controller a state space map of options for which the two-arm system must be prepared.

It is our purpose to illustrate this concept on a simple manufacturing scenario which leads to the known turret game, see [10,12,13].

2 Pick-and-Place Robot

Consider two single link robotic arms, shown in Figure 1, in the horizontal plane with inertial reference frame (ξ, η). Arm 1 has a rigid link of length r and an end-effector e_1, rotating about the base B_1 fixed at $(0, 0)$. Arm 2 has a rigid link of length r and end-effector e_2 rotating about the base B_2, which itself is fixed to a conveyer turn-table rotating about B_1 with angular speed $\dot{\beta}(t)$. The radius of the table is r. The rotation angles of the arms are $\theta_\sigma(t)$, $t \geq 0$, $\sigma = 1, 2$.

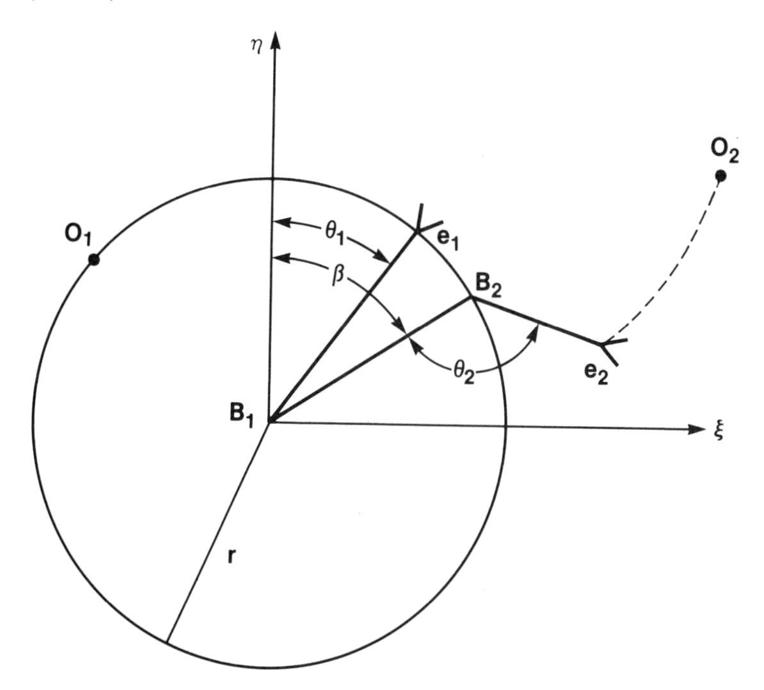

Figure 1. Two-arm robot.

The gripper e_2 is supposed to pick-up an object at some point O_2 in inertial space outside the conveyer and deliver it to location B_1, by controlling the rotation of the turn-table and the rotation of gripper e_2 relative to it. Simultaneously and independently, gripper e_1 is supposed to pick-up an object at some point O_1 in inertial space and deliver it to the conveyor at location B_2. To prevent collision of the arms, both grippers must deposit their objects with zero relative velocity.

Our goal is to seek strategies or control programs that guarantee successful task completion of one arm, despite the actions of the other. Specifically, we seek two pairs of mutually dependent control programs. The first pair secures reaching the target point B_1 by arm 2 while avoiding collision with arm 1; if such a pair of programs exists at a point of the work space, we say that the system is controllable for arm 2 at that point. Conversely, the second pair of programs secures reaching the target point B_2 by arm 1 while avoiding collision with arm 2; if such strategies exist, we have controllability of the system by arm 1. It is clear that at a given point of the work space there are four, and only four possibilities [13]:

(1) System is controllable for only arm 1,

(2) System is controllable for only arm 2,

(3) System is controllable for both arms 1 and 2 and

(4) System is controllable for neither arm 1 nor 2.

Thus the work space is partitioned into four mutually exclusive regions by the controllability properties of the arms. Determining such regions and such pairs of programs is the purpose of the theory of differential games. Specifying the constraints and thus the regions is the goal of qualitative game analysis and specifying the (optimal) program is the goal of quantitative game analysis.

If all system elements are rigid, the system depicted in Figure 1 has three degrees of freedom. The kinetic equations of motion of such a system will be three dynamically coupled, nonlinear second order differential equations; if motors are used to drive the various angular motions, the motor torques would appear as the control variables. In order to avoid this complexity, non-essential for our purpose, we will consider a kinematic version of the problem in which the angular rates themselves, subject to bounds, are the control variables. This corresponds to the limiting case of large torque-to-inertia ratios. Kinetic corrections to the kinematic model, if desired, may be formulated as boundary layers according to singular perturbation theory [13].

It will be convenient to choose the state variables in terms of relative coordinates, with origin at B_1. We let $x_1 \triangleq \beta - \theta_1$, $x_2 \triangleq \theta_2$ with initial conditions $x_i^0 = x_i(0)$, $i = 1,2$, and define the input angular velocities as $u^1 = \dot{\theta}_1$, $u^{21} = \dot{\beta}$ and $u^{22} = -\dot{\theta}_2$. Then the kinematic equation are

$$\left.\begin{aligned} \dot{x}_1 &= u^{21} - u^1, \\ \dot{x}_1 &= u^{22}. \end{aligned}\right\} \tag{1}$$

Since the problem is symmetric, the state work space, i.e. the playing region for the game, is reduced to

$$\Delta = \{(x_1, x_2) | x_i \epsilon [o, \pi],\ i = 1, 2\} \tag{2}$$

The scalar control function $u^1(t)$ is obtained from the actuator of arm 1 and is constrained in some closed interval $u^1 : 0 \leq u^1 \leq \hat{u}$. The control vector $\bar{u}^2(t) = (u^{21}(t),\ u^{22}(t))^T$ is generated by the actuator of arm 2 with values bounded in a closed and bounded set $u^2 \in \mathcal{R}^2$ defined by $u^{21}, u^{22} > 0$, and

$$\frac{u^{21}}{\hat{u}^{21}} - \frac{u^{22}}{\hat{u}^{22}} \leq 1. \tag{3}$$

The first component u^{21} represents coordination control and influences the motion of the conveyor, the second component u^{22} moves arm 2 relative to the conveyor. Constraint (3) is a statement that arm 2 must allocate a fixed amount of control power between control components u^{21} and u^{22}.

According to the work scenerio described, the target for the arm 1 is $\tau^1 = \{(x_1, x_2) | x_1 \leq \epsilon_1\}$, with $\epsilon_1 R > 0$ determining a small neighborhood of B_2

where arm 1 must deposit its object. On the other hand $\tau^2 = \{(x_1, x_2)|x_2 \leq \epsilon_2\}$ about B_1 is the target set for arm 2 aiming at depositing its object at B_1 within a tolerance of $\epsilon_1 R > 0$. The targets in the playing space are shown in Figure 2. Letting $\mathcal{A}^\sigma$ be the subset of the playing space which the arm σ must avoid, it is clear that in our case $\mathcal{A}^1 = \tau^2$ and $\mathcal{A}^2 = \tau^1$. We will later call $\mathcal{A}^\sigma$ the configuration avoidance set for arm σ. In summary, each of the arms must be controlled to reach its target while avoiding the target of the other.

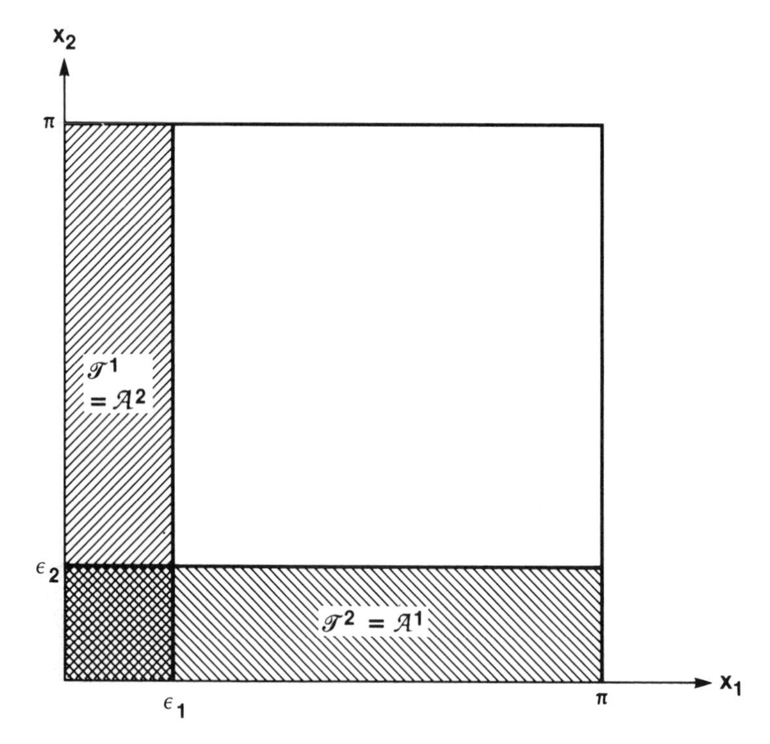

Figure 2. Targets in the playing space for two-arm robot.

In differential games with two (or more) targets, a technical problem arises concerning closure of the target sets. Since τ^1 and τ^2 are closed, so is their intersection, $\tau^1 \cap \tau^2$. Hence the complement of the intersection in τ^1, $\tau^{1*} = \tau^1/(\tau^1 \cap \tau^2)$, and the set τ^{2*}, similarly defined, are not closed. This means that a convergent sequence of guaranteed safe trajections for arm 1, for example, terminating in τ^{1*}, need not converge to a trajectory that terminates to a point in τ^{1*}. To circumvent this difficulty, it is necessary to place an open neighborhood around $\tau^1 \cap \tau^2$ and consider the compliment of this open neighborhood in the two target sets [13,15]. In the sequel, we will assume that this open neighborhood is negligibly small.

3 The Coordination Game

For convenience let us write the state equations (1) in the general format

$$\dot{x}_i = f_i(\bar{x}, \bar{u}^1, \bar{u}^2) \tag{4}$$

where $\bar{x} = (x_1, x_2)^T \epsilon \Delta$. The coordination controls are generally specified by feedback programs, see [9]:

$$\bar{u}^\sigma \triangleq \bar{\mathcal{P}}^\sigma(\bar{x}, u^{\nu\sigma}), \qquad \sigma, \nu = 1, 2, \qquad \sigma \neq \nu,$$

In particular $\bar{u}^\sigma = (u^{\sigma\sigma}, u^{\sigma\nu})^T$, $u^{\sigma\sigma} = \bar{\mathcal{P}}^{\sigma\sigma}(\bar{x})$, $u^{\sigma\nu} = \bar{\mathcal{P}}^{\sigma\nu}(\bar{x}, u^{\nu\sigma})$. Equation (4) under suitable conditions generates unique solutions, called trajectories in Δ.

The semi-game for the arm σ will require knowledge of all the options of the other arm and thus may be expressed by the contingent vector equation in the format

$$\dot{\bar{x}} \epsilon \{ f(\bar{x}, \bar{u}^1, \bar{u}^2) | \bar{u}^\sigma \epsilon \bar{\mathcal{P}}^\sigma(\bar{x}, u^{\nu\sigma}), \bar{u}^\nu \epsilon U^\nu \} \tag{5}$$

with solutions represented by the trajectories of (4) $k(\bar{x}^0, t), t \geq 0$, forming at each $\bar{x}^0 \epsilon \Delta$ a class $\mathcal{K}(x^0)$, subject to required sufficient conditions, see [15].

Given the targets τ^σ, $\sigma = 1, 2$ and the antitargets (avoidance sets) $\mathcal{A}^\sigma$, $\sigma = 1, 2$ with $\mathcal{A}^\sigma = \tau^\nu$, $\sigma \neq \nu$, all in the work envelope Δ, the objective of the semi-game in favor of the arm σ, briefly the σ-game, reduces to the following.

σ-OBJECTIVE

(1) *Stabilization*: A given set $\Delta_0 \subset \Delta$ is positively invariant under (5) if given $\bar{u}^1$, $\bar{u}^2$ in (4), $\bar{x}^0 \epsilon \Delta$ implies $k(x^0,\ t)\epsilon\Delta_0$, $t \geq 0$.

(2) *Reaching*: There is a time interval $t_c^\sigma < \infty$, possibly stipulated, such that for the above $\bar{u}^1, \bar{u}^2, \bar{x}^0 \epsilon \Delta_0$ implies $k(x^0, t)\epsilon\tau^\sigma$, $\forall t > t_c^\sigma$.

(3) *Avoiding collision*: There is a time interval t_A^σ, $t_A^\sigma \geq t_c^\sigma$ such that for the above $\bar{u}^1, \bar{u}^2$, $x^0 \epsilon \Delta_0$ implies $k(x^0, t) \cap \mathcal{A}^\sigma = \phi$, $\forall t \leq t_A^\sigma$, while (2) holds.

Definition 1: The σ-game is strongly controllable at $\bar{x}^0$ for the σ-objective, if simultaneously

(a) there is a control program $\bar{\mathcal{P}}^\sigma(\bar{x}, \bar{u}^\nu)$ such that for any $k(\cdot)\epsilon\mathcal{K}(\bar{x}^0)$ the subobjective (3) holds.

(b) given some $\bar{u}^\nu(t)$, the program in (a) secures the subobjectives (1), (2).

The set Δ_D^σ of all such $\bar{x}^0$ forms the region of strong controllability for the σ-objective. Any subset of such a region is strongly controllable for the said objective. Now let $\Delta_\epsilon^\sigma(\bar{x}^0)$ be the closure of an open subset of Δ such that $\Delta_\epsilon^\sigma \supset \mathcal{A}^\sigma$ and $\partial\Delta_\epsilon^\sigma \cap \partial\mathcal{A}^\sigma = \phi$. Given $\bar{x}^0$ we term $\Delta_A^\sigma(\bar{x}^0) = \Delta_\epsilon^\sigma(\bar{x}^0)/\mathcal{A}^\sigma$ the safety "slow down" zone about $\mathcal{A}^\sigma$. Moreover let $\Delta_0 \subset \Delta$ be a set on which we want the σ-objective to hold. We abbreviate the strong controllability for σ-objective to "strong σ-controllability".

Theorem 1. A given set Δ_0 is strongly σ-controllable, if there is a safety zone $\Delta_A^\sigma(\bar{x}^0)$, two controllers $\bar{\mathcal{P}}^\sigma(\bar{x}, \bar{u}^\nu)$, $\bar{\mathcal{P}}^\nu(\bar{x}, \bar{u}^\sigma)$, and two C^1-functions $V^\sigma(\cdot)$: $\Delta_0 \to \mathcal{R}$, $V_A^\sigma(\cdot) : \Delta_A^\sigma \to \mathcal{R}$ such that

i. $a^\sigma \leq V^\sigma(\bar{x}) \leq b^\sigma$,

where

$a^\sigma = \inf V^\sigma(\bar{x}) | \bar{x}\epsilon\partial\tau^\sigma$,

$b^\sigma = \inf V^\sigma(\bar{x}) | \bar{x}\epsilon\partial\Delta_0$.

ii. $0 < a_A^\sigma < b_A^\sigma < \infty$

where

$a_A^\sigma = \sup V_A^\sigma(\bar{x}) | \bar{x} \epsilon \partial \mathcal{A}^\sigma,$

$b_A^\sigma = \inf V_A^\sigma(\bar{x}) | \bar{x} \epsilon \partial \Delta_A^\sigma.$

iii. for each $(\bar{u}^\sigma, \bar{u}^\nu) \epsilon \bar{\mathcal{P}}^{\sigma\sigma} \times \bar{\mathcal{P}}^{\nu\sigma}$ there is

$c^\sigma = \text{const} > 0$ such that

$$\nabla V^\sigma(\bar{x})^T \cdot \bar{f}(\bar{x}, \bar{u}^1, \bar{u}^2) \leq -c^\sigma, \qquad (6)$$

$\sigma, \nu = 1, 2, \sigma \neq \nu.$

iv. for each $\bar{u}^\sigma \epsilon \bar{\mathcal{P}}^\sigma(\bar{x}, \bar{u}^\nu)$ there is

a constant $c_A^\sigma < \left[\dfrac{b_A^\sigma - a_A^\sigma}{b^\sigma - a^\sigma}\right] c^\sigma$, such that

$$\nabla V_A^\sigma(\bar{x})^T \cdot \bar{f}(\bar{x}, \bar{u}^1, \bar{u}^2) \geq -c_A^\sigma \qquad (7)$$

for all $\bar{u}^\nu \epsilon U^\nu$, $\sigma, \nu = 1, 2, \sigma \neq \nu$.

For the proof see [14 and 16].

When t_c^σ is stipulated we impose

$$c_\sigma = \frac{b^\sigma - a^\sigma}{t_c^\sigma}, \qquad c_A^\sigma = \frac{b_A^\sigma - a_A^\sigma}{t_A^\sigma}, \qquad t_A^\sigma > t_c^\sigma \qquad (8)$$

When $t_A^\sigma \to \infty$ we secure permanent avoidance of $\mathcal{A}^\sigma$. The control programs may be found from the following corollary. Let

$$\mathcal{L}^\sigma(\bar{x}, \bar{u}^1, \bar{u}^2) = \nabla V^\sigma(\bar{x})^T \cdot \bar{f}(\bar{x}, \bar{u}^1, \bar{u}^2)$$

$$\mathcal{L}_A^\sigma(\bar{x}, \bar{u}^1, \bar{u}^2) = \nabla V_A^\sigma(\bar{x})^T \cdot \bar{f}(\bar{x}, \bar{u}^1, \bar{u}^2)$$

Corollary 1: Given $x^0 \epsilon \Delta_0$ if there are u_*^1, u_*^2 such that

$$\mathcal{L}_A^\sigma(\bar{x}, u_*^1, u_*^2) = \min_{u^\sigma} \max_{u^\nu} \mathcal{L}^\sigma(\bar{x}, \bar{u}^1, \bar{u}^2) \leq \frac{b^\sigma - a^\sigma}{t_A^\sigma} \qquad (9)$$

$$\mathcal{L}_A^\sigma(\bar{x}, u_*^1, u_*^2) = \max_{u^\sigma} \min_{u^\nu} \mathcal{L}^\sigma(\bar{x}, \bar{u}^1, \bar{u}^2) \geq \frac{b_A^\sigma - a_A^\sigma}{t_A^\sigma} \qquad (10)$$

then conditions (iii), (iv) are met with $u^\sigma = u_*^\sigma$. The region Δ_D^σ, maximal Δ_0, may be found from

$$\Delta_D^\sigma : \frac{V^\sigma(\bar{x}) - a^\sigma}{c^\sigma} < \frac{V_A^\sigma(\bar{x}) - a_A^\sigma}{c_A^\sigma} \tag{11}$$

with c^σ, c_A^σ given as estimates of $\dot{V}^\sigma(\bar{x})$, $\dot{V}_A^\sigma(\bar{x})$ or from (8).

We are especially interested in the interface between the two semi-games. In particular we went to determine regions in the state space dexterous for the arms, i.e. Δ_D^1, Δ_D^2, which will lead to the state map of options mentioned in the introduction. Introduce the set $C\Delta_D^\sigma = \Delta/\Delta_D^\sigma$, the compliment of Δ_D^σ in Δ, called semi-neutral. It is covered by points where the strong σ-controllability is contradicted. Then introduce a surface $\sum^\sigma$ subdividing Δ into two disjoint sets: $\Delta^\sigma \supset \Delta_D^\sigma$ called interior and $C\Delta^\sigma = \Delta/\Delta^\sigma$ called exterior with the property that for $\bar{x}^0 \epsilon \sum^\sigma$ there is $\bar{\mathcal{P}}^\nu(\bar{x}, \bar{u}^\sigma)$, $\nu \neq \sigma$, such that no $k(\bar{x}^0, t)$, $t \geq 0$ enters the interior. We call $\sum^\sigma$ nonpermeable for arm σ, briefly σ-nonpermeable.

The following theorem was proved in [17].

Theorem 2. A surface S subdividing Δ into disjoint Δ^σ, $C\Delta^\sigma$ is $\sum^\sigma$, if there are $\bar{\mathcal{P}}^\nu(\cdot)$ and a C^1-function $V_B^\sigma(\cdot) : D \to \mathcal{R}$, $D(\text{open}) \supset S$, such that for all $\bar{x} \epsilon \Delta^\sigma$

(i) $V_B^\sigma(\bar{x}) < V_B^\sigma(\bar{\zeta}), \bar{\zeta} \epsilon S;$

(ii) for each $\bar{u}^\nu \epsilon\, \bar{\mathcal{P}}^\nu(\bar{x}, u^\sigma)$,

$$\nabla V_B^\sigma(\bar{x})^T \cdot f(x, u^1, u^2) \geq 0, \forall u^\sigma \epsilon U_\sigma, \tag{12}$$

Colollary 2: Given $\bar{x} \epsilon \sum^\sigma$, if there is an admissible pair u_*^1, u_*^2 such that

$$\mathcal{L}_B(x, u_*^1, u_*^2, t) = \max_{\bar{u}^\nu} \min_{\bar{u}^\sigma} \mathcal{L}_B(\bar{x}, \bar{u}^1, \bar{u}^2) \geq 0 \tag{13}$$

then condition (ii) is met with $u_*^\nu \epsilon \bar{\mathcal{P}}^\nu(\bar{x}, u_*^\sigma)$, making it possible to deduce $\bar{\mathcal{P}}^\nu(\cdot)$ from (12).

Obviously $\sum^{\sigma} \subset C\Delta_D^{\sigma}$, and there may be many of them. We are interested in $\sum^{\sigma}$ which is closest to Δ_D^{σ}, thus if $\partial\Delta^{\sigma}$ is not well defined, we choose the $\sum^{\sigma}$ closest to the target $\mathcal{T}^{\sigma}$ which is always given and obviously in Δ_D^{σ}. We call such $\sum^{\sigma}$ the σ-semibarrier, denoted B^{σ}. In particular, when $\partial\Delta_c^{\sigma}$ is defined and satisfies Theorem 2 i.e. is some $\sum^{\sigma}$, then $\partial\Delta_D^{\sigma} = B^{\sigma}$.

We may now attempt to interface both strong controllabilities on the work envelope Δ. Let us define the neutral zone $\Delta_N = \Delta/(\Delta_D^1 \cup \Delta_D^2)$, closed if Δ_D^1, Δ_D^2 are open, possibly empty and also possibly embedded in $\Delta_D^1 \cup \Delta_D^2$. Then we define the *barrier* $B = B^1 \cap B^2$ obviously in Δ_{ν} and separating Δ_D^1, Δ_D^2, if not empty. However B will not, in general, divide Δ into two disjoint sets. In view of the above, any candidate for B is confirmed by using Theorem 2 twice, i.e. by finding two functions $V_b^1(\cdot)$, $V_B^2(\cdot)$ each satisfying conditions (i), (ii) on Δ^1, Δ^2, respectively. It is convenient to choose $V^2(\bar{x}) = \text{const} - V^1(\bar{x})$. The candidate for B may be conveniently provided from necessary conditions, be a Isaacs' barrier, or be an educated guess. In the first case, which in fact overlaps with the second, we reason as follows. Suppose we have a smooth B^{σ} with the gradient $\bar{n}^{\sigma} = (n_1^{\sigma}, \ldots, n_{2n}^{\sigma})$ directed toward $C\Delta_D^{\sigma}$. We conclude from (14) that for all $x^0 \epsilon B^{\sigma}$ there is u_*^{ν} such that $(n^{\sigma})^T \cdot f(x, u^{\sigma}, u_*^{\nu}) \geq 0$, holding for all $u^{\sigma} \epsilon U_{\sigma}$. On B both above conditions apply with $\bar{n}^1 = -\bar{n}^2$, so that we have the necessary condition

$$(n^{\sigma})^T \cdot f(x, u^{\sigma}, u_*^{\nu}) = 0 \tag{14}$$

which thus specifies the grid condidate for B to be checked by the twice used Theorem 2.

In general we do not have to have Δ_D^1, Δ_D^2 disjoint: $\Delta_D^1 \cap \Delta_D^2 \neq \emptyset$. The sets $D^1 = \Delta_D^1/\Delta_D^2$ and $D^2 = \Delta_D^2/\Delta_D^1$ will be called the "winning" sets, while $D^{12} = \Delta_D^1 \cap \Delta_D^2$ is the set of guaranteed dextrerity, which is of our primary interest when two manipulators operate in the same workspace, see Figure 3.

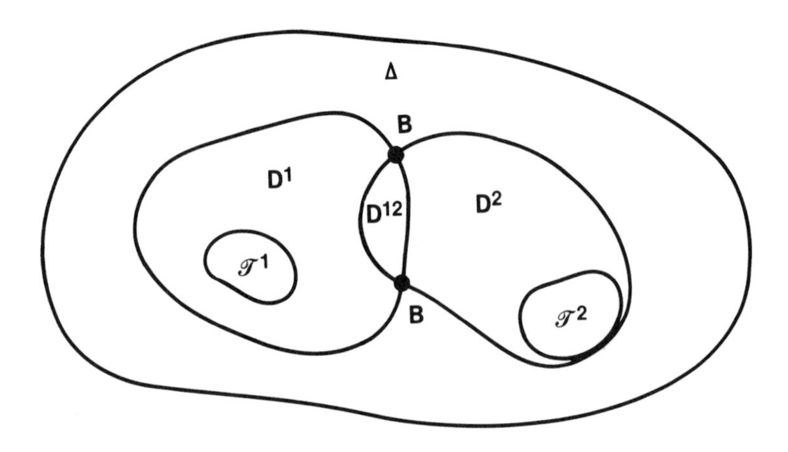

Figure 3. Regions in the workspace.

4 The Turret Game

We return now to our case study of the turret game, and consider first the strong 1-game with the strong 1-controllability. To use Theorem 1, we take $V^1 = x_1$, $V_A^1 = x_2$ with $C_1 = \epsilon_1$, $C_2 = \epsilon_2$ where

$$\begin{aligned} \mathcal{T}^1 \supset \Delta_1 &= \{(x_1, x_2) | V^1 \leq C_1\} \\ \mathcal{A}^1 \subset \Delta_2 &= \{(x_1, x_2) | V_A^1 \leq C_2\} \end{aligned}$$

We now need to find constants c^1, c_A^1, $c_1 > 0$ such that for each

$$x \epsilon \Delta_D^1 = \left\{ (x_1, x_2) \left| \frac{c^1}{V^1(x) - C_1} > \frac{c_A^1}{V_A^1(x) - C_2} \right. \right\},$$

there exists $\bar{\mathcal{P}}^1(\cdot)$, yielding for every $u_*^1 = \bar{\mathcal{P}}^1(\bar{x}, u^2)$

$$\sup_{\bar{u}^2} \nabla V^1(\bar{x})^T \cdot f(\bar{x}, u_*^1, u^2) \leq -c^1 < 0 \tag{15}$$

$$\inf_{\bar{u}^2} \nabla V_A^1(\bar{x})^T \cdot f(\bar{x}, u_*^1, u^2) \geq -c_A^1, \tag{16}$$

See (11). From (15) we obtain

$$\hat{u}^{21} - u^1 \leq -c^1 < 0 \tag{17}$$

For positive c^1 to exist, this implies that

$\hat{u}^{21} - \hat{u}^1 < 0$, that is $\gamma_1 = \hat{u}^{21}/\hat{u}^1 < 1$. Select the control program for arm 1 to be $u_*^1 = \hat{u}^1$ for all $t \geq 0$. Then the largest c^1 compatible with (17), independent of the state, is $c^1 = \hat{u}^1 - \hat{u}^{21}$. From (16) we obtain

$$c_A^1 \geq \hat{u}^{22} \tag{18}$$

The smallest c_A^1 compatible with (18) independent of state, is $c_A^1 = \hat{u}^{22}$.

So for $\gamma_1 < 1$, setting $\gamma_2 = \hat{u}^{22}/\hat{u}^1$, we have

$$\begin{aligned} D^1 &= \left\{ (x_1, x_2) \epsilon \Delta \left| \frac{\hat{u}^1 - \hat{u}^{21}}{x_1 - \epsilon_1} > \frac{\hat{u}^{22}}{x_2 - \epsilon_2} \right. \right\} \\ &= \left\{ (x_1, x_2) \epsilon \Delta \left| x_2 - \epsilon_2 > \frac{\gamma_2}{1 - \gamma_1}(x_1 - \epsilon_1) \right. \right\} \subset \Delta_D^1. \end{aligned}$$

D^1 is a subset of Δ_D^1 for it is easily shown by integrating the state equations that if player 1 always selects control u^* he will win from all initial conditions satisfying $(x_2 - \epsilon_2) > \gamma_2(x_1 - \epsilon_1)$. (The greatest effect player 2 can have on the outcome is when his control selection is $u^{21} = 0$ and $u^{22} = \hat{u}^{22}$ to give the greatest rate of decrease of x_2 towards ϵ_2.) So D^1 is an underestimate of Δ_D^1. We have

$$\Delta_D^1 = \{(x_1, x_2) | (x_2 - \epsilon_2) > \gamma_2(x_1 - \epsilon_1)\}.$$

For $\gamma_1 \geq 1$, clearly $\Delta_D^1 = \phi$.

Consider now the strong game for arm 2, the 2-game. Take $v^2 = x_2$ and $v_A^2 = x_1$ with $C_1 = \epsilon_2$ and $C_2 = \epsilon_1$ where $\mathcal{T}_1^2 \supset \Delta_1$ and $\mathcal{T}_2^2 \subset \Delta_2$. Then the corresponding equations to (3) and (4) are

$$\sup_{u^1} \nabla V_1 \cdot f(x, u^1, u_*^{21}, u_*^{22}) \ \leq \ -c^2 < 0 \tag{19}$$

$$\inf_{u^1} \nabla V_2 \cdot f(x, u^1, u_*^{21}, u_*^{22}) \ \geq \ -c_A^2 \tag{20}$$

where the program of selection of u_*^{21}, u_*^{22} for arm 2 is to be determined. From (20), we obtain

$$-u^{22} \leq -c^2 < 0. \tag{21}$$

This necessarily requires the selection of v_2 to be nonzero. From (20), we obtain

$$c_A^2 \geq \hat{u}^1 - u^{21}. \tag{22}$$

Equations (21) and (22) are independent of the state variables. From (21), a maximum value of c^2 is given by $c^2 = \hat{u}^{22}/(1+\delta)$ for a selection of $u_*^{22} = \hat{u}^{22}/(1+\delta)$ with $\delta\epsilon[0,\infty)$. This means that acceptable values for u^{21} satisfy

$$0 \leq u^{21} \leq \left(1 - \frac{1}{1+\delta}\right)\hat{u}^{21} = \frac{\delta}{1+\delta}\hat{u}^{21}.$$

Hence we can write, selecting the smallest k_2 in (22), given a value of u^{21},

$$\begin{aligned} \frac{c_A^2}{c^2} &= \frac{\hat{u}^1 - \beta\delta\hat{u}^{21}/(1+\delta)}{\hat{u}^{22}/(1+\delta)} \\ &= \frac{1+\delta(1-\beta\gamma_1)}{\gamma_2} \qquad \text{with} \quad \beta\epsilon[0,1]. \end{aligned}$$

Now, provided $\gamma_1 \leq 1$, that is $\hat{u}^{21} \leq \hat{u}^1$, Figure 4(a) and (b), the smallest value of c_A^2/c_2 is obtained when $\delta = 0$. So $c_A^2/c_2 = 1/\gamma_2$ when $u_*^{22} = \hat{u}^{22}$ and $u_*^{21} = 0$. The maximum winning region for arm 2, considering the result for Δ_D^1, is

$$\Delta_D^2 = \{(x_1, x_2)\epsilon\Delta |\ x_2 - \epsilon_2 < \gamma_2(x_1 - \epsilon_1)\}.$$

For $\gamma_1 > 1$, Figure 4(c), it is possible with $\beta = 1$ to select δ sufficiently large, thus defining the selection of u_*^{21} and u_*^{22} to make the above expression for c_A^2/c_2 negative. Equivalently, for $\gamma_1 > 1$ we can select a $\beta\epsilon(0,1)$ such that $\beta\hat{u}^{21} = \hat{u}^1$. Then we require that player 2 play $u_*^{21} = \beta\hat{u}^{21}$, $u_*^{22} = \hat{u}^{22}(1-\beta)$. In this case then, Δ_D^2 must be all of Δ.

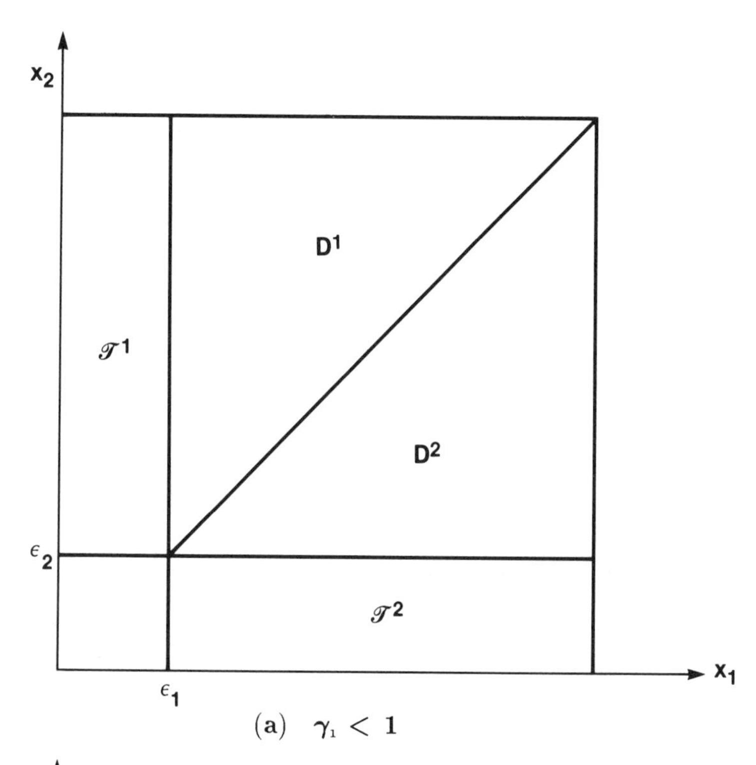

(a) $\gamma_1 < 1$

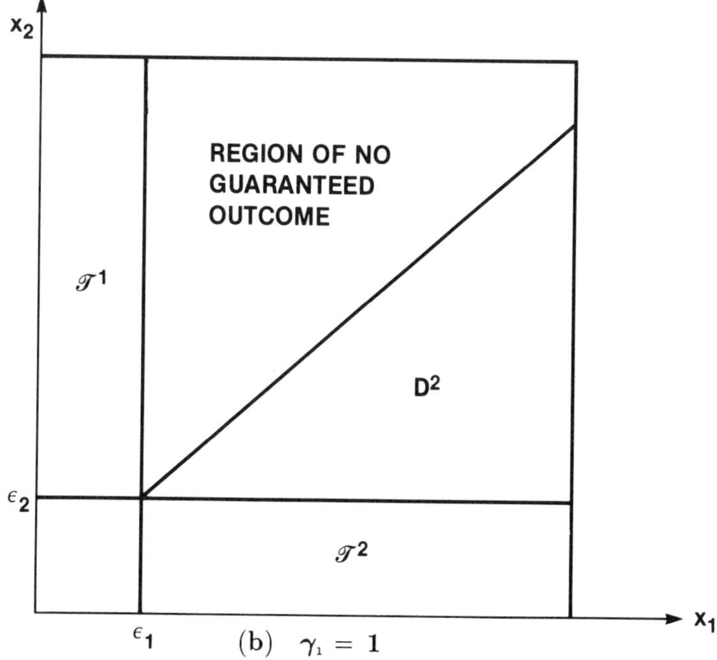

(b) $\gamma_1 = 1$

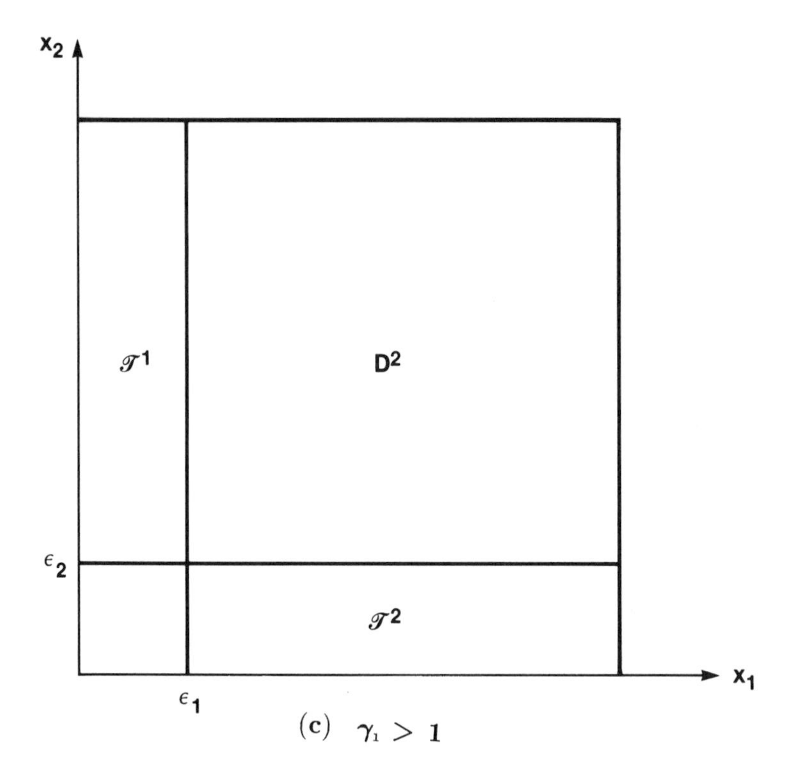

(c) $\gamma_1 > 1$

Figure 4. Region in the playing space for two-arm robots.

In the case concerned, it is seen that $\Delta_D^1 \cap \Delta_D^2 = \phi$.

Even though the qualitative objective for arm 1 (similarly for arm 2) is defined in terms of reachability of $\mathcal{T}^1$, Theorem 1 ensures that there is penetration of the target set. In the case $\gamma_1 > 1$, arm 2 is guaranteed of a win in all of Δ with 2-objective; 1 cannot hope to win unless arm 2 does not play one of his winning strategies. We shall discuss now the case $\gamma_1 < 1$. It may be shown, see [17], that the barrier

$$B = \{(x_1, x_2)\epsilon\Delta | x_2 - \epsilon_2 = \gamma_2(x_1 - \epsilon_1)\},$$

is nonpermeable for arm 1 and arm 2. Consider first the 1 nonpermeability.

Take $D^1 \subset \Delta^1 = \{(x_1, x_2)\epsilon\Delta | x_2 - \epsilon_2 > \gamma_2(x_1 - \epsilon_1)\}$

$D^2 = \Delta^2$

$V^1(x_1, x_2) = 1 - (x_2 - \epsilon_2) + \gamma_2(x_1 - \epsilon_1).$

$V^1(\cdot)$ is a C^1-function and $V^1 \equiv 1$ on B with $V^1(x_1, x_2) < 1$ for all $(x_1, x_2)\epsilon\Delta^1$, so the condition (iii) of Theorem 1 holds:

$$\begin{aligned} \nabla V^1(x_1, x_2)^T \cdot \bar{f}(x_1, x_2, u^1, u^2) &= u^{22} + \gamma_2(u^{21} - u^1) \\ &= u^{22} + \frac{\hat{u}^{22}}{\hat{u}^1} u^{21} - \frac{\hat{u}^{22}}{\hat{u}^1} u^1. \end{aligned}$$

Maximization of $\nabla V^1 \cdot \bar{f}$ with respect to $\bar{u}^2$, given the constraints $u^{21}/\hat{u}^{21} + u^{22}/\hat{u}^{22} \leq 1$, u^{21}, $u^{22} \geq 0$, is a simple linear programming problem with maximization given by $u^{22} = \hat{u}^{22}$ and $u^{21} = 0$. Then $\nabla V^1 \cdot f = \hat{u}^{22}(1 - \hat{u}/\hat{u}^1)$.

We see that $\nabla V^1 \cdot f \geq 0$ for all $0 \leq u^1 \leq \hat{u}^1$, so condition (ii) of Theorem 2 is satisfied for this selection of strategy of arm 2. Hence B is 1-nonpermeable by Theorem 2.

Let us turn now to the 2-nonpermeability.

Take

$$\Delta^2 = D^2,$$

$$\Delta^1 = \{(x_1, x_2)\epsilon\Delta | x_2 - \epsilon_2 > \gamma_2(x_1 - \epsilon_1)\},$$

$$V^2(x_1, x_2) = 1 + x_2 - \epsilon_2 - \gamma_2(x_1 - \epsilon_1).$$

V^2 is a C^1 function and $V^2 \equiv 1$ on B with $V^2(x_1, x_2) < 1$ for all $(x_1, x_2)\epsilon\Delta^2$. Now

$$\begin{aligned} \nabla V^2 \cdot f &= -\gamma_2(u^{21} - u^1) \\ &= -\frac{\hat{u}^{22}}{\hat{u}^1} u^{21} - u^{22} + \frac{\hat{u}^{22}}{\hat{u}^1} u^1. \end{aligned}$$

The maximum $\nabla V^2 \cdot f$ with respect to u^1 is attainable when $u^1 = \hat{u}^1$. So using $u^1 = \hat{u}^1$,

$$\begin{aligned} \nabla V^2 \cdot f &= \hat{u}^{22} - u^{22} - \frac{\hat{u}^{22}}{\hat{u}^1} \\ &= \hat{u}^{22}\left[1 - \frac{u^{21}}{\hat{u}^1} - \frac{u^{22}}{\hat{u}^{22}}\right]. \end{aligned}$$

Now for $\hat{u}^{21} \leq \hat{u}^1$ we have

$$0 \leq \frac{u^{21}}{\hat{u}^1} + \frac{u^{22}}{\hat{u}^{22}} \leq \frac{u^{21}}{\hat{u}^{21}} + \frac{u^{22}}{\hat{u}^{22}} \leq 1.$$

This means that $\nabla V^2 \cdot f \geq 0$ for all admissible u^{22}, u^{21}. The conditions of Theorem 2 are fulfilled, so B is 2-nonpermeable.

5 Conclusions

The obtained D^1, D^2 separated by a given Barrier B give the dexterous regions of operation for the two arms in our pick-and-place scenerio. As $D^{12} \neq \phi$, there is no quaranteed dexterity region when both grippers operate together – which is in agreement with the required manufacturing scenerio.

References

[1] Luh, J. Y. S., Zheng, Y. F., *Asumption of input generalized focus for robots with closed kinematic chains*, IEEE J. Robotics, RA1, 1985, 95–103.

[2] Hemami H., *Kinetics of two arm robots*, IEEE J. Robotics, RA2, 1986, 275–228.

[3] Vukobratovic, M., Potkonjak, V., *Dynamics of Manipulation Robots*, Springer, 1982.

[4] Tan, T. J., Bejczy, A. K., Yun, X. *Nonlinear feedback control of multiple robot arms. Proc. Workshop on Space Telerobotics*, J.P.L. Pub. 87–13, 1987, Vol. 3, 179–192.

[5] Hayati, S. A., *Dynamics and control of coordinated multiple manipulators, Proc. Workshop on Space Telerobotics*, J.P.L. Publ. 87–13, 1987 Vol. 3, 193–204.

[6] Koivo, J. J., *Adaptive position-velocity-force control of two manipulators*, Proc. 24th IEEE CDC, Ft. Lauderdale, 1985, 1529–1532.

[7] Seragi, H. *Adaptive control of dual arm robots*, Proc. Workshop on Space Telerobotics, J.P.L. Publ 87–13, 1987, Vol. 3, 159–170.

[8] Luh, J. Y. S., Zheng, Y. F., *Motion coordination and programmable teleseparation between two industrial robots*, Proc. Workshop on Space Telerobotics, J.P.L. Publ. 87–13, 1987, Vol. 2, 325–334.

[9] Skowronski, J. M. *Control Theory of Robotic Systems*, World Scientific Publ., N. Jersey-London-Singapore, 1989.

[10] Heymann, M. Ardema, M. D., Rajan N., *A formulation and analysis of combat games*, NASA Rep. TP 2487, 1987.

[11] Ardema, M. D., Heymann, M., Rajan J., *Combat games*, J. Opt. Th. Appl. Vol 46, 1985, 391–398.

[12] Ardema, M. D., Heymann, M., and Rajan, N. *Analysis of a Combat Problem: The Turret Game*, J. Opt. Th. Appl. Vol. 54, 1987, 23–42.

[13] Ardema, M. D., ed., *Singular Perturbations in Systems and Control, International Centre for Mechanical Sciences Courses and Lectures*, Vol. 280, Springer, 1983.

[14] Getz, W. M., Leitmann G., *Qualitative differential games with two targets*, J. Math. Aml. Applic., Vol. 68, 1979, 421–430.

[15] Filippov, A. F., *Classical solutions of differential equations with multivalued right hand sides*, SIAM J. Control, Vol. 5, 1967, 609–621.

[16] Skowronski, J. M., *Coordination control of independent two robot arms on moving platform*, AIAA Paper No. 89–0584, 1989.

[17] Skowronski, J. M., Storier, R. J., *Two person qualitative differential games with two objectives*, Comp. Math. & Applic., Vol. 18, 1989, 133–150.

CONTROL AND DYNAMIC SYSTEMS, VOL. 35

MRAC TECHNIQUES WITH APPLICATION TO ROBOT MANIPULATORS

R.J. STONIER

Department of Mathematics & Computing
Capricornia Institute of Advanced Education
Rockhampton, Qld., Australia 4702

C.N. WHEELER

Tarong Power Station
Nanango, Qld., Australia 4610

I. INTRODUCTION

The model reference adaptive control (MRAC) technique has become a widely used tool in the search for control methods, in particular to control robot manipulators. For a review, see Skowronski [1] and Craig [2].

Its application has been predominantly associated with the design of control to make the output state of a nonlinear plant (the robot system) track the output state of a linear reference model via input control modification. Skowronski in [1] points out that in many applications, however, the desired output response may not be associated with the output of a linear reference model. Even if a linearised plant model is used as a reference model, the analysis is only valid locally. Indeed, a disadvantage of such methods is that a linear reference model with its single equilibrium may be incompatible with the generally multi-equilibria manipulator to the extent of rendering the tracking ineffective.

Skowronski in recent papers [3], [4], [5] and [6], has abandoned the error equation analysis and examined the nonlinear model tracking by robot manipulators by securing stability of the diagonal set in the Cartesian product of the state parameter spaces of the plant and model using the Liapunov direct method.

In this paper we look at the nonlinear model tracking by robot manipulators using an error equation analysis via the Liapunov direct method. The results are illustrated on the RP-manipulator model used by Skowronski in his papers.

II. THE MANIPULATOR AND NONLINEAR REFERENCE MODELS

We consider the manipulator with n DOF's whose dynamics are determined by the joint coordinates $q_1,\ldots,q_n$ and their time derivatives $\dot{q}_1,\ldots,\dot{q}_n$, in terms of the n-vector general type Lagrangian equation

$$A(\underset{\sim}{q},\underset{\sim}{w})\ddot{\underset{\sim}{q}} + \bar{\underset{\sim}{\Phi}}(\underset{\sim}{q},\dot{\underset{\sim}{q}},\underset{\sim}{\lambda}) + \bar{\underset{\sim}{\Pi}}(\underset{\sim}{q},\underset{\sim}{\lambda},\underset{\sim}{w}) = \bar{\underset{\sim}{F}}(\underset{\sim}{q},\dot{\underset{\sim}{q}},\underset{\sim}{u}) \tag{1}$$

where $\underset{\sim}{q}(t) = [q_1 \cdots q_n]^T \in \Delta_q \subset R^n$ is the configuration vector at time $t \geq t_0 = 0$, $\underset{\sim}{\lambda}(t) \in \Delta \subset R^{\ell}$ is the vector of all adjustable system parameters, $\underset{\sim}{w}(t) \in W \subset R^k$ is the vector of uncertain parameters, $\underset{\sim}{u}(t) \in U \subset R^m$ is the vector of actuator controls at each joint. Here A is the nonsingular inertia matrix, $\bar{\underset{\sim}{\Phi}}$ is a vector comprising all the internal nonconservative forces such as Coriolis and damping both structural in links and Coulomb in joints, $\bar{\underset{\sim}{\Pi}}$ is the vector of potential forces such as gravity and spring forces and $\bar{\underset{\sim}{F}}$ is the vector of input forces, actuator transmission and control.

We shall restrict our discussion to models that may be written in the form

$$\ddot{\underset{\sim}{q}} + \underset{\sim}{\Phi}(\underset{\sim}{q},\dot{\underset{\sim}{q}},\underset{\sim}{\lambda}) + \underset{\sim}{\Pi}(\underset{\sim}{q}) = \underset{\sim}{F}(\underset{\sim}{q},\dot{\underset{\sim}{q}}) + \bar{\underset{\sim}{u}} , \tag{2}$$

assuming no noise $\underset{\sim}{w}$, which have a finite number of isolated equilibria.

The selected nonlinear reference model is also assumed to be given in this form

$$\ddot{\underset{\sim}{q}}_n + \underset{\sim}{\Phi}_m(\underset{\sim}{q}_m, \dot{\underset{\sim}{q}}_m, \underset{\sim}{\lambda}_m) + \underset{\sim}{\Pi}_m(\underset{\sim}{q}_m) = \underset{\sim}{F}_m(\underset{\sim}{q}_m, \dot{\underset{\sim}{q}}_m) + \bar{\underset{\sim}{u}}_m \tag{3}$$

with the appropriate notation. It is required that the equilibria of this reference model coincide with the isolated equilibria of the manipulator Eq. (2). Furthermore, the control vector $\underset{\sim}{u}_m$ is selected to generate a desired state space performance and it is assumed that the reference model is overall Lagrange stable.

Both the manipulator and the reference models may be written in state format in the usual manner.

Let

$$y = [q_1\ \dot{q}_1\ q_2\ \dot{q}_2 \cdots q_n\ \dot{q}_n]^T \in R^{2n}$$

and

$$\underset{\sim}{y}_m = [q_{m1}\ \dot{q}_{m1}\ q_{m2}\ \dot{q}_{m2} \cdots q_{mn}\ \dot{q}_{mn}]^T \in R^{2n} .$$

Then

$$\dot{y}_i = y_{2i}$$

$$\dot{y}_{2i} = -\Phi_i - \Pi_i + F_i + \bar{u}_i , \quad i = 1,\ldots,n , \tag{4}$$

$$\dot{y}_{mi} = y_{m2i}$$

$$\dot{y}_{m2i} = -\Phi_{mi} - \Pi_{mi} + F_{mi} + \bar{u}_{mi} , \quad i = 1,\ldots,n , \tag{5}$$

or in vector format

$$\dot{\underset{\sim}{y}} = \underset{\sim}{f}(\underset{\sim}{y}, \bar{\underset{\sim}{u}}, \underset{\sim}{\lambda}) \tag{6}$$

and

$$\dot{\underset{\sim}{y}}_m = \underset{\sim}{f}_m(\underset{\sim}{y}_m, \bar{\underset{\sim}{u}}_m, \underset{\sim}{\lambda}_m) . \tag{7}$$

Sufficient conditions on the relevant functions in each model are assumed to apply to ensure the existence of solution trajectories at given $\underset{\sim}{y}^0, \underset{\sim}{y}_m^0 \in \Delta$.

III. CONTROL ALGORITHMS FOR MODEL REFERENCE TRACKING

For control input modification, we shall assume that the parameter vectors λ and λ_m are known constant vectors. Our objective is simply to design a control law determining u which will obtain the model reference tracking of the output states of the manipulator and reference models.

The development of control algorithms will be based upon the Liapunov sufficiency results of asymptotic stability. Writing $e = y - y_m$ the state equations defining the evolution of the error vector e are

$$\dot{e}_i = e_{2i}$$

$$\dot{e}_{2i} = -\chi_i - Z_i + \tilde{u}_i , \qquad i = 1,2,\ldots,n , \tag{8}$$

where

$$\chi_i = \Phi_i - \Phi_{mi}$$

$$Z_i = \Pi_i - \Pi_{mi}$$

and

$$\tilde{u}_i = \bar{u}_i - \bar{u}_{mi} , \qquad i = 1,2,\ldots,n .$$

Here χ_i defines the misalignment in nonpotential force terms between manipulator and reference model, Z_i defines the misalignment in conservative force terms and $\tilde{u}_i$ the misalignment in control.

Our construction infers that the origin $e = 0$ is an equilibrium state of Eq. (8) when control $\tilde{u} = 0$.

Let us begin with the positive definite Liapunov function

$$V = e^T e / 2 .$$

Its derivative is determined as

$$\dot{V} = \sum_{i=1}^{n} e_i^T \dot{e}_i$$

$$= \sum_{i=1}^{n} e_{2i}[e_i - \chi_i - Z_i + \tilde{u}_i] .$$

The straightforward selection of the control law

$$\tilde{u}_i = -e_i + \chi_i + Z_i - e_{2i} , \quad i = 1,\ldots,n , \tag{9}$$

makes $\dot{V}$ negative semi-definite:

$$\dot{V} = -\sum_{k=1}^{n} e_{2k}^2 .$$

Substituting the control Eq. (9) into Eq. (8), we find $\dot{V}(t) = 0$ only for $\underset{\sim}{e}(t) = \underset{\sim}{0}$ because of our construction, so by La Salle's Theorem, the origin is in fact asymptotically stable.

To improve the system response for faster convergence, the following control law

$$\tilde{u}_i = \begin{cases} -e_i + \chi_i + Z_i - \beta_{2i}e_{2i} - \beta_i e_i^2/e_{2i} , & |e_{2i}| > \delta_i , \\ -e_i + \chi_i + Z_i - \beta_{2i}e_{2i} , & |e_{2i}| \le \delta_i , \end{cases} \tag{10}$$

where δ_i, $i = 1,\ldots,n$ and β_j, $j = 1,\ldots,2n$ are selected positive constants, yields

$$\dot{V} = -\sum_{k=1}^{2n} \beta_k e_k^2 ,$$

in the region defined by $|e_{2i}| > \delta_i$, $i = 1,\ldots,n$. By increasing the β_k, the speed of convergence of each component e_j towards zero through this region may be improved.

It has been assumed that there is sufficient control power available to implement the nonlinear control laws Eqs. (9) and (10) which remove the nonlinearity in order to obtain convergence. In general we need to take into account that in a given physical problem control power is bounded. However, these control laws will be not as 'strong' as those required to obtain control to follow a linear reference model.

Let us consider now an alternative approach to develop a control law to obtain asymptotic stability using the following construction. We introduce parameters k_i, $i = 1,\ldots,2n$, and write Eqs. (8) in the following form:

$$\dot{e}_i = e_{2i}$$
$$\dot{e}_{2i} = k_i e_i + k_{2i} e_{2i} - \chi_i - Z_i + \tilde{u}_i - k_i e_i - k_{2i} e_{2i}, \quad i = 1,\ldots,n,$$

or in matrix form:

$$\dot{\underset{\sim}{e}} = A\underset{\sim}{e} - S(\underset{\sim}{\chi} + \underset{\sim}{Z} - \underset{\sim}{\tilde{u}}) + R\underset{\sim}{e} \tag{11}$$

where

$$A = \begin{bmatrix} 0 & 1 & & & & \\ k_1 & k_2 & & & 0 & \\ & & \cdot & & & \\ & & & \cdot & & \\ & & & & \cdot & \\ & 0 & & & 0 & 1 \\ & & & & k_{2n-1} & k_{2n} \end{bmatrix}_{2n\times 2n}$$

$$S = \begin{bmatrix} 0 & 0 & 0 & 0 & \cdots & 0 \\ 1 & 0 & 0 & 0 & \cdots & 0 \\ 0 & 0 & 0 & 0 & \cdots & 0 \\ 0 & 0 & 1 & 0 & \cdots & 0 \\ & & & \cdot & & \\ & & & \cdot & & \\ & & & \cdot & & \\ 0 & 0 & 0 & 0 & \cdots & 1 \end{bmatrix}_{2n\times n}$$

$$R = \begin{bmatrix} 0 & 0 & \cdots & 0 & 0 \\ -k_1 & -k_2 & \cdots & 0 & 0 \\ & \cdot & & & \\ & \cdot & & & \\ & \cdot & & & \\ 0 & 0 & \cdots & -k_{2n-1} & -k_{2n} \end{bmatrix}_{2n\times 2n}$$

and

$$\underset{\sim}{\chi} = [\chi_1 \ \chi_2 \ \cdots \ \chi_n]^T$$
$$\underset{\sim}{Z} = [Z_1 \ Z_2 \ \cdots \ Z_n]^T$$

$$\tilde{u} = [\tilde{u}_1 \ \tilde{u}_2 \ \cdots \ \tilde{u}_n]^T$$

$$e = [e_1 \ e_2 \ \cdots \ e_n]^T .$$

It is clear that the k_i can be selected to make A a stability matrix. A positive definite symmetric matrix P then exists to the Liapunov equation

$$A^T P + PA = -Q \tag{12}$$

given a positive definite matrix Q of our choice. Consider now the positive definite quadratic form

$$V = e^T P e$$

as Liapunov function for system (11). We find

$$\dot{V} = \dot{e}^T P e + e^T P \dot{e}$$

$$= -e^T Q e + 2 e^T P[Re - S(\chi + Z - \tilde{u})] .$$

Assuming sufficient control, the selection of the control law to satisfy

$$S(\chi + Z - \tilde{u}) = Re$$

will make $\dot{V}$ negative definite. This selection yields

$$\tilde{u}_i = \chi_i + Z_i + k_i e_i + k_{2i} e_{2i} , \qquad i = 1,\ldots,n , \tag{13}$$

which may be compared with that given by Eqs. (9) and (10). The difference here is that $\tilde{u}_i$ defined by Eq. (13) ensures for this selection of V that its derivative is negative definite. Now the system response is enhanced by the appropriate selection of the parameters k_i.

We recognize k_i and k_{2i} as proportional and derivative gains. Our development of control law Eq. (10) for the model reference tracking of the nonlinear model may be compared with the linearising and decoupling control laws discussed in Chapter 8 of Craig [7]. The difference between the two theories is that the analysis given here is based upon the use of Liapunov functions to obtain the nonlinear control laws.

IV. ADAPTIVE CONTROL FOR MODEL REFERENCE TRACKING

For the purpose of developing an adaptive control algorithm, let us assume that some or all of the parameters $\underset{\sim}{\lambda}$ of the plant are directly adjustable. The scheme for input modification and direct adaption is shown in Figure 1.

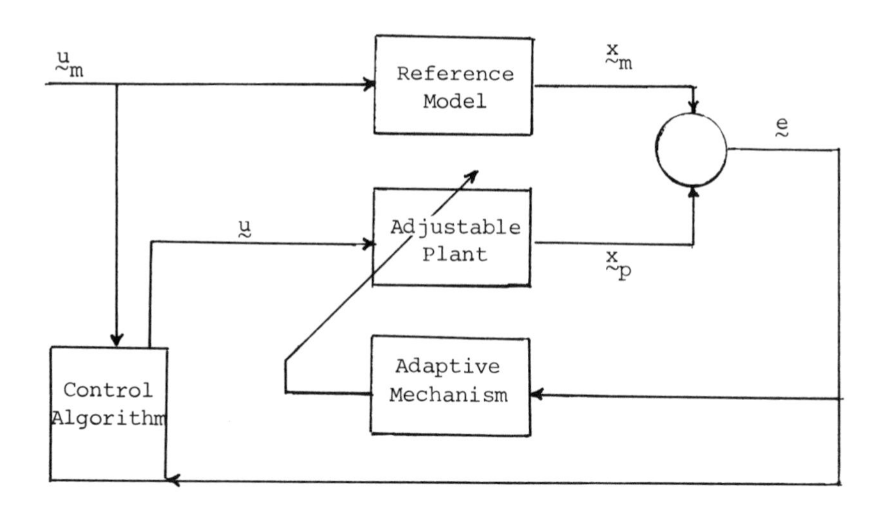

FIGURE 1: Adaptive Control

Following the approach taken in the classical Linear Model Reference Theory, we form the Liapunov function

$$V(\underset{\sim}{e},\underset{\sim}{\chi}) = \underset{\sim}{e}^T P \underset{\sim}{e} + \underset{\sim}{\chi}^T \underset{\sim}{\chi} ,$$

with error equation (11), and P defined by the Liapunov equation (12). We find

$$\dot{V} = -\underset{\sim}{e}^T Q \underset{\sim}{e} + 2\underset{\sim}{e}^T P[R\underset{\sim}{e} - S(\underset{\sim}{z} - \tilde{u})] + 2[\dot{\underset{\sim}{\chi}}^T - \underset{\sim}{e}^T P S]\underset{\sim}{\chi} .$$

The adaptive law is defined by

$$\dot{\underset{\sim}{\chi}}^T = \underset{\sim}{e}^T P S ,$$

or

$$\dot{\underset{\sim}{\chi}} = S^T P \underset{\sim}{e} . \tag{14}$$

The adaptation mechanism given by Eq. (14) determines the change in $\underset{\sim}{\chi}$ and consequently the variable λ_i, in terms of the error and the matrix P whose coefficients are determined by our selection of Q and the matrix A through the selection of k_i.

The control law for input modification is defined by

$$S(\underset{\sim}{Z} - \underset{\sim}{\tilde{u}}) = R\underset{\sim}{e} . \tag{15}$$

This reduces to the formal equations

$$\tilde{u}_i = Z_i + k_i e_i + k_{2i} e_{2i} , \qquad i = 1,\ldots,n .$$

With the implementation of these algorithms,

$$\dot{V} = -\underset{\sim}{e}^T Q \underset{\sim}{e}$$

is negative definite. For bounded input control in the reference model, we conclude that

$$\lim_{t \to \infty} \dot{V}(\underset{\sim}{e},\underset{\sim}{\chi}) = 0 , \quad \text{and so} \quad \lim_{t \to \infty} \underset{\sim}{e} = \underset{\sim}{0} .$$

Hence the adaptive scheme is stable. The analysis we have given here introducing control gains k_i may be compared to the nonlinear based torque control discussed by Craig in Chapter 5 of [2]. Each pair k_i and k_{2i} may be chosen to place the closed loop poles associated with each joint.

In our analysis above, we have assumed all λ_i are variable. If some are not available, these may be mathematically separated out and included in $\underset{\sim}{Z}$ without affecting the analysis.

A numerical simulation of an RP - manipulator with two degrees of freedom is given in the following example.

V. EXAMPLE

We consider the RP - manipulator shown in Figure 2 below. It is assumed that the mass of each arm m_1 and m_2 are situated at the prismatic joint and gripper as shown, and that the arms are light inextensible rods. The prismatic joint allows only for sliding motion of the second arm relative to the first - there is no rotation of the second arm relative to the first. Motion is assumed in the vertical plane and the external gravity force is assumed constant.

Two springs are included into the physical setup to stabilise the system. A nonlinear rotational spring is at the base and a nonlinear elastic spring at the prismatic joint.

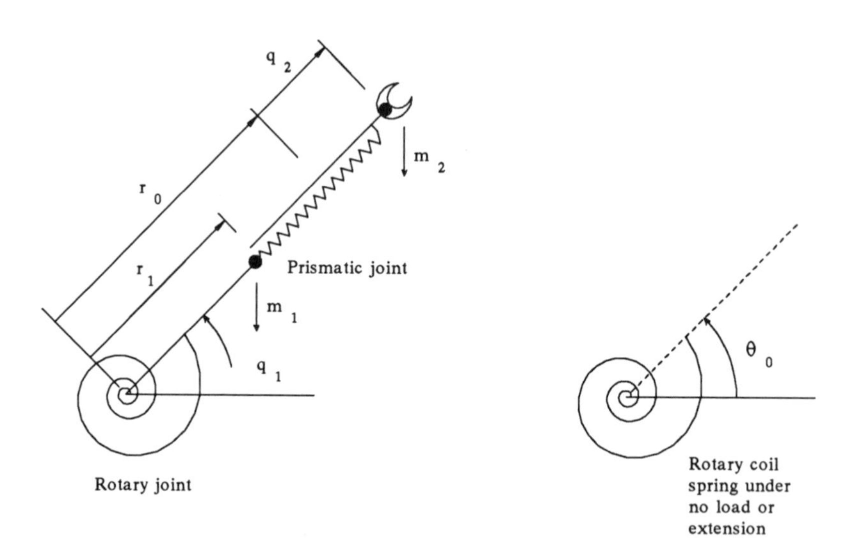

FIGURE 2: RP - manipulator

In developing the dynamic model, we choose generalised coordinates q_1 defining the angular displacement of the rotational coil spring, and q_2 the linear extension in the elastic spring between the prismatic joint and gripper.

Using the Lagrangian approach, we obtain the dynamics as per Eq. (4) where

$$\Phi_1 = [2m_2(r_0 + y_3)y_4 + \lambda_1 |y_2|]y_2 / [m_1 r_1^2 + m_2(r_0 + y_3)^2]$$

$$\Pi_1 = [g(m_1 r_1 + m_2(r_0 + Y_3)) \cos(\theta_0 + y_1) + a_1 y_1 + b_1 y_1^2 + c_1 y_1^3] / [m_1 r_1^2 + m_2(r_0 + y_3)^2]$$

$$\Phi_2 = -(r_0 + y_3)y_2^2 + \lambda_2 y_4 / m_2$$

$$\Pi_2 = g \sin(\theta_0 + y_1) + [a_2 y_3 + b_2 y_3^3] / m_2$$

$$\bar{u}_1 = u_1 / [m_1 r_1^2 + m_2(r_0 + y_3)^2]$$

$$\bar{u}_2 = u_2 / m_2$$

$$F_i = 0 , \quad i = 1,2 .$$

Here:

- a_1, b_1 and c_1 are the constants defining the nonlinear characteristics of the spring at the base;
- a_2 and b_2 are the constants defining the nonlinear characteristics of the spring at the prismatic joint;
- $\lambda_1 |y_2| y_2 = \lambda_1 |\dot{q}_1| \dot{q}_1$ defines the damping force in the rotary joint (λ_1 a positive constant);
- $\lambda_2 y_4 = \lambda_4 \dot{q}_2$ defines the damping force in the prismatic joint (λ_2 a positive constant);
- θ_0 is the natural angular displacement of rotary coil spring as shown in Figure 2.

External disturbance forces F_i due to the environment are assumed to be zero.

To establish the reference model, we take the Φ_i and Π_i terms of Eq. (5) to be

$$\Phi_{m1} = \lambda_{m1} |y_{m2}| y_{m2} / [m_1 r_1^2 + m_2(r_0 + y_{m3})^2]$$

$$\Pi_{m1} = \frac{g[m_1 r_1 + m_2 (r_0 + y_{m3})] \cos(\theta_0 + y_{m1}) + a_1 y_{m1} + b_1 y_{m1}^2 + c_1 y_{m1}^3}{[m_1 r_1^2 + m_2 (r_0 + y_{m3})^2]}$$

$$\Phi_{m2} = \lambda_{m2} y_{m4} / m_2$$

$$\Pi_{m2} = g \sin(\theta_0 + y_{m1}) + [a_2 y_{m3} + b_2 y_{m3}^3] / m_2$$

$$\bar{u}_{m1} = u_{m1} / [m_1 r_1^2 + m_2 (r_0 + y_{m3})^2]$$

$$\bar{u}_{m2} = u_{m2} / m_2$$

$$F_{mi} = 0 , \quad i = 1,2 .$$

In this nonlinear reference model, it is assumed that the masses m_1, m_2, and the characteristic constants $a_1, b_1, c_1, a_2, b_2, r_0, r_1$ and θ_0 are the same, Φ_{m1}, Φ_{m2} have been defined from Φ_1 and Φ_2 by the deletion of a term. With this construction, it is easily verified that an equilibrium state of the reference model $\underset{\sim}{y}^*_m$ will be identical to the manipulator model $\underset{\sim}{y}^*$:

$$\underset{\sim}{y}^* = [y_1^* \; 0 \; y_3^* \; 0]^T$$

$$y_m^* = [y_{m1}^* \; 0 \; y_{m3}^* \; 0]^T$$

where the pairs y_1^*, y_3^* and y_{m1}^*, y_{m3}^* are solutions of

$$g[m_1 r_1 + m_2 (r_0 + z_3)] \cos(\theta_0 + z_1) + a_1 z_1 + b_1 z_1^2 + c_1 z_1^3 = 0$$

and

$$m_2 g \sin(\theta_0 + z_1) + a_2 z_3 + b_2 z_3^3 = 0 .$$

The constants selected for the numerical simulation are

$g = 9.81$ m/sec^2

$m_1 = 10.0$ kg	$m_2 = 5.0$ kg
$\lambda_1 = 6.0$ kg/m^2	$\lambda_2 = 3.0$ kg/sec
$a_1 = 500.0$ kg m^2/sec^2	$b_1 = 3.0$ kg m^2/sec^2
$c_1 = 200.0$ kg m^2/sec^2	$a_2 = 200.0$ kg/sec^2
$b_2 = 210.0$ kg/m^2 sec^2	
$\lambda_{m1} = 5.0$ kg m^2	$\lambda_{m2} = 2.0$ kg/sec

$r_0 = 1.5$ m $\qquad\qquad r_1 = 0.66$ m .

Given these values, there is only one equilibrium state $\underset{\sim}{y}_m^*$:

$$y_{m_1}^* = y_1^* = q_{m_1} \approx -0.1247 \text{ radians}$$

$$y_{m_2}^* = y_2^* = \dot{q}_{m_1} = 0.0 \text{ rad/sec}$$

$$y_{m_3}^* = y_3^* = q_{m_2} \approx -0.2049 \text{ m}$$

$$y_{m_4}^* = y_4^* = \dot{q}_{m_2} = 0.0 \text{ rad/sec} .$$

The equilibrium value $y_1^* = -0.1247$ radians shows the rotary coil is under compression and the elevation of the arm relative to the horizontal at equilibrium is 1.0603 radians or 64.4°.

The length of the gripper from the base rotary joint is $r_0 + y_{m_3}^* = 1.2591$ metres.

For the reference model, this equilibrium state is stable. Given $E_m(y_m, \dot{y}_m)$ represents the positive definite, total energy of the system, the instantaneous change of total energy, given no input control, can be shown to be

$$\dot{E}_m(\underset{\sim}{y}_m, \underset{\sim}{\dot{y}}_m) = -\lambda_{m_1}|y_{m_2}|y_{m_2}^2 - \lambda_{m_2} y_{m_4}^2 ,$$

which is negative semi-definite, for positive constants λ_{m_1} and λ_{m_2}. Since given $y_{m_2}(t) = y_{m_4}(t) = 0$ for all $t \geq 0$ implies that $\dot{E} = 0$ only for those states associated with the simple equilibrium, we conclude that the equilibrium state $\underset{\sim}{y}_m^*$ is in fact asymptotically stable.

The asymptotic nature of this equilibrium state is shown in Figure 3 which shows the reference output behaviour with no input control $\underset{\sim}{u}_m = \underset{\sim}{0}$. The numerical integration of the state equations (7) was performed using a Runge Kutta variable step algorithm.

Assuming sinusoidal reference input control $u_{m_1} = 6 \sin t$, $u_{m_2} = 6 \cos t$, a computer simulation using control law Eq. (10) with

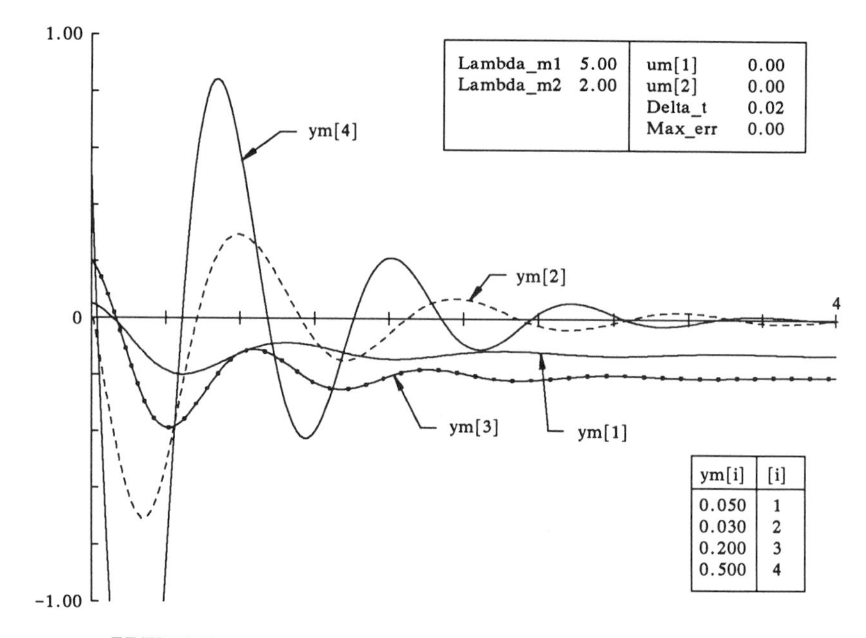

FIGURE 3: Asymptotic behaviour of reference model

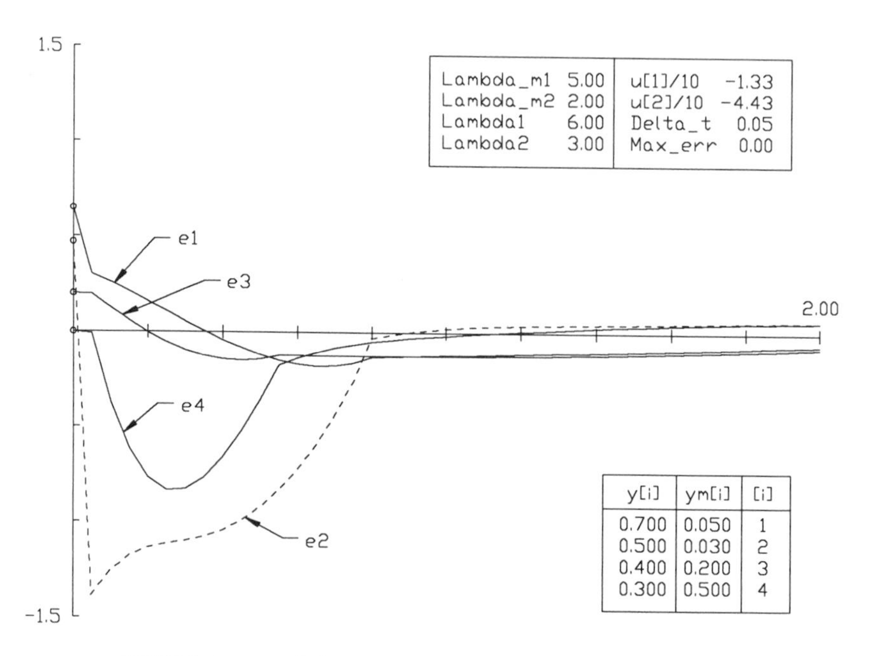

FIGURE 4: Model reference control of manipulator using control law (10)

$\beta_1 = 40$, $\beta_2 = 1$, $\beta_3 = 40$, $\beta_4 = 1$,

$\delta_1 = 0.2$, $\delta_2 = 0.2$, step size = 0.05 sec ,

over an interval of 10 seconds resulted in good convergence as shown in Figure 4 above. Initial conditions for $\underset{\sim}{y}$ and $\underset{\sim}{y}_m$ are tabulated in each diagram.

To discuss adaptive control, assume both λ_1 and λ_2 are variable. Following the construction in the previous analysis, select

$k_1 = -1$, $k_2 = -1$, $k_3 = -1$, $k_4 = -2$,

so that

$$A = \begin{bmatrix} 0 & 1 & 0 & 0 \\ -1 & -1 & 0 & 0 \\ 0 & 0 & 0 & 0 \\ 0 & 0 & -1 & -2 \end{bmatrix} .$$

Matrix A has eigenvalues $(-1 \pm i\sqrt{3})/2$ and -1 (repeated). Taking the positive definite matrix,

$$Q = \begin{bmatrix} 4 & 0 & 0 & 0 \\ 0 & 2 & 0 & 0 \\ 0 & 0 & 2 & 0 \\ 0 & 0 & 0 & 6 \end{bmatrix} ,$$

the solution P of the matrix equation is determined as

$$P = \begin{bmatrix} 5 & 2 & 0 & 0 \\ 2 & 3 & 0 & 0 \\ 0 & 0 & 4 & 1 \\ 0 & 0 & 1 & 2 \end{bmatrix} .$$

The adaptive control algorithm Eq. (14) in this case translates to the equations

$$\dot{\chi}_1 = -2e_1 - 3e_2$$

$$\dot{\chi}_2 = -e_3 - 2e_4 \ , \tag{16}$$

and the control algorithm Eq. (15)

$$u_1 = -e_1 - e_2 + Z_1$$

$$u_2 = -e_3 - 2e_4 + Z_2 \ . \tag{17}$$

To implement the equations (16), we use the forward Euler approximation

$$\chi_1(t+\delta t) = \chi_1(t) - (2e_1(t) + 3e_2(t))\delta t$$

$$\chi_2(t+\delta t) = \chi_2(t) - (e_3(t) + 2e_4(t))\delta t \ , \tag{18}$$

and require a specification of an initial estimation of $\underset{\sim}{\chi}(0)$, or $\phi_i(0)$, $i = 1$ and 2 .

Assuming again the same sinusoidal reference input control $u_{m1} = 6\sin t$ and $u_{m2} = 6\cos t$, the results of a computer simulation using Eqs. (17) and (18) with $\phi_1(0) = 0.5$ and $\phi_2(0) = 1.0$, and $\delta(t) = 0.05$ over an interval of 8 seconds is shown in Figure 5.

The unknown λ_1 and λ_2 calculated by formula from χ_1 and χ_2 were found to be 278.8 and 1.92 at the end of the iteration. (Result on the graph for λ_1 is actually λ_1 divided by 10.)

It is observed from the graph that under the above conditions and parameter selections, the misalignment terms χ_i tended to zero. Indeed this was found to be true for numerical simulations with different initial states and initial values of $\phi_i(0)$ yielding initial estimates of λ_i . In this paper we have not attempted a general theoretical examination of parameter error convergence.

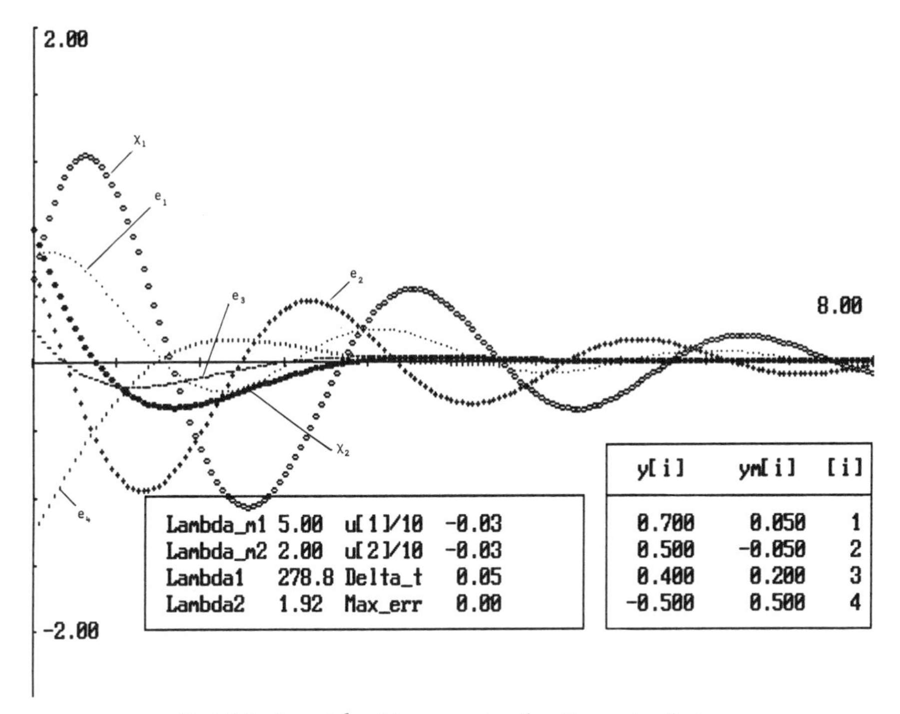

FIGURE 5: Adaptive control of manipulator using equations (17) and (18)

In this paper we have examined model reference control techniques using Liapunov techniques to control, in particular, a nonlinear n DOF manipulator to follow a prescribed or selected nonlinear model that has desired characteristics. It is assumed that the selected nonlinear model has equilibria compatible with those of the manipulator.

The control law development for asymptotic stability of the error and for adaptive control using Liapunov theory, takes into account the specific structural form of the manipulator dynamics. The performance of the system error convergence can be determined by selection of the given parameters and the coefficients of Q in the matrix Liapunov equations.

The inclusion of bounded disturbances into the mathematical models and robustness has not been considered.

References

1. J.M. Skowronski, "Control Dynamics of Robot Manipulators," Academic Press, 1986.

2. J.J. Craig, "Adaptive Control of Mechanical Manipulators," Addison-Wesley, 1988.

3. J.M. Skowronski, "Nonlinear Model Tracking by Robot Manipulators," *in* Proceedings of ASME, Boston, Massachusetts, December, 1987.

4. J.M. Skowronski, "Algorithm for Adaptive Control of Two Arm Flexible Manipulators under Uncertainty," *IEEE Trans. Aerospace and Electronic Systems 24*, No. 5, 1988.

5. J.M. Skowronski, "Adaptive Control of Robotic Manipulators under Uncertain Payload," *in* "Advances in Robotics," M.H. Namza (ed.), Acta Press, Anaheim, 1985, pp. 40-44.

6. J.M. Skowronski, "Model Reference Adaptive Control under Uncertainty of Nonlinear Flexible Manipulators," *in* Proc. AIAA Guidance, Navigation and Control Conference, Williamsburg, VA, 1986.

7. J.J. Craig, "Introduction to Robotics, Mechanics and Control," Addison-Wesley, 1986.

CONTROL AND DYNAMIC SYSTEMS, VOL. 35

Optimal Preview Controllers Based upon Explicit Trajectory Models

H. Ali Pak and **Rowmau Shieh**
Department of Mechanical Engineering
University of Southern California
Los Angeles, CA 90089-1453

I Introduction

Preview controllers are a class of servo controllers in which the future information about the reference trajectory is used in generating the instantaneous control effort. The preview portion of the control law is a feedforward action which is derived from the future states of the desired trajectory. This feedforward action when augmented with some form of feedback control (usually state feedback) constitutes the total control effort. The realizability of preview control is contingent upon the availability of the future reference trajectory information at any instance in time. In applications where this information is available the use of preview control has been shown to drastically improve the tracking performance relative to the use of state feedback alone.

Examples of preview control applications include a discrete time disturbance rejection controller for a process plant [1]. Also several input tracking controllers have been designed using preview control. These include computer controlled trajectory tracking in robotic and machine tool systems [2-4].

In the above examples the dynamics and the gains of the controllers were determined by formulating, and subsequently solving, a specific form of linear quadratic optimal control problem. This formulation is known as the *finite length preview control problem* [5]. In finite length preview control a critical prerequisite is the prior selection of the length of time for which the desired trajectory is to be previewed. For discrete time control the above prerequisite translates to the need for the preselection of the number of preview points for any given sampling period. Since no exact relationship exists for the determination of the finite preview length, various intuitive selection rules have been proposed. For example, as a rough rule of thumb, a preview length equal to three times the longest time constant of a closed loop plant has been suggested [6]. Another intuitive suggestion states that a minimum preview length should, at least, equal the longest open loop plant time constant [3].

This paper focuses on the application of optimal preview control for a class of motion control problems in which the reference trajectories may be explicitly generated as outputs of linear time invariant sampled-data systems. (eg. polynomials, sinusoids, cycloids etc.). For discrete time control it will be shown that the explicit use of the desired trajectory model implicitly determines the exact preview length. In the next section the general *explicit trajectory model based optimal preview* controller will be presented for linear time invariant systems. This will be followed by an example of its application to a positioning servo. In the subsequent sections the tracking performance of the controller is assessed in comparison with the finite length preview controller and the more common inverse dynamic feedforward controller designed using frequency domain techniques.

II Controller Design Method

Consider a time invariant linear SISO discretized plant described by equations:

$$\mathbf{x}(k+1) = A_p\mathbf{x}(k) + B_p u(k) \tag{1}$$

$$y(k) = C_p\mathbf{x}(k) \tag{2}$$

The dimensions of A_p, B_p, and C_p are $(n \times n), (n \times 1)$, and $(1 \times n)$ respectively, and the pair (A_p, B_p) is controllable and the pair (A_p, C_p) is observable. Let y_r be the reference trajectory which the output y is required to follow. We assume that y_r can be explicitly generated using a linear time invariant model of the form

$$\tilde{\mathbf{x}}_r(k+1) = \tilde{A}_r\tilde{\mathbf{x}}_r(k)$$

$$y_r(k) = \tilde{C}_r \tilde{\mathbf{x}}_r(k)$$

where the dimensions of $\tilde{A}_r$ and $\tilde{C}_r$ are $(m \times m)$ and $(1 \times m)$ respectively. By definition the generator must be observable, hence a transformation can always be found to convert $(\tilde{A}_r, \tilde{C}_r)$ to the following observability canonical form

$$\mathbf{x}_r(k+1) = A_r \mathbf{x}_r(k) \tag{3}$$

$$y_r(k) = C_r \mathbf{x}_r(k) \tag{4}$$

$$A_r = \begin{bmatrix} 0 & 1 & 0 & \cdots & 0 \\ 0 & 0 & 1 & \cdots & 0 \\ & & & \ddots & \vdots \\ & & & & 1 \\ \alpha_1 & \alpha_2 & \alpha_3 & \cdots & \alpha_m \end{bmatrix}, \; C_r = [\, 1 \;\; 0 \;\; 0 \;\; \ldots \;\; 0 \,].$$

Note that in this form each state appears as one-step ahead of its previous state, i.e. $x_{r_m}(k) = x_{r_{m-1}}(k+1), \ldots, x_{r_2}(k) = x_{r_1}(k+1)$. Also the first state is the desired trajectory at the kth sampling instant, i.e. $y_r(k) = x_{r_1}(k)$. Furthermore, using the above explicit trajectory model, the future reference points may be automatically generated recursively as long as the generator is run $m-1$ samples ahead of the controller. Combining Eqs. (1) to (4) the following augmented model is formed:

$$\mathbf{w}(k+1) = A\mathbf{w}(k) + Bu(k) \tag{5}$$

$$e(k) = C\mathbf{w}(k) \tag{6}$$

where

$$\mathbf{w}(k) = \begin{bmatrix} \mathbf{x}(k) \\ \mathbf{x}_r(k) \end{bmatrix}, \; A = \begin{bmatrix} A_p & 0 \\ 0 & A_r \end{bmatrix}, \; B = \begin{bmatrix} B_p \\ 0 \end{bmatrix}, \; C^T = \begin{bmatrix} C_p \\ -C_r \end{bmatrix}$$

The quadratic cost function to be minimized is defined by

$$J_k = \sum_{j=k}^{\infty} \left[e^2(j) + Ru^2(j) \right] \tag{7}$$

where $e(j) = y(j) - y_r(j)$ is the tracking error. The augmented plant (A, B, C) and the above cost function constitute an optimal linear quadratic regulator (LQR) problem. Using the standard LQR technique, the optimal control law $u^o(k)$ is given by

$$u^o(k) = -G\mathbf{w}(k) \tag{8}$$

where $G = (R + B^T KB)^{-1} B^T KA$ is the optimal gain vector. And the Riccati gain matrix K can be determined by solving the algebraic Riccati equation (ARE),

$$K = A^T KA + C^T C - A^T KB \left[R + B^T KB\right]^{-1} B^T KA. \tag{9}$$

To clarify the role of the feedback and the preview gains we may partition the Riccati gain matrix as

$$K = \begin{bmatrix} K_1 & K_2 \\ K_2^T & K_3 \end{bmatrix} \tag{10}$$

where the partitioned matrices K_1, K_2 and K_3 are $(n \times n), (m \times n)$ and $(m \times m)$ respectively. Now, using Eqs. (8), (9) and (10) the gain vector G is given by

$$G = \begin{bmatrix} G_{fb} & G_{pr} \end{bmatrix} \tag{11}$$

where the feedback gain vector $G_{fb} = MB_p^T K_1 A_p$ and the preview gain vector $G_{pr} = MB_p^T K_2 A_r$ with $M = (R + B_p^T K_1 B_p)^{-1}$. Thus the relevant Riccati gain matrices are

$$K_1 = C_p^T C_p + A_p^T K_1 A_p - A_p^T K_1 B_p M B_p^T K_1 A_p \tag{12}$$

and

$$K_2 = -C_p^T C_r + A_c K_1 A_r \tag{13}$$

where $A_c = A_p - B_p M B_p^T K_1 A_p$.

From Eq. (12) it is apparent that the elements of the feedback gain vector G_{fb} are independent of the reference trajectory model and may be evaluated using the standard optimal LQR method. In contrast, from Eq. (13), the elements of the preview gain vector G_{pr} depend both on the optimal closed loop dynamics and the reference trajectory model. Furthermore, using the observability canonical model of the reference trajectory has ensured that the feedforward states are the preview points along the reference trajectory. Hence, the optimal control input of Eq. (8) may be rewritten as

$$u^o(k) = -\sum_{i=1}^{n} g_{fb}(i)x_i(k) + \sum_{l=0}^{m-1} g_{pr}(l)y_r(k+l) \tag{14}$$

where $g_{fb}(i)$ and $g_{pr}(l)$ are the elements of the optimal feedback gains and the preview gains of Eq. (11). The block diagram in Fig. 1 shows the structure of the controller of Eq. (14) acting on the discretized plant described by Eqs. (1) and (2).

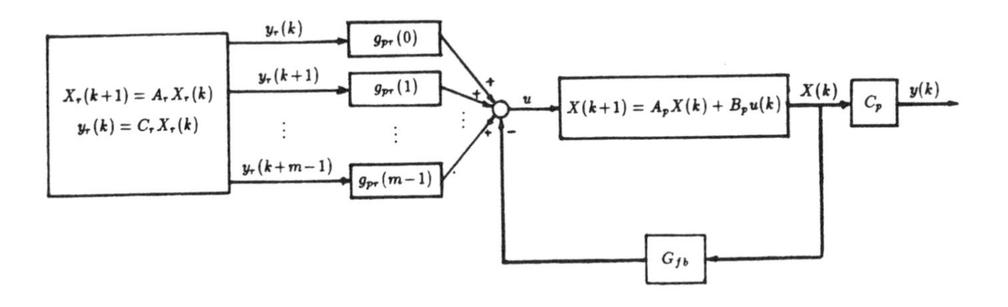

Fig. 1 General Structure of Optimal Preview Control System based on Explicit Trajectory Model

In connection with the optimal controller of Eq. (14) three points merit further discussion. First, the trajectory generator dynamics A_r is only used for the off-line determination of the optimal feedforward gains. For implementation purposes, it is not required to have the generator run in real-time. All that is needed is availability of $m-1$ future reference points along the trajectory at any sampling instant. Second, the tracking controller does not require an on-line implementation of the *adjoint* of the closed loop plant which is required in the standard optimal tracking controller [7]. Third, since the number of preview points is implicitly determined by the order of the trajectory model, the need to intuitively preselect the preview length is also avoided. However, to maintain optimality, the reference trajectory must be generated $m-1$ samples past its final point. Furthermore, each time a reference trajectory's state matrix, A_r, is changed new preview gains are required. In brief, this formulation of the tracking controller is particularly suitable for applications where the trajectory generator models are prespecified. Two examples of such applications are CNC machine tool contouring and cam profile replication by servo systems.

III Application in Motion Control

Consider a positioning system actuated by a direct current (dc) servo motor with the following open loop transfer function

$$\frac{\theta(s)}{u(s)} = \frac{K_t}{s(\tau_m s + 1)(\tau_e s + 1)} \tag{15}$$

Here K_t is the open loop gain, and τ_m and τ_e are the mechanical and the electrical time constants respectively. For the purpose of computer control, the plant model is preceded by a zero order hold and a sampler. This gives a discrete time state space representation of the form

$$\mathbf{x}(k+1) = A_p\mathbf{x}(k) + B_p u(k) \tag{16}$$

$$\theta(k) = C_p\mathbf{x}(k) \tag{17}$$

where

$$\mathbf{x}(k) = \begin{bmatrix} \theta(k) \\ \omega(k) \\ \alpha(k) \end{bmatrix}, \quad C_p^T = \begin{bmatrix} 1 \\ 0 \\ 0 \end{bmatrix},$$

$$A_p = \frac{1}{\tau_m - \tau_e}\begin{bmatrix} \tau_m - \tau_e & \tau_m^2(1-E_m) - \tau_e^2(1-E_e) & \tau_m^2\tau_e(1-E_m) - \tau_e^2\tau_m(1-E_e) \\ 0 & \tau_m E_m - \tau_e E_e & \tau_e\tau_m(E_m - E_e) \\ 0 & E_e - E_m & \tau_m E_e - \tau_e E_m \end{bmatrix}$$

$$B_p = \frac{K_t}{\tau_m - \tau_e}\begin{bmatrix} T_s(\tau_m - \tau_e) + \tau_e^2(1-E_e) - \tau_m^2(1-E_m) \\ \tau_m(1-E_m) - \tau_e(1-E_e) \\ E_m - E_e \end{bmatrix}$$

Here $E_m = exp(-T_s/\tau_m), E_e = exp(-T_s/\tau_e)$, and T_s is the sampling period. The state variables $\theta(k)$, $\omega(k)$, and $\alpha(k)$ are the angular position, velocity, and acceleration respectively. Assume that the positioning system is required to replicate a predefined motion trajectory generated by a model of the form given by Eqs. (3)

and (4). Equations (11) through (14) are used to give the following optimal preview tracking controller

$$u^o(k) = -(g_\theta \theta(k) + g_\omega \omega(k) + g_\alpha \alpha(k)) + \sum_{l=0}^{m-1} g_{pr}(l)\theta_r(k+l) \tag{18}$$

The control structure of Eq. (18) is shown in the block diagram of Fig. 2.

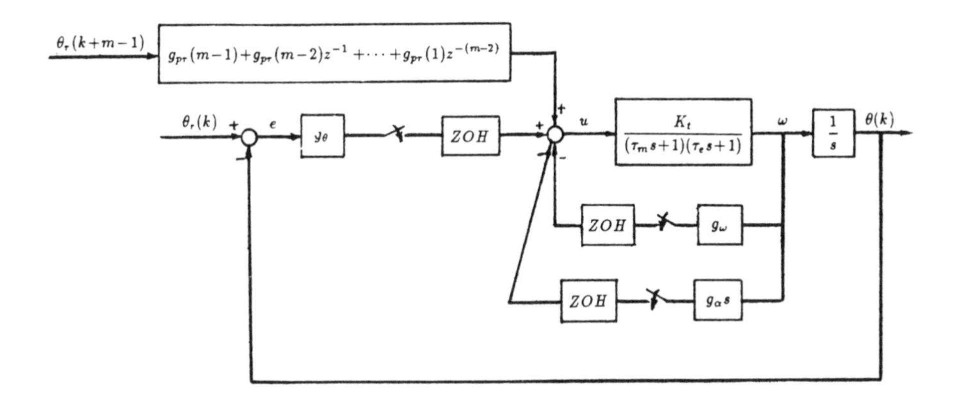

Fig. 2 Structure of Optimal Preview Control Design in Motion Control System

In practice, the controller of Eq. (18) is implemented by restructuring such that

$$u^o(k) = g_\theta(\theta_r(k) - \theta(k)) - g_\omega \omega(k) - g_\alpha \alpha(k) + \sum_{l=1}^{m-1} g_{pr}(l)(\theta_r(k+l) - \theta_r(k)) \tag{19}$$

Note that for Eq. (19) to correspond to Eq. (18), the following condition must hold

$$g_{pr}(0) = g_\theta - \sum_{l=1}^{m-1} g_{pr}(l) \tag{20}$$

This condition implies that the future positional information are only effective as differences with respect to the current instantaneous reference point θ_r. It can be numerically shown that the use of the matrix Riccati Eqs. (12) and (13) will always result in a set of numerical values for the feedback and the preview gains which satisfy Eq. (20).

In practice, an encoder and a tachogenerator may be used as cost effective position and velocity sensors. However, direct acceleration feedback is often economically unjustified. An estimate of the angular acceleration may be obtained through a reduced order observer. Figure 3 shows the block diagram of the optimal controller with the implementation where only position and velocity sensors are used. The acceleration signal is estimated using a discrete time reduced order observer which is derived in the appendix. The modified control law including the observer is given by

$$u^o(k) = u_r(k) + \sum_{l=1}^{m-1} g_{pr}(l)(\theta_r(k+l) - \theta_r(k)) \tag{21}$$

where u_r is the regulator action using only the current and the previous position and velocity signals (see Eq. (A.5)).

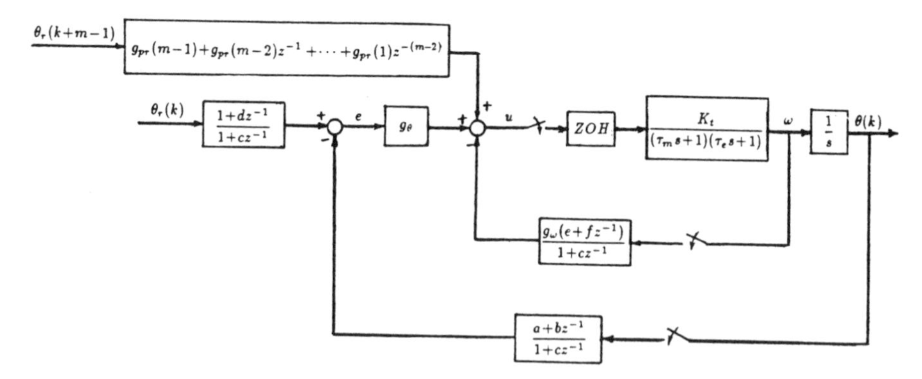

Fig. 3 Implementation Block Diagram of Motion Control System

To illustrate the tracking performance of the controller two computer generated trajectories were considered: a sinusoidal profile $\theta_r(k) = 1000\sin(2\pi k\frac{T_s}{T_c})$ and a cycloidal profile $\theta_r(k) = 1000(\frac{kT_s}{T_c} - \frac{1}{2\pi}(\sin(2\pi k\frac{T_s}{T_c}))$, where T_c is the cycle time. The trajectories' state space representation in the required observability canonical form are:

$$A_{r(sinusoid)} = \begin{bmatrix} 0 & 1 \\ -1 & 2\rho \end{bmatrix} \tag{22}$$

$$A_{r(cycloid)} = \begin{bmatrix} 0 & 1 & 0 & 0 \\ 0 & 0 & 1 & 0 \\ 0 & 0 & 0 & 1 \\ -1 & 2(1+\rho) & -2(1+2\rho) & 2(1+\rho) \end{bmatrix} \tag{23}$$

where $\rho = \cos(2\pi\frac{T_s}{T_c})$. The numerical values for the actuator parameters were chosen to represent a high performance servo actuator characteristic used in high speed cam profile replication with $K_t = 40s^{-1}, \tau_m = 4ms$ and $\tau_e = 1ms$. The value of control effort penalty R in the quadratic cost function of Eq. (7) was chosen to be .005. This gives the first break frequency at approximately 50 Hz for the regulated system without any feedforward action (see the frequency response plots in Fig. 8).

Figure 4 shows the response of the controlled system to the sinusoidal input of period $T_c = 240ms$. In comparison to the use of the regulator u_r alone, the combined controller of Eq. (21) appears to considerably improve the tracking performance. The response to the cycloidal input is shown in Fig. 5. Again, the combined controller, including the preview action, appears to follow the desired trajectory with a negligible amount of tracking error. These results indicate that substantial tracking error reductions may be expected when the preview controller is added to the standard feedback controller. In next two sections we assess the response of the system using the controller of Eq. (21) in comparison with the finite length preview controller and the more common inverse dynamic feedforward controller.

IV Comparison with Finite Length Preview Control

In the formulation of the discrete time finite length optimal preview controller the number of preview points, N_p, is chosen by the designer prior to the design of the control law [5]. The designer has a choice of selecting an arbitrary value of N_p provided that his choice is realizable in terms of future input availability at every sampling instant. However, in contrast to the controller derived in section 2, the explicit formulation of the trajectory dynamics are not used in the design of the feedforward controller. Instead, it is assumed that all of the future reference points after the final preview point, $y_r(k + N_p + 1)$ to $y_r(\infty)$, may be extrapolated linearly from the final preview point $y_r(k+N_p)$. This assumption allows a finite dimensional state space representation of the preview points in the form

$$\mathbf{x}_r^*(k+1) = A_r^*\mathbf{x}_r^*(k) \quad (24)$$

$$y_r(k) = C_r^*\mathbf{x}_r^*(k) \quad (25)$$

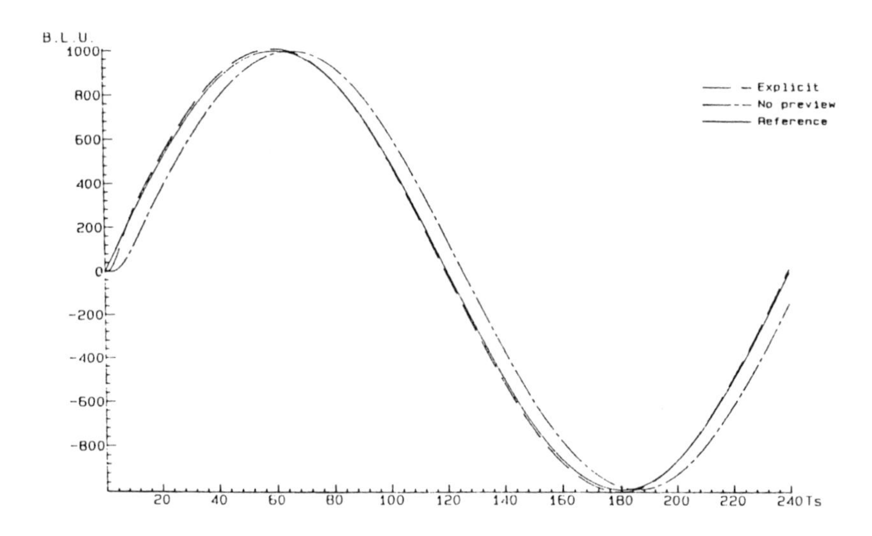

Fig. 4 Tracking Performance for the Sinusoidal Input

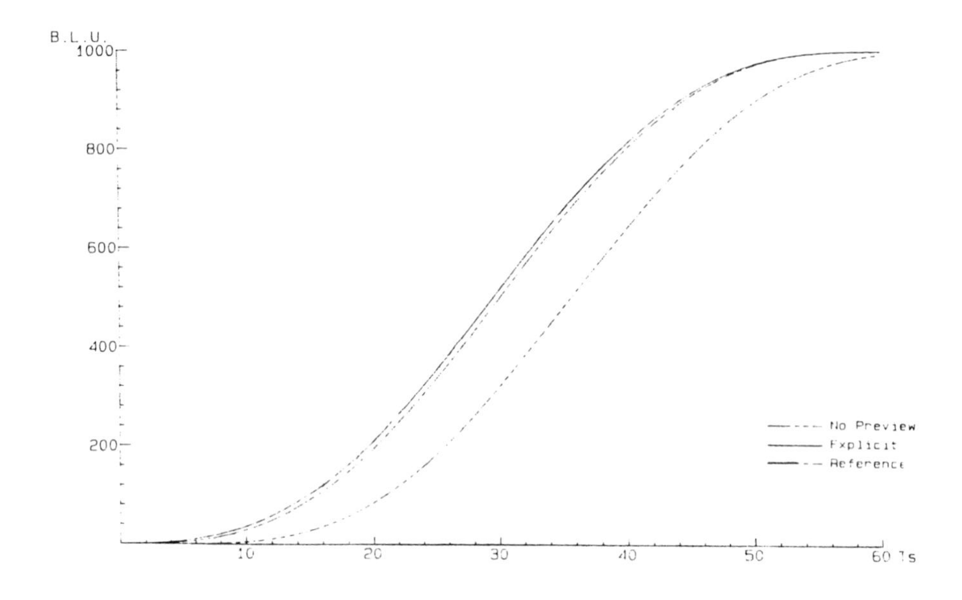

Fig. 5 Tracking Performance for the Cycloidal Input

where

$$\mathbf{x}_r^*(k) = \begin{bmatrix} y_r(k) \\ y_r(k+1) \\ \vdots \\ y_r(k+N_p) \end{bmatrix}, \quad C_r^* = \begin{bmatrix} 1 \\ 0 \\ \vdots \\ 0 \end{bmatrix}^T,$$

$$A_r = \begin{bmatrix} 0 & 1 & 0 & \cdots & 0 \\ & 0 & 1 & & \\ & & & \ddots & \vdots \\ & & & 0 & 1 \\ 0 & 0 & \cdots & -1 & 2 \end{bmatrix}$$

It is noted that the above representation of the preview states is completely independent of any particular trajectory generator model.

For the positioning system of Eq. (15), to design an optimal preview tracking controller based on N_p preview states, Eqs. (24) and (16) are augmented and the quadratic cost function of Eq. (7) is minimized according to standard LQR technique. Using a similar derivation as in [3], but including the electrical motor transients, the following optimal preview controller is obtained

$$u^{*o}(k) = g_\theta(\theta_r(k) - \theta(k)) - g_\omega \omega(k) - g_\alpha \alpha(k) + \sum_{l=1}^{N_p} g_{pr}^*(l)(\theta_r(k+l) - \theta_r(k)) \quad (26)$$

The feedback gains g_θ, g_ω and g_α are identical to the corresponding gains of the controller of Eq. (19), and can be determined using Eq. (12). The preview gains $g_{pr}^*(l)$ are different from those of Eq. (19) and their values are given by

$$g_{pr}^*(l) = -MB_p^T k_r(l-1), \quad l = 1, 2, \cdots, N_p - 2$$

$$g_{pr}^*(N_p - 1) = -MB_p^T(k_r(N_p - 2) - k_r(N_p))$$

$$g_{pr}^*(N_p) = g_\theta - \sum_{l=1}^{N_p} g_{pr}^*(l)$$

The vectors $k_r(l)$ can be evaluated using the following recursive equations

$$k_r(0) = \begin{bmatrix} -1 \\ 0 \\ 0 \end{bmatrix}$$

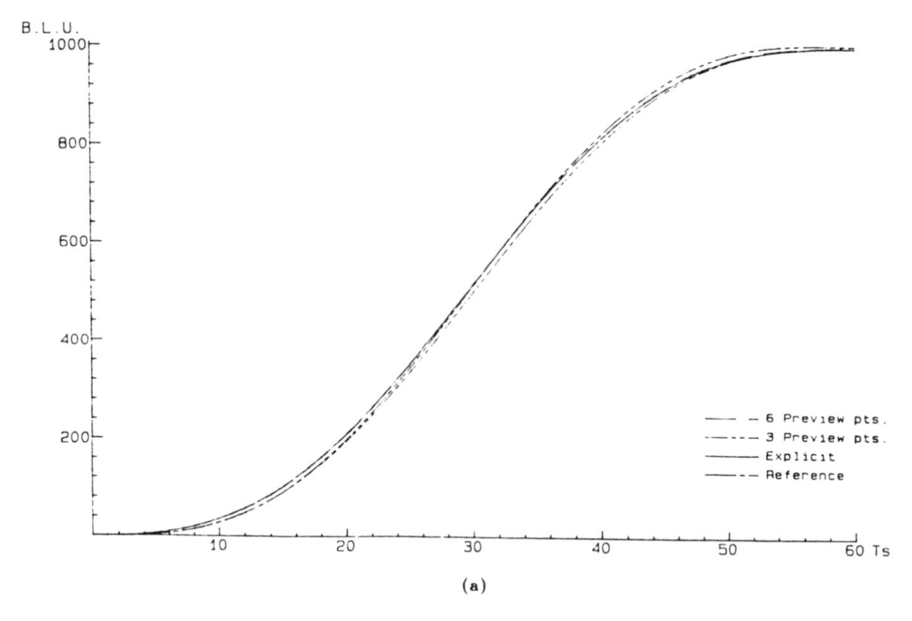

(a)

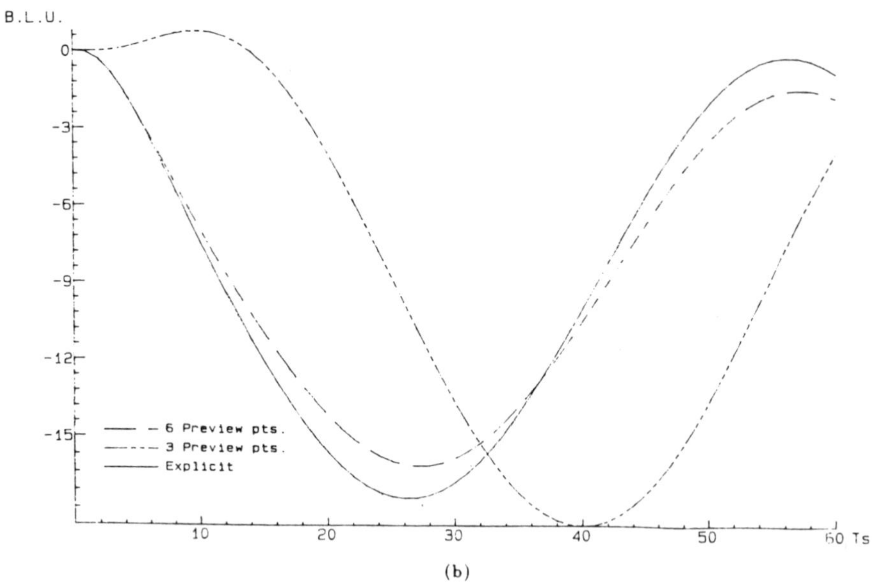

(b)

Fig. 6 Comparison with Finite Length Preview Control, (a) Output Response (b) Tracking Error

$$k_r(l) = A_c^T k_r(l-1), \quad l = 1, 2, \cdots, N_p - 2$$

$$k_r(N_p - 1) = (1 - A_c^T)^{-1} A_c^T (k_1 + \sum_{l=0}^{N_p-1} k_r(1) + k_r(N_p - 2))$$

$$k_r(N_p) = -k_1 - \sum_{l=0}^{N_p-1} k_r(l)$$

and

$$A_c = A_p - B_p \begin{bmatrix} g_\theta & g_\omega & g_\alpha \end{bmatrix},$$

k_1 = 1st column of the Riccati gain matrix K_1 of Eq. (12).

For the purpose of comparison of the performance of the controllers of Eqs. (21) and (26) on the positioning system of Eq. (16), the complete state feedback action is replaced by the same regulator u_r, giving

$$u^{*o}(k) = u_r(k) + \sum_{l=1}^{N_P} g_l^*(l)(\theta_r(k+l) - \theta_r(k)). \tag{27}$$

Figure 6 shows the tracking performance of the positioning system to the same cycloidal profile under the control of Eq. (27) with two different preview lengths. The response to the controller of Eq. (21) is also included for comparison purposes. From Fig. 6 it is evident that the three responses are almost identical taking into account numerical quantization effects which may cause a few basic length units (BLU) of error. Furthermore, it can be concluded that for the cycloidal trajectory, with $m = 4$, using more than 3 preview points, $(m-1)$, does not provide any significant tracking improvement. In addition,when the computational time for the generation of the feedforward control action is a critical bottleneck, using more than three preview points may adversely affect the tracking performance as the sampling period would have to be increased.

V Comparison with Inverse Dynamic Feedforward Control

For the positioning system of Fig. 3 a *perfect tracking feedforward* controller may be designed based on the feedforward of inverse dynamics [8]. As long as the sampling zeros are within the unit circle, there are two alternatives available for the structure of the feedforward controller. The first structure uses the inverse of the open loop plant, and the resulting feedforward action attempts to cancel the open loop dynamics [9]. The complete controller also includes an independently designed feedback regulator. The second structure for a perfect tracking feedforward controller is based on the inverse of the closed loop regulated plant. For the purpose of comparison with the tracking controller of Eq. (21), the second alternative will be used in this section. From the block diagram of Fig. 3, the closed loop z-transform function of the regulated positioning system including the reduced order observer is given by

$$\frac{\theta(z)}{\theta_r(z)} = \frac{K' z^{-1}(1 + z_1 z^{-1})(1 + z_2 z^{-1})(1 + dz^{-1})}{D(z^{-1})} \tag{28}$$

where K' is a constant gain which is a function of sampling period, the actuator parameters, and the observer gains. The observer zero at $z = d$ is within the unit circle. The closed loop denominator polynomial $D(z^{-1})$ is of order 5 and includes the two observer poles. Of the two sampling zeros, z_1 and z_2, z_1 is equal to -2.78 when the sampling period T_s is chosen to be $1ms$. This zero is outside the unit circle implying that the transfer function of the closed loop system cannot be inverted to provide a perfect tracking feedforward controller. This is a direct result of a generic property of any sampled-data controlled continuous plant when the pole excess of the continuous plant transfer function is greater than 2, and the sampling rate is sufficiently high [10].

To generate a *near perfect* feedforward controller for transfer functions with unstable or oscillatory inverses a novel approach has been proposed by Tomizuka [11]. The resulting feedforward controller is known as the *zero phase error tracking controller* (ZPETC) and has the property that the phase error between the desired reference θ_r and the actual trajectory $\theta(k)$ is zero for sinusoidal profiles. Using the ZPETC design technique, the feedforward controller's transfer function is given by

$$\frac{u_{zpetc}(z)}{\theta_r(z)} = \frac{z^2 F(z^{-1})}{K'(1 + z_1)^2(1 + z_2 z^{-1})(1 + dz^{-1})} \tag{29}$$

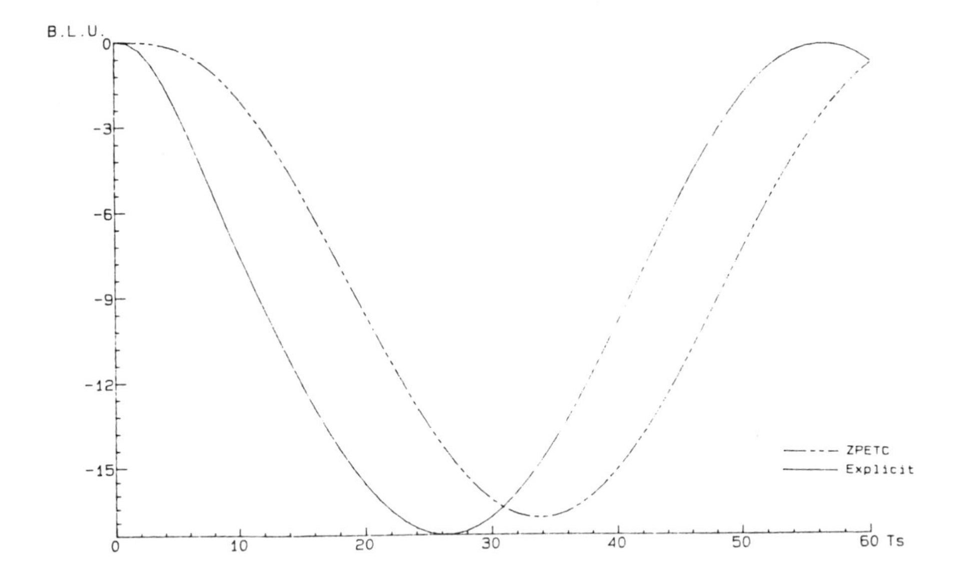

Fig. 7 Tracking Error Comparison with ZPETC

where $F(z^{-1}) = D(z^{-1})(z^{-1} + z_1)$. From Eq. (29) the algorithm for the discrete time $u_{zpetc}(k)$ may be written as

$$u_{zpetc}(k) = \frac{1}{K'(1+z_1)^2} F(q^{-1})\theta_r(k+2) - (d+z_2)u_{zpetc}(k-1) - dz_2 u_{zpetc}(k-2) \quad (30)$$

where q^{-1} is a unit delay operator. Here $F(q^{-1})$ is a stable sixth order monic polynomial containing the poles of the closed loop system (including the obsever poles), as well as, a phase concelling pole at $z = z_1^{-1}$, where z_1 is the zero outside the unit circle. The combined regulation and tracking controller is given by

$$u(k) = u_r(k) + u_{zpetc}(k) \quad (31)$$

Here the regulation controller u_r is identical to that used in Eq. (21).

Comparing the control law of Eq. (31) to the preview controller of Eq. (21), it is noted that the ZPETC is always two-step ahead irrespective of the trajectory to be followed. Furthermore, the feedforward action u_{zpetc} is now recursive due to the existence of the two poles in the transfer function of Eq. (29). The recursive nature of the ZPETC can make its performance more sensitive to coefficient quantization effects particularly when the pole z_2 is in the vicinity of -1 point. Such inverse dynamic poles, although stable, should in fact be regarded as unacceptable, and treated as uncancellable [11]. In addition, since $F(q^{-1})$ is a sixth order polynomial and a second order recursion exists, the computational load in executing the control law of Eq. (31) would be typically higher for trajectory models of order $m \leq 8$. Figure 7 shows the tracking errors resulting from the use of the control laws of Eqs. (31) and (21) for the same cycloidal trajectory. Despite the extra complexity of u_{zpetc}, from the graphs it is evident that the two responses are almost identical.

VI Conclusion

In this paper a class of optimal feedforward tracking controllers have been proposed for linear time invariant systems tracking trajectories which have prespecified dynamic models. It is shown that, when the controller's design is based upon the use of the observability canonical form of the trajectory model, the feedforward states are implicitly given as the future reference points. Furthermore, the proposed formulation of the tracking problem automatically determines the required preview length for a given trajectory model. This characteristic results from the fact that knowledge about the autonomous signal generator (Eqs. (3) and (4)) and its current state is equivalent to the knowledge about its future outputs. In this sense the preview problem with the explicit trajectory generator model is an "infinite" preview problem. The continuous time equivalent of this problem requires either the feedforward of all the states of the signal generator or $m - 1$ derivatives of its output [12].

The frequency response plots in Fig. 8 provide further insight into the cause of the superior tracking performance of the preview controller in comparison with a purely regulatory control law. It is apparent that the preview action reduces the phase error substantially. Yet, it does not generate extra amplitude error in the low frequency range (up to 10 percent of the closed loop bandwidth). As noted by Tomizuka [11], a phase error of ϕ at frequency ω implies a time delay of ϕ/ω in the tracking response. It should be pointed out that for a sinusoidal reference trajectory the ZPETC would provide a superior tracking performance since, in theory, it eliminates the phase error completely. However, for trajectories other than sinusoids, for example cycloids, our simulation studies indicate that the preview controller is equally effective in reducing the tracking error.

As a concluding remark, it is stressed that the preview action simply generates an auxiliary feedforward signal. As such it does not affect the stability of any closed loop system upon which it acts. Furthermore, each time the reference trajectory model is changed, only the number and the value of the preview gains need to be altered while the feedback controller remains unaffected. This is in contrast to the use of trajectory models within feedback loops based on the internal model principle of control [13]. Tracking controllers designed using the internal model principle increase the order of the closed loop system by the order of the trajectory model. Consequently, careful pole replacement procedures must be used in order to preserve stability and transient performance characteristics each time the reference

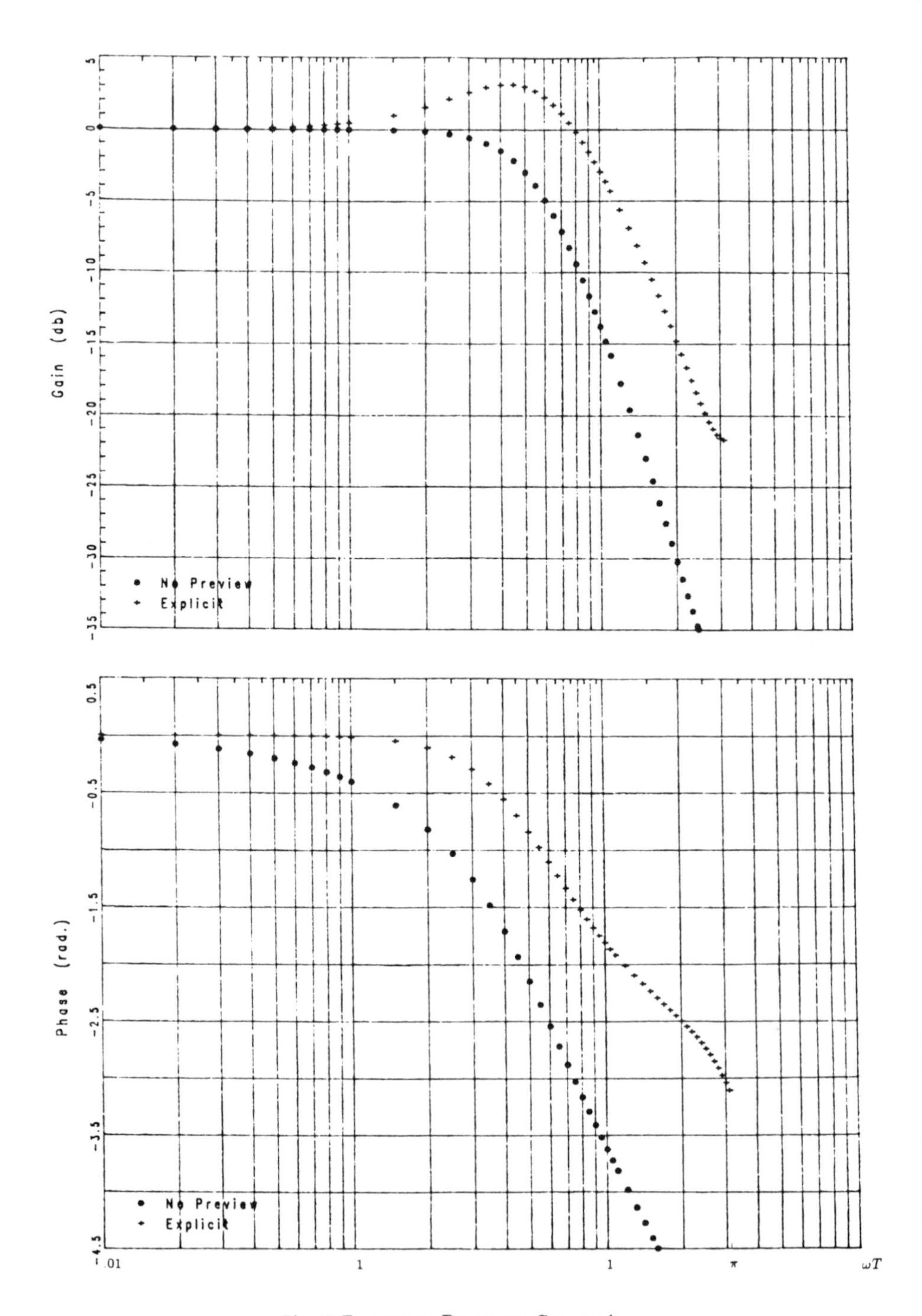

Fig. 8 Frequency Response Comparisons

trajectory model is changed.

VII Appendix A

Derivation of the Regulator with a Reduced Order Observer

In motion control systems usually only position and velocity signals are measured. A reduced order observer is developed here to provide an estimate of the acceleration $\hat{\alpha}(k)$, given the position and the velocity signals $\theta(k)$ and $\omega(k)$. Let the plant state equation be partitioned as

$$\begin{bmatrix} \theta(k+1) \\ \omega(k+1) \\ ---- \\ \alpha(k+1) \end{bmatrix} = \begin{bmatrix} A_1 & A_2 \\ & \\ -- & -- \\ A_3 & A_4 \end{bmatrix} \begin{bmatrix} \theta(k) \\ \omega(k) \\ --- \\ \alpha(k) \end{bmatrix} + \begin{bmatrix} B_1 \\ \\ -- \\ B_2 \end{bmatrix} u(k)$$

$$y(k) = \begin{bmatrix} I & 0 \end{bmatrix} \begin{bmatrix} \theta(k) \\ \omega(k) \\ --- \\ \alpha(k) \end{bmatrix}$$

Using the standard reduced order observer design technique, we can obtain the observer state equation as

$$\eta(k+1) = (A_4 - LA_2)\eta(k) + (B_2 - LB_1)u_r(k) + (A_3 - LA_1 + LA_4 - LA_2L)y(k) \tag{A.1}$$

where L is the observer gain vector and $\eta(k)$ is the state of the observer defined as $\eta(k) = \hat{\alpha}(k) - L[\theta(k)\ \omega(k)]^T$ with $\hat{\alpha}(k)$ as the estimated acceleration.

Consider the plant given in the section 3, which has the state equations given by Eqs. (16) and (17). Then we have partitioned matrices as

$$A_1 = \begin{bmatrix} 1 & a_{12} \\ 0 & a_{22} \end{bmatrix}, \quad A_2 = \begin{bmatrix} a_{13} \\ a_{23} \end{bmatrix}, \quad A_3 = \begin{bmatrix} 0 & a_{32} \end{bmatrix}, \quad A_4 = [a_{33}],$$

$$B_1 = \begin{bmatrix} b_1 \\ b_2 \end{bmatrix}, \quad B_2 = [b_3],$$

where a_{ij} and b_i are denoted as the elements of state matrices A_p and B_p. Substitute the above partitioned matrices into Eq. (A.1) and let $L = [l_1 \ l_2]$, then we obtain

$$\eta(k+1) = (a_{33} - l_2a_{23} - l_1a_{13})\eta(k) + (b_3 - l_2b_2 - l_1b_1)u_r(k) + (l_1a_{33} - l_1l_2a_{23} - l_1^2a_{13} - l_1)\theta(k) + (l_2a_{33} + a_{32} - l_2^2a_{23} - l_2a_{22} - l_1l_2a_{13} - l_1a_{12})\omega(k).$$

The state of the observer can now be rewritten such that

$$\eta(k) = \frac{1}{1+dz^{-1}}((b_3 - l_2b_2 - l_1b_1)u_r(k-1) + (l_1a_{33} - l_1l_2a_{23} - l_1^2a_{13} - l_1)\theta(k-1) + (l_2a_{33} + a_{32} - l_2^2a_{23} - l_2a_{22} - l_1l_2a_{13} - l_1a_{12})\omega(k-1)),$$

where $d = l_1a_{13} + l_2a_{23} - a_{33}$. Since $\hat{\alpha}(k) = \eta(k) + l_1\theta(k) + l_2\omega(k)$, substitute with $\eta(k)$; then we have

$$\hat{\alpha}(k) = \frac{1}{1+dq^{-1}}((b_3 - l_2b_2 - l_1b_1)q^{-1}u_r(k) + l_1(1-q^{-1})\theta(k) + (l_2 + (a_{32} - l_2a_{22} - l_1a_{12})q^{-1})\omega(k). \quad (A.2)$$

The optimal control input for regulator feedback loop is given by

$$u_r = g_\theta(\theta_r(k) - \theta(k)) - g_\omega\omega(k) - g_\alpha\hat{\alpha}(k). \quad (A.3)$$

Substitute $\hat{\alpha}(k)$ into Eq. (A.3), we obtain

$$u_r = \frac{1}{1+cq^{-1}}(g_\theta((1+dq^{-1})\theta_r(k) - (a+bq^{-1})\theta(k)) - g_\omega(e+fz^{-1})\omega(k)), \quad (A.4)$$

where

$$a = 1 + \frac{g_\alpha}{g_\theta}l_1, \quad b = d - \frac{g_\alpha}{g_\theta}l_1, \quad c = d + g_\alpha(b_3 - b_2l_2 - b_1l_1)$$
$$e = 1 + \frac{g_\alpha}{g_\omega}l_2, \quad f = d - \frac{g_\alpha}{g_\omega}(l_1a_{12} + l_2a_{22} - a_{32}).$$

A DARMA representation of this control input can be written as

$$u_r(k) = g_\theta(\theta_r(k) + d\theta_r(k-1) - a\theta(k) - b\theta(k)) - g_\omega(e\omega(k) + f\omega(k-1)) - cu_r(k-1) \quad (A.5)$$

References

1. M. TOMIZUKA and D. E. ROSENTHAL, "On the Optimal Digital State Vector Feedback Controller with Integral and Preview Actions," *ASME Journal of Dynamic System, Measurement and Control,* 101, 172–178, (1979).

2. M. TOMIZUKA, D. DORNFELD, X.-Q. BIAN and H.-G. CAI, "Experimental Evaluation of the Preview Servo Scheme for a Two Axis Positioning Systems," *ASME Journal of Dynamic System, Measurement and Control,* 106, 1–5, (1984).

3. H. A. PAK and P. J. TURNER, "Optimal Tracking Controller Design for Invariant Dynamics Direct-Drive Arms," *ASME Journal of Dynamic System, Measurement and Control,* 108, 360–365, (1986).

4. H. A. PAK , "Adaptive Maching of Feed Drive Dynamics within a Non-Adaptive Tracking Controller," *Proceedings of the USA-Japan Symposium on Flexible Automation,* Minneapolis, Minnesota, July 18–20, (1988).

5. M. TOMIZUKA, "Optimal Continuous Finite Preview Problem," *IEEE Trans. on Automatic Control,* AC-20, 362–365, (1975).

6. M. TOMIZUKA and D. E. WHITNEY, "Optimal Discrete Finite Preview Problems (Why and How is Future Information Important?)," *ASME Journal of Dynamic System, Measurement and Control,* 97, 319–325, (1975).

7. F. L. LEWIS, Optimal Control, John Wiley & Sons, New York, 1986.

8. H. A. PAK, "Issues in Precision Motion Control of High Speed Production Devices," *ASME, PED,* 21, 321–332, (1986).

9. H. SERAJI, "Design of Feedforward Controllers for Multi-variable Plants," *Int. J. Control,* 46, 1633–1651, (1987).

10. K. J. ÅSTRÖM and B. WITTENMARK, Computer Controlled Systems: Theorem and Design, Prentice-Hall, Inc., Englewood Cliffs, New Jersey, 1984.

11. M. TOMIZUKA, "Zero Phase Error Tracking Algorithm for Digital Control," *ASME Journal of Dynamic System, Measurement and Control,* 109, 65–68, (1987).

12. E. KOEINDLER, "On the Linear Servo Problem," *Int. J. Control,* 9, 4, 465–472, (1969).

13. B. A. FRANCIS and W. M. WONHAM, "The Internal Model Principle of Control Theory," *Automatica,* 12, 457–465, (1976).

ANALYSIS OF A ROBOT DRAGLINE FOR MINING

G F Shannon BE PhD

Associate Professor of Electrical Engineering
University of Queensland
ST LUCIA Q AUSTRALIA

1.0 INTRODUCTION

Mining operations are essentially materials handling and in the case of open cut coal mining there are two materials involved. The main aim is to mine the product, coal, but to gain access to it, the material lying of top of the coal needs to be removed. The total tonnage of this overburden often greatly exceeds the coal tonnage and hence this becomes the dominant activity in the materials handling process.

The item of machinery that is most commonly used to move the overburden is the drag-line. This has a large bucket (20 to 50 cubic metres capacity) which is attached to two ropes; one, the hoist rope, passes over a pulley at the end of a long boom and then onto a winding drum in the base of the drag-line; the second, the drag rope passes over a guide and onto a winding drum in the base of the dragline. The base, boom and attached ropes and bucket can rotate about the platform on which the drag-line rests. This platform is normally capable of "walking" to move itself to a new worksite. Figure 1 shows a schematic drawing of a dragline and typical physical details are listed in Table 1.

In normal use, the bucket is lowered into the work area, dragged up the sloping face of the overburden, filling in the process, then lifted and swung around and the bucket emptied over a dump zone. This is a more or less repetitive action occurring with a cycle time of about 90 seconds. It is carried out under the control of skilled operators in a noisy and dirty environment.

In as much as repetitive operations in adverse environments are often considered as possible areas for the application of Robots, a feasibility study has commenced.

The physical form of the drag-line does not fit the conventional coordinate axes classification but it is seen to be a system in which

(1) two of the linkages (ropes) are flexible
(2) one linkage (the boom) is non rigid
(3) the mass of material being moved is variable.

A control system which could be used on the drag line must have some means of determining the weight of the bucket and contents and be able to work with the flexible link systems.

2.0 DRAG-LINE MODELLING

Whilst it may be convenient to make measurements to determine values of variables to describe the state of the drag-line, consideration will be restricted to the information which is already available.

This information is:-

(1) Physical description of the dragline, including
  (a) boom length
  (b) inclination of boom from vertical
  (c) winding drum diameters
  (d) motor/winding drum inertias
  (e) Inertia of dragline in swing mode
  (f) mass per unit length of drag and hoist ropes

(2) Paid out length of hoist and drag ropes

(3) Electrical current and voltage delivered to drive motors

Figure 1

8050 WALKING DRAGLINE

WARD LEONARD - STATIC CONTROL - INDEPENDENT PROPEL

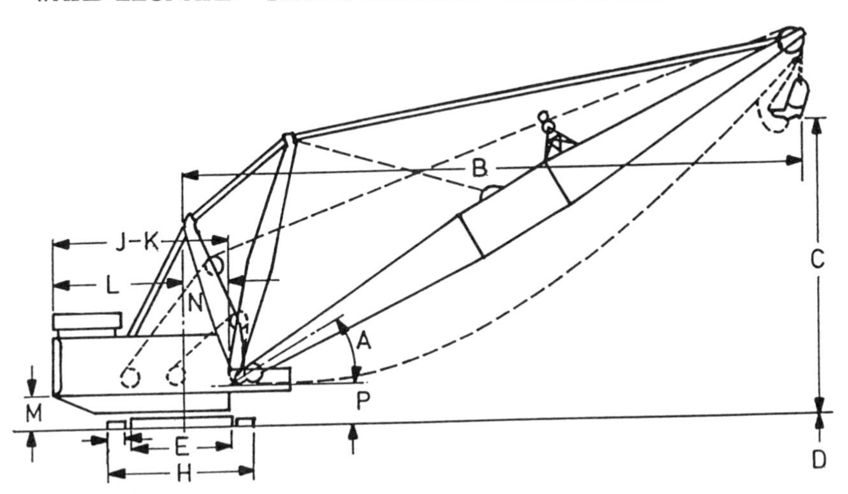

SPECIFICATION 8050-38 **TABLE I**

Boom Length (Centreline)	(99 m)
A - Boom Angle (Approx.)	(38°)
B - Dump Radius	(86.3 m)
C - Dump Height	(45.7 m)
D - Depth	(54.9 m)
E - Tub Nominal Outside Diameter	(17.7 m)
Rail Circle Mean Diameter	(14.2 m)
Main Swing Gear Pitch Diameter	(12.4 m)
F - Width of Shoe	(3.35 m)
G - Length of Shoe	(17.7 m)
H - Width over Both Shoes	(25.3 m)
Length of Step (Approx.)	(1.9 m)
J - Rear End Width	(20.5 m)
K - Length of Rotating Frame	(28.8 m)

TABLE I (cont)

L - Rear End Clearance Radius............................	(21.8 m)
M - Clearance Under Rotating Frame........................	(2.2 m)
N - Centre Rotation to Boom Foot..........................	(6.6 m)
P - Ground to Boom Foot..................................	(3.9 m)
Working Weight.......................................	(2,862,216 kg)
Ballast...	(408,240 kg)
Maximum Allowable Load...............................	(136,987 kg)
Ropes...	(30.5 kg/m)

The paid out lengths of hoist and drag ropes are measured via angular position measurements on the respective winding drums: electrical voltages and currents for each of the drive motors are also measured.

The weights of full and empty dragline buckets must be measured on each and every swing of the dragline. Further, these measurements are to be made whilst the dragline is being used in its normal working environment. Despite this, it is assumed that some time would be available to conduct periodic tests to evaluate key parameters of the dragline and to check zero references and calibration.

3.0 PRIMARY ANALYSIS OF DRAG LINE BUCKET MOVEMENT.

For this primary analysis, it will be assumed that the ropes lie on a straight line between support pulleys and the bucket.

3.1 Establishment of a 'Frame of Reference'

The actual position of the bucket in space, if the flexibility of the ropes is ignored, may be determined from knowledge of the three angles

θ_S, θ_D and θ_H. {i.e., angles of swing, drag and hoist motors - see Appendix I - list of symbols.}

Of these, θ_S provides a suitable variable to use in a reference frame as it defines the angular position of the boom. It is convenient to consider that the bucket lies in the vertical plane passing through the boom. Errors introduced by the assumption are studied later.

The position of the bucket in this plane may be expressed in terms of its (x, y) coordinates or horizontal position from the swing axis and vertical height above the dragline base.

For purposes of this analysis, the bucket position will be expressed in terms of this frame of reference and it will have coordinates

$$(\theta_S, x, y).$$

3.2 Relationships between input variables and bucket coordinates.

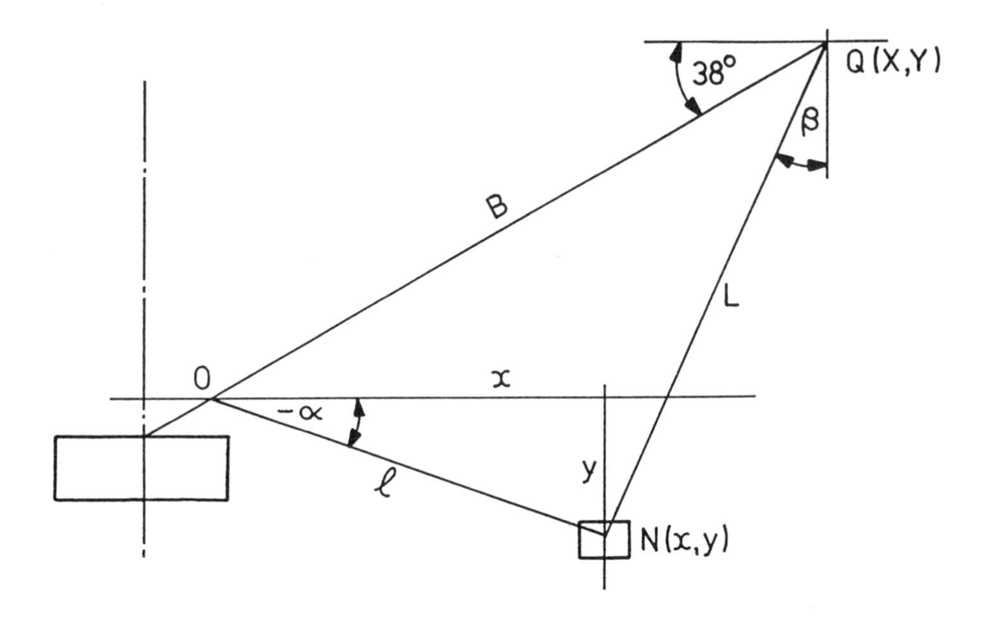

Figure 2 - Key Geometry of Dragline.

$$\ell = \ell_C + r_D \{\theta_D(mT) - \theta_D(0)\} \tag{1}$$

where $\theta_D(o)$ is zero position of drag winding drum

$$L = L_C + r_H \{\theta_H(mT) - \theta_H(o)\} \tag{2}$$

$$x = X - L \sin \beta \tag{3}$$

$$y = Y - L \cos \beta \tag{4}$$

To enable (3) and (4) to involve input variables alone, the value of β must be obtained.

3.3 Basic Geometry of ropes and Bucket

Because of the mass of the ropes, the ropes themselves will not be straight and the distance between boom ends and bucket will differ from the paid out lengths " ℓ " and "L".

As a first order approximation however, the 'rope sag' will be ignored when finding the position of the bucket.

Within the triangle created by the boom and the drag and hoist ropes, the following expressions are true,

$$\cos (38^\circ - \alpha) = \frac{B^2 + \ell^2 - L^2}{2B\ell} \tag{5}$$

$$\cos (52^\circ - \beta) = \frac{B^2 + L^2 - \ell^2}{2BL} \tag{6}$$

From which α and β may be found.

3.4 Velocity of the Bucket

The instantaneous velocity of the bucket may be expressed as the (vector) sum of its three components

v_x, v_y and v_z

where v_z is in a direction mutually perpendicular to v_x and v_y.
From the geometry of the system

$$\nu_z = \frac{d}{dt} (r \, \phi_s)$$

$$\nu_z = r \, \omega_s \tag{7}$$

$$\nu_x = \frac{dx}{dt}$$

$$\nu_y = \frac{dy}{dt}$$

Substituting from (3) and (4),

$$\nu_x = \frac{dx}{dt} = 0 - \dot{L} \sin \beta - L \, \dot{\beta} \cos \beta \tag{8}$$

$$\nu_y = \frac{dy}{dt} = 0 - \dot{L} \cos \beta + L \, \dot{\beta} \sin \beta \tag{9}$$

and from (6).

$$\dot{\beta} \sin (5\dot{2}-\beta) = \tfrac{1}{2} \left[\frac{2BL(L\dot{L} - \ell\dot{\ell}) - B\dot{L}(B^2 + L^2 - \ell^2)}{B^2L^2} \right] \tag{10}$$

and from (2)

$$\dot{L} = r_H \, \omega_H \tag{11}$$

$$\dot{\ell} = r_D \, \omega_D \tag{12}$$

4.0 ENERGY CONSIDERATIONS

4.1 The bucket energy may be expressed as the sum of Kinetic and potential energies and the work done in moving the bucket from one position (1) to another (2) is equal to this change in energy.

$$\text{ie } \Delta \text{ Energy} = K_{E2} - K_{E1} + K_{P2} - K_{P1}$$

$$= \tfrac{1}{2} M \{v_2^2 - v_1^2\} + Mg \{y_2 - y_1\} \tag{13}$$

The instantaneous velocities v_1 and v_2 may be expressed in terms of the resolved parts

$$\text{ie} \quad v_1^2 = v_{1x}^2 + v_{1y}^2 + v_{1z}^2 \tag{14}$$

For the moment attention will be restricted to motion within the vertical plane passing through the boom.

Motion in this plane produces energy changes which are caused by energy inputs to the drag and hoist motors.

$$\text{Hence} \quad v_1^2 = v_{1x}^2 + v_{1y}^2 \tag{15}$$

From (8) and (9)

$$v^2{}_x = \dot{L}^2 \sin^2 \beta + 2 L\dot{L} \dot{\beta} \sin \beta \cos \beta + L^2\dot{\beta}^2\cos^2\beta \tag{16}$$

$$v^2{}_y = \dot{L}^2 \cos^2 \beta - 2 L\dot{L}\dot{\beta} \sin \beta \cos \beta + L^2\dot{\beta}^2\sin^2\beta \tag{17}$$

Adding (16) and (17),

$$v^2 = v_x^2 + v_y^2 = \dot{L}^2 + L^2\dot{\beta}^2 \tag{18}$$

Substituting (18) into (13) the change in energy may be written as,

$$\Delta\text{Energy} = M \; g \left[y_2 - y_1 \right] + \tfrac{1}{2} \left[\dot{L}_2^2 - \dot{L}_1^2 + L_2^2 \dot{\beta}_2^2 - L_1^2 \dot{\beta}_1^2 \right] \tag{19}$$

4.2 Input Electrical Energy

The way that the electrical input to the motor is utilised is expressed as,

$$\text{Energy in} = \text{Elec loss} + \text{Mech loss} + \text{Energy stored} \tag{20}$$

$$\text{Energy in} = \int_{t_1}^{t_2} VI_a \, dt$$

$$\text{Elec losses} = \int_{t_1}^{t_2} I_a^2 R_a \, dt \tag{22}$$

where V is voltage at input to DC drive motor.

$$\text{Mech losses} = \int_{\Theta_1}^{\Theta_2} (\tau_F + B_F\omega) \, d\Theta \tag{23}$$

Energy stored = change in kinetic energy of winding drums and drive motor of winding drums and drive motor + Δ Energy (24)

Change in winding drum energy

$$= \tfrac{1}{2} J_D \{\omega_{2D}^2 - \omega_{1D}^2\} + \tfrac{1}{2} J_H \{\omega_{2H}^2 - \omega_{1H}^2\} \quad (25)$$

The expressions (21), (22) and (23) involve quantities which are not constant and the integral is most readily calculated as a summation

$$\text{Energy in} = \Sigma_{n=0}^{k} V(nT)\, I_a(nT)\, T \quad (26)$$

$$\text{Elec losses} = \Sigma_{n=0}^{k} R_a I_a^2(nT)\, T \quad (27)$$

$$\text{Mech losses} = \int_{\Theta_1}^{\Theta_2} (\tau_F + B_F\,\omega)\, d\Theta$$

$$= \int_{t_1}^{t_2} (\tau_F + B_F\,\omega)\, \omega dt$$

$$= \Sigma_{n=0}^{k} \{\tau_F(nT) + B_F\omega(nT)\}\, \omega(nT)T \quad (28)$$

on substituting (19), (25), (26), (27) and (28) into (20) yields an expression from which M may be evaluated.

In doing this, V, I_a and ω need to be sampled regularly to enable the other quantities to be calculated.

5.0 ANALYSIS OF THE SWING MODE.

The swing motor produces a torque which must overcome

(1) friction in the system
(2) accelerate the dragline itself
(3) accelerate the bucket
(4) overcome the coriolis forces

These may be expressed as follows.

$$\text{Friction Torque} = \tau_{FS} + B_F\,\omega_S \quad (29)$$

$$\text{Acceleration torque} = J_S \dot{\omega}_S + Mr^2\dot{\omega}_S \tag{30}$$

$$\text{Coriolis torque} = 2\ M\ r^2\omega_S\ v_X \tag{31}$$

where J_S is moment of inertia of dragline about swing axis
r = radius of bucket about swing axis
v_X = radial velocity of bucket
ω = angular velocity about swing axis

Thus

$$\tau_S = \tau_{FS} + B_F\omega_S + J_S\dot{\omega}_S + Mr^2\dot{\omega}_S + 2\ Mr^2\omega_S v_X \tag{32}$$

From this M can be evaluated.

6.0 EMPTY BUCKET DETECTION

A general solution to the detection of the time at which a bucket that is dumping, is empty, is very complicated.

The drag rope is slack and hence triangulation to determine bucket position is not possible, nor is the calculation of the angle of the hoist rope. As the bucket is swinging almost free, various forces could be involved and until such time as a more rigorous analysis can be studied the following approximations will be made to enable a "starting point" solution to be advanced.

Assume that at the moment the bucket is almost empty, the angle of the hoist rope away from vertical is small.

Under there conditions the rope tension may be regarded as acting vertically.

From this

$$F = M(g+a) \tag{33}$$

where a is acceleration of the hoist rope

$$a = r_H \dot{\omega}_H \tag{34}$$

If the weight of the empty bucket is M_E then an indication of bucket empty is given by

$$F < (M_E + 500)(g + a) \tag{35}$$

where 500 is some margin of mass to account for small disturbances.

For the hoist motor, the power balance equation is

Power into = rope power + power to mech loss + power to accelerate inertias ie.

$$\begin{aligned} I_a E &= F_H \omega_H r_H + (B_{FH} \omega_H + \tau_{FH}) \omega_H + J \omega_H \dot{\omega}_H \\ &= I_a (V - I_a R) \end{aligned} \tag{36}$$

To use equation (36) as it stands to calculate F would require division by ω_H which could be equal to zero. Hence check $\omega_H \neq 0$ before using.

Alternatively, if the field of the motor is known, the back emf

$$E = I_f k_m \omega_H \tag{37}$$

Hence when $\omega_H = 0$, $E = 0$.

Substituting (37) into (36) yields,

$$F = 1/r_H \{I_f k_m I_a - J\dot{\omega}_H - \tau_{FH} - B_{FH} \omega_H\} \tag{38}$$

Thus, find F from (36) or (38) and use (35) to verify the bucket is empty.

$$\text{Then} \quad M_E = \frac{F}{g + r_H \dot{\omega}_H} \tag{39}$$

7.0 SECONDARY MODELLING

The assumption that the ropes are linear is obviously erroneous, but a rigorous finite element analysis of the system is unlikely to be feasible in the time available. Consequently an alternative is sought.

Figure 3 shows the typical real life situation.

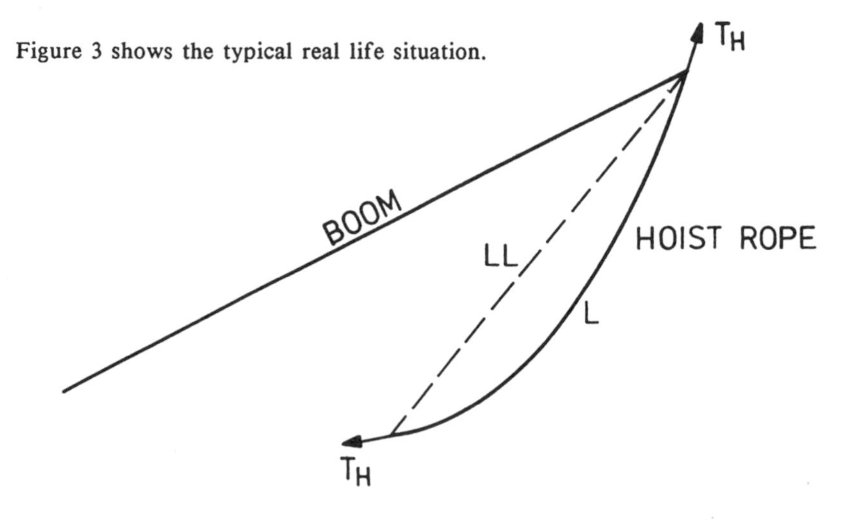

Fig. 3

The primary analysis considered that the linear length LL is equal to the paid out length L.

The actual situation is that a rope with distributed mass m kg/unit length and subject to a tension T_H will fall into an arc such that

$$LL = f(L, \beta, T_H, m_H) \tag{40}$$

In this expression, there is no difference between dynamic and static behaviour if the small contribution of the non vertical acceleration forces on the rope are neglected.

Hence, rather than attempt to formally determine f it is convenient to experimentally determine LL as a function of the variable in a static case and build up a Table. This will enable LL to be found for given values of L, T_H, and β with an interpolation routine being used to extend the information in the table. Alternatively, as the shape is a catenary with a known analytical form, the analytical expression may be used.

7.1 Dumping

The bucket is dragged in position shown in Figure 4(a), lifted as in Figure 4(b).

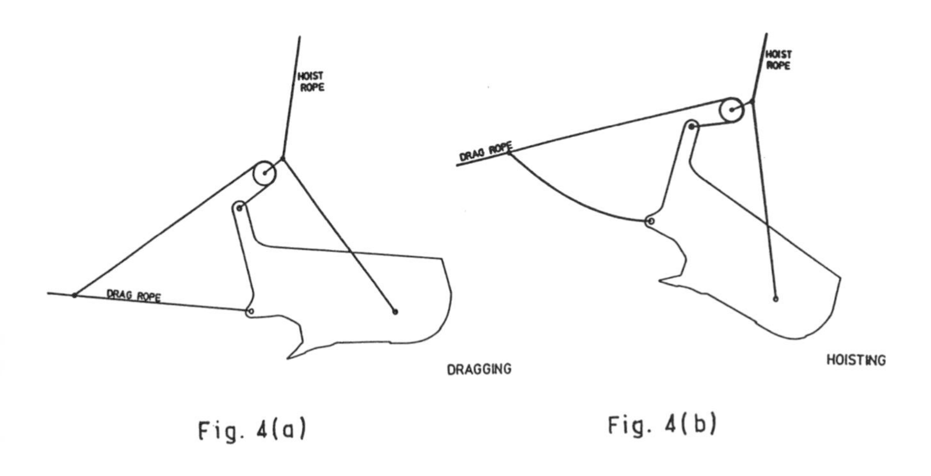

Fig. 4(a) Fig. 4(b)

In position (b) both hoist and drag ropes tension are required to maintain this orientation. As the drag rope tension is slackened, the overturning moment provided by gravity tips the bucket and its contents spill on the ground below.

For the spill to be on target, dynamic analysis of the tipping must be studied. This is the subject of on going studies.

8.0 CONCLUSIONS

The above analysis presents a set of equations from which the required bucket weights may be found. Several simplifying assumptions have beeen made to realise these solutions, but practical tests have revealed that bucket weithts are not in error by more than 10%.

The accuracy may be determined by one of two ways.

(1) by comparing measured values using the above empressions with the results of analysis by quantity surveys.

(2) by simulation studies where the model based on assumptions has its performance compared to that of a model based on rigorous analysis of the dragline.

As this detailed, rigorous model does not exist at the moment, the first method has been relied upon.

REFERENCES

1. I F Turner, "Boom Protection for Draglines", IEAust Q Div Tech Papers, QBG 1756, March 84.

2. I F Turner, "Microprocessors in Mining Industry", ME Thesis, University of Queensland, 1989.

3. G F Shannon, J M Skowronski, "Dynamic Weighing of Contents of a Dragline Bucket", University of Queensland, Report EE 87/18, April 87.

APPENDIX 1

LIST OF SYMBOLS

Θ_S = angle of swing
Θ_D = angle of drag winding drum
Θ_H = angle of hoist winding drum
ω_S = Θ_S = angular velocity of swing
ω_D = Θ_D = angular velocity of drag drum
ω_H = Θ_H = angular velocity of hoist drum

ℓ = paid out length of drag rope
L = paid out length of hoist rope
M_B = Mass of bucket
M_C = Mass of contents of bucket
M = Mass of bucket and contents
m = mass of winding rope (kg/m)
B = length of Boom (m)
y = height of boom top above horizontal plane of dragline base
x = horizontal distance of boom tip from dragline swing axis
α = angle between a 'straight line' drag rope and the horizontal
β = angle between a 'straight line' hoist rope and the vertical
τ_S = output torque developed by the swing motors
τ_D = output torque developed by the drag motors
τ_H = output torque developed by the hoist motors
τ_{FS} = static friction torque of swing motion
τ_{FD} = static friction torque of drag motion
τ_{FH} = static friction torque of hoist motion
B_{FS} = velocity dependent friction torque coefficient in swing motion
B_{FD} = velocity dependent friction torque cooefficient in drag motion
B_{FH} = velocity dependent friction torque coefficient in hoist motion
x = horizontal distance of bucket centre of mass from the centre line swing axis. (metres)
y = vertical height of the bucket centre of mass from the base of the drag line. (metres)
r = x
v_x = $\frac{dx}{dt}$ (ms^{-1})
v_y = $\frac{dy}{dt}$ (ms^{-1})
T = time interval between samples
n,k = number of samples
t = time (seconds)
K_E = Kinetic Energy (Nm)
K_p = Potential Energy (Nm)
P = Power ($J = Nm\ S_{-1}$)
I = electrical current

I_a = armature current

R_a = armature resistance

V = electrical voltage

E = induced emf

r_D = radius of drag winding drum

r_H = radius of hoist winding drum

ℓ_c = length of drag rope when bucket wound in to closest position determined by limits.

L_c = length of hoist rope when bucket wound in to closest position determined by limits.

INDEX

D

O

P

Q

R

T

U

V

W

Y

Z